Improve your Remote Viewing Accuracy Techniques using Quantum Microtubules

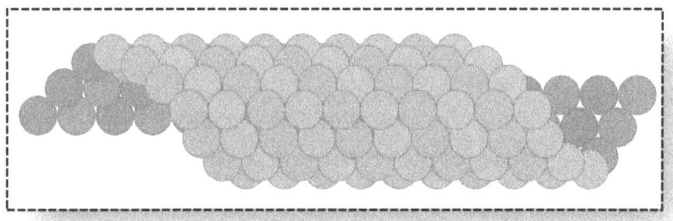

Published by the Institute for Solar Studies.
Santa Monica, CA.

Scott Rauvers

Read the First 3 Chapters of this book FREE at

www.mightyz.com/faqs.html

View the Associative Remote Viewing Dow Jones Project at

www.ez3dbiz.com/dow_project_research_summary.html

This third edition is also printed in workbook style with extra wide margins and headers on each page for the ease of writing in notes and details for students of remote viewing. If you prefer the non-workbook edition, this title has been reprinted for your convenience under the name **Secret Gems Foods & Essential Oils for Intuition & Associative Remote Viewing** (ISBN-10:1979771464), which has been printed in a non-workbook format.

Improve your Remote Viewing Accuracy Techniques
using Quantum Microtubules

Scott Rauvers

While it is impossible to outline every single detail necessary for a successful associative remote viewing session, this book's purpose is to serve as a guide, an instrument if you will, that will greatly increase your associate remote viewing skills. If this book serves that purpose, then I am satisfied that this book will have done its job

Improve your Remote Viewing Accuracy Techniques using Quantum Microtubules

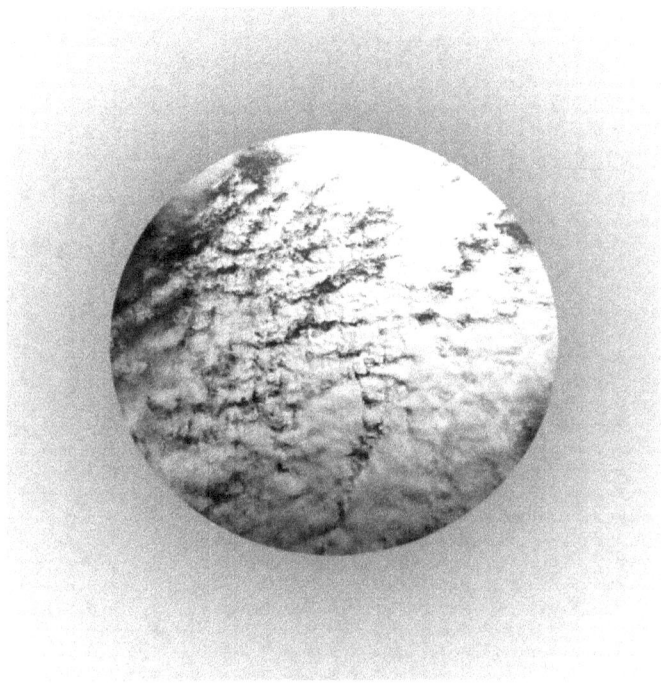

Scott Rauvers

Copyright © **November** 2017
All rights reserved. **The Solar Institute**

ISBN-10:1978254644

Improve your Remote Viewing Accuracy Techniques
using Quantum Microtubules

Other great titles published by the Institute for Solar Studies on Behavior and Human Health

- Sunspot Secrets. Jesus as the Sun. Technology, Prophecy and Human Evolution

- The 2018 Tao Nutrition Planetary Almanac and Intuitive Astrology Organizer

- New Millennium Millionaire Secrets to Fortune, Prosperity and Happiness

- Unique Anti-Aging Techniques to Live Beyond 100 years

- Avoid Root Canals. 101 Homeopathic Nutrition Remedies to Stop Tooth Cavities

- How to Make Tinctures, Extracts, Flower Essences and Homeopathic Remedies

You may preview the first 3 chapters of any of these books by visiting:

www.mightyz.com/faqs.html

Scott Rauvers

DEDICATION

This book is dedicated to the founders **of** **Quantum Mechanics:** Erwin Schrödinger, Werner Heisenberg, Max Born John von Neumann, Paul Dirac, Max Planck, Niels Bohr, Werner Heisenberg, Louis de Broglie, Arthur Compton, Albert Einstein, Enrico Fermi, Wolfgang Pauli, Max von Laue, Freeman Dyson, Satyendra Nath Bose, Arnold Sommerfeld, David Hilbert, Wilhelm Wien, and others.

We thank them for their contributions to science which has changed the way we live today. The last 4 years (2014 to 2017) has seen a revolution in quantum physics, especially in the area of quantum chemistry, quantum biology and quantum photosynthesis. This book merges these breakthroughs into the associative remote viewing framework to dramatically enhance the accuracy of associative remote viewing.

A special thanks goes out to the many scientists and researchers who have not only published some brilliant papers on quantum biology and quantum chemistry in the last few years, but for their courage and fearless research. Our **ARV** (*associative remote viewing*) technology would not have been made possible without their contributions.

It is interesting to note that Michael J. Fox who was has Parkinson's played the role of Marty in the Back to the Future Science Fiction Movie Trilogy. Parkinson's is associated with a loss of dopamine in the brain and dopamine enhances anticipatory effects (*Role of dopamine in anticipatory and consummatory aspects of sexual behaviour in the male rat. Pfaus JG and Phillips AG. Oct 1991*). It is the anticipatory effect that is partly responsible for pre-stimulus effects taking place during precognition (*Future directions in precognition research: more research can bridge the gap between skeptics and proponents. Michael S. Franklin et al. Aug 2014*), (*PAA; Mossbridge et al., 2014*). Parkinson's also affects the brain's resonating microtubules which is affected by water moisture and atmospheric water moisture is affected by lunar phases, as we shall clearly show in this book (*microtubules are tubular shaped structures inside the brain's neurons*).

It is the sole aim of this book to reveal methods, techniques and their related technologies that extend the pre-stimulus response time for receiving information during remote viewing from seconds to a maximum of 4 days into the future in order to enhance the success of associative remote viewing sessions.

Recommended Reading
Dopamine-dependent oscillations in frontal cortex index "start-gun" signal in interval timing. June 2015

Further Reading
Anomalous anticipatory response on randomized future conditions. Percept. Mot. Skills 84, 689–690. Bierman D. J., Radin D. I. (1997)

Anomalous anticipatory brain activation preceding exposure to emotional and neutral pictures. Bierman D. J., Scholte H. S. (2002). Paper Presented at The Parapsychological Association, 45th Annual Convention (Paris)

Predicting the unpredictable: critical analysis and practical implications of predictive anticipatory activity. Mossbridge J., Tressoldi P. E., Utts J., Ives J., Radin D., Jonas W. (2014). Front. Hum. Neurosci. 8:146 P 10.3389/fnhum.2014.00146

Attitudes of college professors towards extrasensory perception. Zetetic Scholar 6, 7–17. Wagner M. W., Monnet M. (1979).

Stock market prediction using associative remote viewing by inexperienced remote viewers. Smith C. C., Laham D., Moddel J. (2014). J. Sci. Explor. 28, 7–16

Improve your Remote Viewing Accuracy Techniques using Quantum Microtubules

The Solar Institute's Remote Viewing Series

Our remote viewing sessions over the previous 3 years have totaled more than 70 associative remote viewing sessions involving the future position of the Dow Jones Industrial Average. The best of our research has been spread out over a 3 part workbook series. This is book #3.

CONSTELLATIONS AND REMOTE VIEWING

Book 1 - ***Wormhole Theories, Sunspot Activity and Remote Viewing Stocks.*** Topics Covered: Quantum Tunneling, Herbs for Remote Viewing, 13:30LST, The Star Arcturus, Cosmic Rays and Remote Viewing, Air Pressure, The Human Nervous System and Pre-Stimulus Activity, Frequencies that Enhance the Results of Remote Viewing, Solar and Weather Conditions for Prime Associative Remote Viewing Sessions, Intuitive Biorhythms and Remote Viewing, Magnetic Midnight, the Ophiuchus Constellation, Mayer Waves, Moisture as a Medium for Conveying Information, The Associative Remote Viewing Procedure, Studies Involving Remote Viewing the Markets, Torsion Effects and Time, Magnetic Fields, Paramagnetic Materials, Angular Momentum and the Density of Time and much more!

REMOTE VIEWING HARDWARE AND TECHNOLOGY

Book 2 - ***Remote Viewing. The Complete User's Manual on Experiencing Future Consciousness.*** Topics Covered: Emotions as Sensors for Future

Stimuli, Associative Remote Viewing and power of Expectation, The Maharishi Effect, Remote Viewing the Future of the Dow Jones, Remote Viewing Electronics / Technology, Dealing with Remote Viewing Interference, Schumann Resonance, Heart Math Coherence and Remote Viewing, Humidity as an Emotional Intensifier, Polarized Light, Finding the Ideal Remote Viewing "Sweet Spot", The Key of Time, The Quarter Moon, Neutrinos and the Nervous System, Tungsten and the Electroweak Force, Hydrocarbons, Barometric Air Pressure and Intuition, Maintaining Strong Brainwaves During Remote Viewing Sessions, Triboluminescence, The Color Yellow, Environmental Radiation and Remote Viewing, Biodynamic Gardening Phases and Remote Viewing, Photoelectrics and much more!

THE QUANTUM REALM AND REMOTE VIEWING

Book 3 – *Improve your Remote Viewing Accuracy Techniques using Quantum Microtubules.* Topics Covered: The Quantum Mind, Remote Viewing and Quantum Mechanics, HRV and Remote Viewing, The role Microtubules play in Remote Viewing, Remote Viewing and Non-locality, The Hypothalamus and Remote Viewing, Gems and Minerals that Enhance Remote Viewing, Quantum Coherence, The Hippocampus, Empathy and Psychic Ability, Substances that Enhance Remote Viewing, Linoleic Acid and Quantum Mechanics, Quantum Photosynthesis, Dopamine and Remote

Improve your Remote Viewing Accuracy Techniques using Quantum Microtubules

Viewing, Transthyretin, Neurotransmitters and Remote Viewing, Lithium, Monoterpenes, The Signal to Noise Ratio and Remote Viewing, Essential Oils and Quantum Effects, Anesthetics, Taxol, The Pacific Yew Tree, Bacteria, Monoterpenes and Quantum Photosynthesis, Consciousness and Frequency, Meditation, Brainwave Rhythmus and Remote Viewing, Photons, Alternate Timelines and Parallel Universes, The Zero Point Field, The Best Moon Phases for Remote Viewing, Favorable Environments and Conditions for Remote Reviewing and much more!

You may preview the first 3 chapters of any of these books by visiting:

www.mightyz.com

Scott Rauvers

This third edition includes essential oils and other researched substances, including brainwave frequencies that enhance associative remote viewing. It also includes methods to enhance dopamine levels and keep it circulating in the brain longer to enhance remote viewing. Dopamine is one of the key substances that enhance remote viewing due to its anticipatory actions. This edition also shows in great depth how to utilize substances to clarify the connection to the quantum realm, from which all future information comes from. This book also looks at how the genes in our body play a role in intuition showing the latest scientific studies confirming that genes are connected to intuition.

Scott Rauvers

Scott Rauvers

Table of Contents

Introduction   **Page 35**
The Breakthrough Discovery that Enhanced Associative Remote Viewing___Heart Rate Variability___The Parasympathetic Nervous System and Future Events___ The Parasympathetic Nervous System Effects on Bodily Functions___The Sympathetic Nervous System___ Effects on the Body___Heart Rate Variability and the Nervous System___Heart rate variability (HRV)___The Vagus Nerve___Depression and Sympathetic Dominance____Obesity and Sympathetic Dominance___Symptoms of Sympathetic Dominance___A Research Study Examining the Effects of Moon Phase on Intuition on Roulette___Pupil Dilation and Pre-Stimulus___The Effects of Solar Weather on Heart Rate Variability and the Body's Parasympathetic and Sympathetic Nervous Systems___The Schumann resonance and its Influence on Human Brainwaves

Chapter 1. Solar Activity, HRV and the Nervous System.   **Page 62**
HRV and Magnetic Storms___The Sun's 10.7cm Solar Radio Flux Increases the body's Parasympathetic Nervous System___The KP "Sweet Spot" and the Sun's 10.7 cm Radio Flux___The 10.7 cm radio flux and Anticipatory Reactions___Inflammatory Responses and Cosmic Rays___Magnetic Fields Influence the Human Autonomic Nervous System___ Geomagnetic Storms and Heart Rate Variability

Chapter 2. Essential Oils for a Healthy Parasympathetic Nervous System.   Page 74

Meniki and Hinoki Increase Parasympathetic Nervous System Activity___Lavender's Effect on the Parasympathetic Nervous System___Juniper Essential Oil and the Parasympathetic Nervous System___Rose and Patchouli Essential Oils reduce Sympathetic Nervous System Activity___Bergamot Stimulates the ___Parasympathetic Nervous System___Pepper, Estragon, Fennel and Grapefruit Increase Sympathetic Nervous System Activity___Anxiety During Remote Viewing___Anxiety, St. John's Wort and Valerian

Chapter 3. Lunar Rhythms and Remote Viewing.   Page 83

A Summary of the Full Moon and Its Effects___Gout Attacks and the Full Moon___Insect Flight and the Lunar Cycle___Strokes and the Moon's First Quarter

Chapter 4. Alpha Brain Waves and Performance.   Page 88

Alpha Brainwave Activity during Air Pistol Shooting, Basketball free-throws and Golf Shots___Professional Golfers and Alpha Brainwave States___Alpha-Theta (A/T) Training___Sensory Motor Rhythm___Lavender essential oil and Brainwaves___Nicotine and Precognition___The Hippocampus and Nicotine___Photosynthesis and Quantum Biology___Quantum Photosynthesis and the Human Heart___Why photosynthesis in a remote viewing book?___Microtubules and Consciousness___Water Moisture and

Intuition___Lithium and Moisture___Sap Flow and Season___Seasonal Variation of Photosynthesis

Chapter 5. Microtubules, Resonance and Precognition. Page 109
What is a Microtubule?___Quantum Effects observed at Room Temperature___What is Intuition or Psychic Awareness?___How Fear Can sometimes be Mistaken for Intuition___Using a Computer to Predict Dice Position___Can Computers Help Us Hone our Psi Faculties?

Chapter 6. Remote Viewing and Non-locality. Page 122
The Schuman Resonance and Human Consciousness___How the Brain Receives Information via the Quantum Field During Remote Viewing___What is a Quantum State?___Chemical Reactions caused by Magnetic Fields___What does Non-locality Mean?___Method of Information Transfer during ARV Sessions___Ferromagnetism and Quantum Mechanical Effects___Where does Consciousness Come From?___Quantum Behavior Powers the Sun___Cordless Telephones affect the Brain's Microtubules___
An Experiment to "Stretch Time" to get More Done

Chapter 7. The Hypothalamus and its Sensitivity to Light. Page 141
What is Glutamate?___Nicotine and Glutamate ___Linoleic Acid found in Cyanobacteria___Digoxin and the Hypothalamus___Blue light and the Thalamus___

Light, Alertness and the Hypothalamus___Blue Light Enhances Photosynthesis___Light and Tubulin___Hypothalamus Activity During Midnight___Melatonin and Microtubules___Why the metal Tungsten Enhances Associative Remote Viewing Sessions___The Schottky Diode___Wulfenite___Lead Molybdate___What does TMD stand for?___What is Tungsten Disulfide?___Quantum Dots___Mixing Tungsten diselenide (WSe2) with Gold causes a 20,000-fold Increase in Photoluminescence___Anglesite___Quantum Dots and Stained Glass___Carbon Nano-materials___Carbon Materials___Quantum Dot Size = Frequency___The Selenium Cadmium Photoresistor___What is Cadmium Selenide?___Associative Remote Viewing and Geographical Location / Region___Hexagon Shaped Minerals___Does Carbon Dioxide Increase Photosynthesis?___How Tungsten Diselenide is used to Make an Artificial Leaf Solar Cell___TMDC Catalysts

Chapter 8. Quantum Transitioning from Photons as the carrier of Information during Remote Viewing.   Page 180

What is Quantum Coherence?___Remote viewing and Time___What is The Universal Wave Function?

Chapter 9. The Hippocampus, Empathy and Psychic Ability.   **Page 188**
Where is the Brain's Hippocampus?__Extrasensory Perception and Hippocampus__Hippocampus Empathy and Psychic Ability__What is Telepathy?__The Results of an MRI Study on Telepathy__Where is the Parahippocampal Gyrus?__The Substance Bergamot and the Hippocampus__The Brainwave Frequency of 7hz. The Key Frequency to Remote Viewing?

Chapter 10. Substances that Enhance Remote Viewing.   **Page 198**
Is ATP Exhibiting Quantum Effects?__Linoleic Acid as Quantum Fuel__What is the Cytoskeleton?__What is Linoleic Acid?__Linoleic Acid Synergy__Linoleic Acid Amounts in Some Oils__Gems and Minerals that Enhance Remote Viewing

Chapter 11. Polarized Light.   **Page 210**
Polarized Light and Plant Growth__Left-handed Circularly Polarized Light and its effects on Lentil and Pea plant Growth

Chapter 12. The Mid-Brain Dopamine System.   **Page 214**
What is Dopamine?__What are the Effects of Dopamine?__The Zacks Functional MRI Experiment__What is Parkinson's disease?__The Reward Effect, Dopamine and Enhanced Precognition__Dopamine and Feelings of Satisfaction__Dopamine, Reward and Pre-sentiment__The Immune System and

Dopamine___Foods highest in Tyrosine___Dopamine extends Lifespan in Worms___Linoleic acid Protects against loss of Dopamine___Excess Linoleic Acid Accelerates Aging

Chapter 13. Methods that Enhance Dopamine
Page 228

Tyrosine___The Herb White Peony and Dopamine___Gingko___Geraniol and Dopamine___Cacao Essential Oil___Pistachio___Protecting Dopamine Flow___Lower Lipolysaccharide Levels___Brilliant Blue___Selenium and Dopamine___Garlic and Dopamine Interaction___Dopamine and Onion Powder___Clary Sage___Creatine Boosts Dopamine Levels___The Thyroid and Dopamine Function___Selenium___L-DOPA___Mucuna Pruiens and Dopamine___Transthyretin___Is Transthyretin (TTR) the Psychic Gene?___Nicotine Protects Against Alzheimer's___Additional Substances that increase Transthyretin in the body___Fish Oil and Transthyretin___Transthyretin Synergy___What is IGF-I?___The PON1 Gene___Resveratrol and Fish Oil reduces Catecholamine levels___Aspirin and Salicylate Protect Dopamine___Methods that Enhance the Release of Dopamine___Nicotine___If Plants contain nicotine would that not kill them?___Nicotinamide / NAD___L-DOPA ___Do Nicotine Patches Increase Endurance?___Lithium Enhances Nicotine Sensitivity___Nicotine as a Plant Defense Mechanism and Nicotine in Food__ Jasmonic acid___GABA and Jasmine___Jasmine___Foods that contain

Nicotine___Algae and Quantum Effects___Blue-Green Algae used to make High-Performance Battery Electrodes___Algae Makes Better Lithium Ion Batteries___Excess Nicotine and Parkinson's___A flower that naturally contains Geraniol and Linalool___4-Anisaldehyde___Piperonal___Bumblebees are attracted to Nicotine at Low Concentrations___Lavender and Nicotine

Chapter 14. Substances that Enhance the Brain's Neurotransmitters. Page 271

What is OR?___Linoleic Acid and Neurotransmitter Activity___Exercise and Neurotransmission___Opiates___Myelin___Mantis Shrimp___The Sunstone and Polarized Light___ Substances that Enhance Myelin Growth ___Lithium___ Lithium and Microtubules___Cholesterol___ Lecithin___Foods that promote regeneration of Myelin___Foods Highest in Pyrroloquinoline from highest to lowest___Catecholamines___What are Gibberellins?___What is an MAO?___Foods that enhance Catecholamines___Rhodiola Rosea___Theanine___PTFE___Aspartate and Glutamate___Neurotransmitters and The Spine___What is Glutamic Acid?___What is GABA?___Geraniol is used for Spinal Cord Injuries___ Valerian and Dopamine___Valerian's Calming Effects on the Nervous System___ Anxiety, St. John's Wort and Valerian___Valerian and Dopamine___Valerian's Calming Effects on the Nervous System___Anxiety, St. John's Wort and Valerian___Valerian Root is as effective as a Pharmaceutical___Nardostachys___Why

Improve your Remote Viewing Accuracy Techniques using Quantum Microtubules

Older People May be More Intuitive___The Connection between GABA and Enhanced Intuition, Psychic and Precognition___How to Generate acetylcholine and GABA in the body___ The Russian Telepathic Experiments___Extending the 'Split Second' Retrieval of Information from the Future___The Underlying Mechanism of Telepathy___A list of former USSR PSI Labs___Nicotine Produces Alpha Brainwaves___Herbs for Healthy Neurotransmission___Ginkgo and Brain Circulation___Huperzia Serrata___What are Preganglionic Fibers?___What is Phosphatidylcholine?___ Bergamot Essential Oil___ Monoterpenes___Monoterpenes levels in Essential Oils___Rosemary and Sandalwood___Selenium___ Vitamin B6___What are PAH's?___The Structure of PAH's___Alpha Brain Waves___Theta Brain Waves___ Alpha Brain Waves and Remote Viewing___Properties of Alpha Brain Waves___Alpha waves and Heart Math___Tobacco Enhances Alpha Brainwaves___ Theanine Produces Alpha Brainwaves___10 Hz Current Produces Alpha Brainwave Rhythmus___The Schuman Resonance and Alpha Brainwaves___Nicotine increases Alpha Brainwaves___Tobacco Enhances Alpha Brainwaves___Stochastic Resonance and Alpha Waves___Weak Noise Enhances Neural Synchronization___Alpha and Gamma Enhance Creativity___Sunlight, Opiates and Exercise___Essential Oils for a Healthy Autonomic Nervous System___ Anxiety___Work Productivity___Self Esteem___ Peppermint Oil and Athletic Ability___Exercise and Jasmine Rose and Lavender Essential Oils___Ylang-Ylang Lengthens Processing Speed___Essential Oils for

Scott Rauvers

Enhancing Attention__Topical Application of Ylang Ylang Increases Skin Temperature__Carvone and Limonene are Chiral Fragrances

Chapter 15. Techniques for Controlling the Signal to Noise Ratio during Associative Remote Viewing. Page 329

What does the Signal to Noise Ratio Mean?__A Scientific look at the Signal to Noise Ratio__Radio propagation and Signal to Noise Ratio__The Signal to Noise Ratio and the Pareto Principle__Working with the Signal to Noise during Remote Viewing sessions__The Three Crucial Concepts to Remote Viewing Sessions__Methods to Enhance the Signal and Reduce the Noise

Chapter 16. Using Disruptors to Enhance Quantum Coherence in Microtubules. Page 339

Microtubule Disruption__Monoterpenes and Citrus__The Action of Essential Oils at the Microscopic Scale__Essential Oils and their Effects on Microtubules__What is Carvone?__What is A Monoamine Releasing Agent?__Geraniol__Geraniol is used for Spinal Cord Injuries__Citral__The Effect of Citral on Microtubules__Citral Protects the Liver__Synergy between Citral and Geraniol__ Synthetic Microtubule Disruptors__Anesthetics and Microtubules__Anesthetics and Microtubules__What is an Action Potential?__Katanin and its role in the Nervous System__Kinesin__What is Kinesin?__Lidocaine__ Substances with similar effects to Lidocaine__QX-314__Limonene__Anesthesia for

Stress Relief__ What is Phenoxyethanol?__Synthetic Disruptors__Vincristine as a Microtubule Disruptor__The Vinca Plant__Oryzalin as a Microtubule Disruptor__Griseofulvin as a Microtubule Disruptor__Nocodazole as a Microtubule Disruptor__RH4032__Acetylsalicylic Acid (Aspirin)__Rotenone__Nicotine as a Central Nervous System Antioxidant__Psychotropics and Microtubules__Colchicine__High Pressure and its Effects on Microtubule Functioning__High Pressure and Microtubule Depolymerization__High Pressure and X-rays__Coherence and High Pressure__The Effects of High Pressure on Muscles__Sensory Deprivation and Remote Viewing__The Hippocampus And Extremely Low-Frequency Electromagnetic Fields__Moon Phase and Geomagnetic Activity

Chapter 17. Substances that Strengthen and Enhance the Operation of Microtubules. Page 380

Ashwagandha as a Microtubule Stabilizer__The Pacific Yew, Taxol and its Microtubule Stabilizing Effects__Taxol as a powerful Anti-Cancer Substance__Taxol Stabilizes Microtubules__Where to Find the Pacific Yew Tree__Deuterium and Taxol__Reversing Aging using Quantum Coherence__Linoleic Acid and Lifespan__Acetylation and its Effects on Microtubules__What does Acetylation Mean?__The rare earth element Lanthanum Acetate enhances flexibility of the Arteries__Properties of Lanthanum__Lanthanum Enhances Photosynthesis

Chapter 18. How Plants 'See' Page 393

What are Flagellates?___Eyespot Proteins___What is a Flavoprotein?___E106___Algae, Quantum Effects and Photosynthesis___What are Chalmydonionas___Where are the Eyes located in Green Algae?___How Plants Utilize Quantum Coherence for their Photosynthesis___Detection of Quantum Fluctuations___The Quantum Process of Photosynthesis___Bacteria and Quantum Photosynthesis

Chapter 19. Monoterpenes and Photosynthesis. Page 411

Monoterpenes are Produced by Trees During Photosynthesis___The Monoterpene Linalool___What are Monoamines?___Neurotransmitters for Calm Moods and Emotions___Pirenzepine___Manganese and Copper promote binding of Dopamine to Serotonin___Seasonal Variation of Serotonin___Essential Oils and Neurotransmitters___Molecules that Exhibit Quantum Effects___Low Phenylalanine levels and Dopamine___Where to Obtain Monoterpenes___Monoterpene Synergy___ Menthone___Geraniol___Fenchone___Terpenoids___What is a Terpenoid?___Sesquiterpenes___What Essential oils that have a High Percentage of Sesquiterpenes?___Sesquiterpenes Effect on the Body___Sesquiterpenoids in Algae___The Scent of Burning Incense Induces Alpha Waves___Scents that Enhance Theta Brainwaves___Valerian and Depression___Valeriana wallichii DC and Depression___Could Monoterpenes be assisting Quantum Photosynthesis?___

Chapter 20. Do Certain Essential Oils Exhibit Quantum Effects? Page 436

Some Essential Oils and their Molecular Composition___ Transfer of Information via Quantum Effects is attributed to the Coherent Resonation of Water___Microtubules and hyper-computation___Heart Math and Non-Locality___Coherence and Super fluidity___Super fluidity effects in Nature___A Quantum Computer based on Superposition___What is Coherence?___Polymers and the Quantum Effect___Vanadium___Vanadyl Sulfate___Vanadyl Sulfate Protects the Heart

Chapter 21. Does Consciousness operate at a Measurable Frequency?. Page 450

Geomagnetic Storms and 40Hz___The Ajna Light___GABA and Consciousness___Can Meditation Enhance Superposition?___Quantum Collapse and the Brain's Microtubules

Chapter 22. Types of Meditation and its effect on Brainwave Activity. Page 461

Types of Meditation and Brainwave Patterns___The 10hz Frequency___What are Gamma Brain Waves?___Flickering Light and Brainwave Activity___How to Generate 10Hz and 40Hz Gamma___Methods that Amplify 40Hz Gamma___Alpha and Gamma waves and Creativity___What is the Eye Blink Rate?___Nicotine Enhances Right Brain Functioning___ Phenylacetaldehyde Enhances Photon Emission___What is

Phenylacetaldehyde?__Hemispheric Balancing__Stochastic Resonance__Noise Amplifies Electrical Signals in the Brain__10 Hz Current Induces Alpha Brainwave Rhythmus__Seasonal Variation of Sesquiterpenes in the Essential Oil (Lamiaceae)

Chapter 23. Can Photons Travel Backwards Through Time?.   Page 479

Photon Emissions from Living Organisms__Thomas Edison, Luminescence and Silver Sterling Mine__Zincite__Manganite__Todorokite__Birnessite__The Time Travelling Photon Experiment__Quantum Superposition and Travel to the Past

Chapter 24. Remote Viewing and Alternate Timelines.   Page 490

Parallel Worlds and the Biophysical Field

Chapter 25. Neutrinos and Parallel Universes.   Page 495

Hydrogen and Alternate Universes__Do Neutrinos Behave like Quantum Waves?__Tachyons__Neutrinos Travel Faster than Light via Super-Luminosity__Why Aren't Parallel Universes "bumping" into one Another? __Parallel Universes and Healing__Changing the Past to Change the Future

Chapter 26. Microtubules and The Quantum Brain.   Page 509

How many Neurons does the brain contain?__Why Meditation may strengthen the connection with the Liquid Crystalline structures within our body__Are

Microtubules Quantum Computers?___A detailed interior of a Microtubule___Consciousness and Neurons___Microtubules in plants___Piezoelectric Properties of Microtubules___Microtubule Structures found in Basalt and Pumice Stone___A History of Microtubules

Chapter 27. Microtubule and Essential Oils.   **Page 524**

The Action of Essential Oils at the Microscopic Scale___Geraniol___The Terpenes___The Terpenes___Borneo___Essential Oil Synergy___Seasonal Variation of Blood Pressure___Essential Oils that Lower Blood Pressure___The Composition of Ylang Ylang Essential Oil___Barometric Air Pressure and Blood Pressure___Seasonal Variation and Blood Pressure___Bergamot and Blood Pressure___Synergistic Ratios

Chapter 28. Essential Oils and their Effects on Brainwave Activity.   **Page 540**

Linalool and Brainwave Patterns___What is Methylisoborneol?___What is Geosmin?___Algae dissolves Radiation___MSG

Chapter 29. The Thalamus Region of the Brain and Remote Viewing.   **Page 546**

Chapter 30. Tungsten as a Photon Light Emitter   **Page 549**

Transition Metal Dichalcogenide___What is a TMD?___Molybdenum in foods___Lunar Phase and Mung Beans___What is Tungsten Disulfide?___

Microtubules and the van der Waals force__Composition of Rare Earth in Common Minerals

Chapter 31. Microtubules and the Schuman Resonance. **Page 557**
The Schumann Resonance Affects the Parahippocampal gyrus

Chapter 32. How Tobacco, Photosynthesis and Manganese all relate to one another. **Page 562**
Manganese and Photosynthesis__RH4032__Elicitors__Acetylsalicylic Acid__Phenylpropanoids__Reduced Iron Enhances Manganese Retention

Chapter 33. The TXP Formula. **Page 567**

Chapter 34. Favorable Environments and Solar Weather Conditions for Successful Associative Remote Viewing Sessions. **Page 568**
Geomagnetic Activity Levels__Earth's Magnometer__Favorable Gravity Bouguer Environments for Remote Viewing__The Tao__The Properties of Yang__Minerals__The Seven System__Body Properties__Nutritional Properties__Food Properties__Supplements__Voice__Chakras

Chapter 35. The Brain as a Hologram and the Field of Zero-Point Energy. **Page 578**

Chapter 36. The Zero Point Field and Memory
Page 585
Waking Conscious states similar to Microtubules__Are Microtubules Interacting with the Quantum Foam?__The Planck Scale and Quantum Consciousness__Do Birds Utilize The Quantum Realm?__Birds may sense Earth's magnetic field via Quantum Fields__What does Quantum Mechanics Mean?__Quantum Mechanics and Bird Navigation__Birds can Sense Polarized Light

Chapter 37. Variations of Water Moisture Caused by Moon Phases.
Page 597
Scientific Studies of Moon Phase and Water Moisture__Polarized Light and Effects on Microorganism__ A list of Dextrogyre Substances__Theta Brainwaves and Rotating Polarized Light__Clozapine__Fenchyl__Rainfall according to Phase of Moon__Perigee and Apogee Moon and Rainfall__Weather and Trauma__Time Travel, the Sun and Science Fiction__Air Pressure and Moon Phase__Barometric Air Pressure and Moon Distance__Seasonal Variation of Barometric Air Pressure According to Sunspots and Region__High Air Pressure and Births__Biodynamic Gardening and the Influence of the Moon's Forces__Seasonal Variation__Cosmic Rays and Water Moisture__The Moon's Influence on Nature__The Star Arcturus and Remote Viewing

Chapter 38. How to Find Favorable Solar Weather Conditions to Enhance Remote Viewing Accuracy   Page 620

The 0.8 MEV Energetic Particles___An LST Time Clock___LST Seasonal Calendar___How to Use the Calendar___Peak Seasonal Remote Viewing Seasonal LST Accuracy Time Slots___ Reasons for Failed ARV Sessions___A Remote Viewing Financial Markets Template___Why Time Flows at Different Speeds According to its Mass___Closing Remarks / Final Summary___Essential Oils and Creativity___A List Of 6 Tea Recipes That Enhance Intuition___Solar Eclipses and Wind Speed___Heart Rate Variability (HRV) And Exercise___Clinical Trials of Dr. Dardik's Lunar Exercise Routine

Materials References   Page 651

Monoterpenes in Essential Oils___Phenol Levels in Essential Oils___Keytone Levels in Essential Oils___ Monoterpenes in Essential Oils Chart #2___Essential Oils that have the Most Popular Monoterpenes___ A list of Terpene alcohols___A list of Cyclic ketones___A list of Aromatic ketones___Van Der Waals Radius of the Elements___Tellurium Dioxide and Lead Molybdate. Minerals used in Acousto-optic Devices___Piezoelectric Crystals that have Elastic Properties___Ancient Egyptian Healing Rods and the Schumann Resonance___Tuning Forks for the Parasympathetic Nervous System___Other Tuning Forks for the body's Parasympathetic Nervous system___The Otto Tuners___ **Acupressure Points for Influencing Heart Rate Variability (HRV)** ___Research Study Number 2.

Improve your Remote Viewing Accuracy Techniques using Quantum Microtubules

Accupressure Points and Heart Rate Variability (HRV) ___ Acupuncture Points for Influencing Heart Rate Variability (HRV) ___Music and Exercise and its effects on the body's Autonomic Nervous System

Book Index   **Page 704**

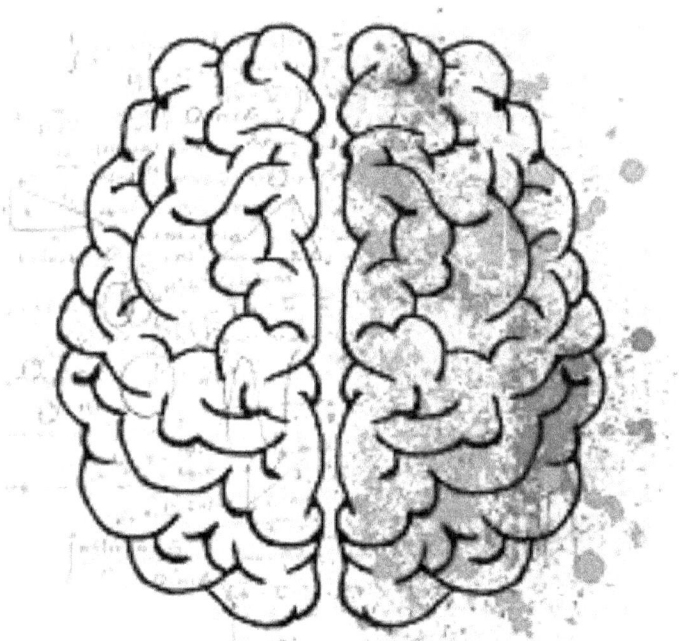

Improve your Remote Viewing Accuracy Techniques using Quantum Microtubules

Introduction

Our previous 2 editions on remote viewing, Wormhole Theories, Sunspot Activity and Remote Viewing Stocks and Remote Viewing. The Complete User's Manual on Experiencing Future Consciousness, laid the groundwork for methods and techniques that enhance associative remote viewing. This third edition ties them all together, including how the body receives the information during remote viewing, both via quantum methods and the nervous system.

The Breakthrough Discovery that Enhanced Associative Remote Viewing

The accuracy of our associative remote viewing sessions dramatically increased after using the device known as the **EMwave 2**. This device tells you when your heart and brain are in coherence with one another using a method known as Heart Math. This creates a clearer connection to activity taking place in the nervous system..

Heart Rate Variability
The most significant changes that occur in the body before an anticipatory event include physiological changes in the skin, cardiopulmonary and nervous system *(Preparation- or intention-to-act, in relation to pre-event potentials recorded at the vertex. B. Libet et al. 1983).* Hence the scientific data shows our bodies have direct access to information about the future, at least short term. Soon we shall show a scientific study involving a roulette wheel putting this principle into action and how Heart Math was used to enhance the odds of winning.

Precognition via our body takes place without our thinking about it. Hence it takes place in our autonomic nervous system. If we did not have our autonomic nervous system, every exhale/inhale of our breath or every beat of the heart would be in our conscious minds every waking moment. Future events would quickly overwhelm our mind. It may be that our autonomic nervous system acts as the main buffer between the conscious mind and precognitive information.

The value lies in being able to select the information you wish to receive using conscious effort and trusting the Parasympathetic Nervous System to clearly communicate with the conscious mind. Hence this clearly explains why long forgotten fears hiding in the subconscious and spontaneous bursts of anger left over from past experiences cause illness to come out of nowhere.

Traumatic incidents and stressful events leave lasting marks in the body's autonomic nervous system. Hence the reverse is also true, future incidents and situations leave subtle marks can be detected via the body's autonomic nervous system.

During Associative Remote Viewing Sessions, the Parasympathetic Nervous System can recognize precognitive information and directly rely this information to the conscious mind through spontaneous action, thoughts or emotions. Remote Viewing in general relies on the Freudian levels of consciousness levels. The lowest level is the "*unconscious*." This states a that a part of our mental processes we take for granted as "*awareness*" does not have direct access to the unconscious.

This part of the psyche is what first receives signals. Next it is passed to the body's autonomic nervous system (ANS) and as the signal line impinges on the autonomic nervous system, the information becomes converted into a reflexive nervous response. As this takes place, the signal enters the subconscious mind and then into the lower fringes of the conscious mind.

Hence our subconscious and conscious minds communicate through our bodies. This is made possibly due to the fact that the body's nervous system contains 100 trillion synapses and 100 billion neurons, more than enough processing power to communicate precognitive information to the conscious mind.

Further Reading
The Autonomic Nervous System and Future Events

Skin conductance prestimulus response: Analyses, artifacts and a pilot study. Spottiswood J, May E. Journal of Scientific Exploration, in press; 2003

Anomalous anticipatory skin conductance response to acoustic stimuli: experimental results and speculation about a mechanism. E.C May et al. J Altern Complement Med. 2005 Aug; 11(4):695-702.

The Parasympathetic Nervous System and Future Events
Before we look at the roulette study, we first need to understand the specific part of the nervous system sends us messages from the future.

The Hypothalamus and the Nervous System
The hypothalamus regulates a specific part of the nervous system known as the **Autonomic Nervous System**. The Autonomic Nervous System is associated with bodily functions that take place without our knowing about it (*unconsciously*). This includes blood pressure, heart beat, respiration and digestion. Like Neutrinos, the Autonomic Nervous System has consists of two separate flavors.

1 - **Sympathetic**

2 - **Parasympathetic**

The Sympathetic Nervous System is the "*fight or flight*" system which is associated with the neurotransmitter epinephrine (*or adrenaline*).

The Parasympathetic Nervous System Effects on Bodily Functions
Parasympathetic Nervous System activity is accompanied by an increase in high frequency power (HF) of **Heart Rate Variability** (**HRV**) spectrum.

- Relaxation
- Rest
- Healthy Digestion

The Sympathetic Nervous System Effects on the Body

Sympathetic Nervous System activity is accompanied by an increase in low frequency power (LF) of the heart rate variability (HRV) spectrum.

- Pupil dilation
- Accelerated Heart Rate
- Widens bronchial passages
- Decrease movement of the large intestine
- Constricts blood vessels
- Increase peristalsis in the esophagus
- Creates goose bumps and perspiration (sweating)
- Raises blood pressure

When both of these systems are functioning properly, a balance (*tandem*) takes place, creating homeostasis. Overall the Parasympathetic Nervous System is usually slightly more dominant then the Sympathetic Nervous System. However, what happens if the Dominance of the Parasympathetic Nervous System becomes reversed?

If the Sympathetic Nervous System became the dominant force, instead of our body being relaxed with healthy digestion, healthy blood pressure and a strong heart, the body would exist in a perpetually hyped-up state of being. It will always be hungry and never satisfied. You would always feel exhausted and never be able to sleep. The heart will be racing, unable to handle strenuous physical activity. This is what would happen if the Sympathetic Nervous System became dominant.

Heart Rate Variability and the Nervous System
The function of the body's Autonomic Nervous System is measured via what's known as **HRV (Heart Rate Variability)**. This is also called 'Cycle Length Variability,' and is used as a guideline or test to determine what state the Autonomic Nervous System (Parasympathetic vs Sympathetic) is in. To put it simply, Heart Rate Variability is the difference between peaks seen on an EKG (the distance between the peeks) or frequency.

Heart rate variability (HRV)
Heart rate variability (HRV) parameters are commonly used to as a method to accurately indicate the condition of the body's autonomic nervous system (ANS) behavior. The sympathetic and parasympathetic systems work similar to a neuronal network which continually control heart rate.

    The beauty of Heart Rate Variability is that it can very precisely measures a person's ability to rapidly increase or decrease their pulse rate in response to stress or changes in environment. This is one of the main keys to detecting/interpreting future information.

These measurements are measured in milliseconds. This is much like the bioavailability of a mineral. Nano-sized minerals are rapidly absorbed, whereas standard sized minerals take much longer to absorb into the body.

The faster your body can vary your rate or its high variability the easier it is for the nervous system to be parasympathetic dominant.
However low variability shows that the body's sympathetic nervous system is dominant, which over the long term leads

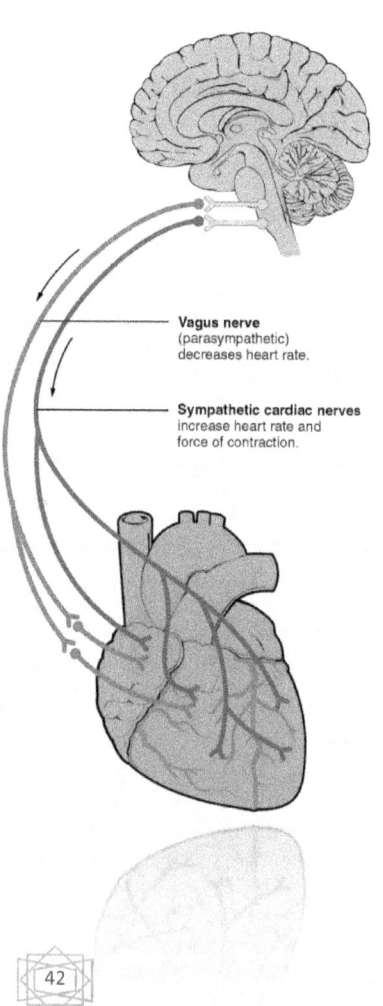

to health problems.

The Vagus Nerve
The Vagus nerve is directly connected to the brain and is one of the **major conduits that control parasympathetic nervous system responses**. Another region that is connected with the body's parasympathetic nervous system is the brainstem region (*lower brain*) and the sacrum (*tailbone*) area. responses by the sympathetic nervous system comes mostly from the thoracic spine region.

Depression and Sympathetic Dominance
The January 2015 issue of Progress in Neuropsychopharmacology & Biological Psychiatry (*Reactive Heart Rate Variability in Male Patients with First-Episode Major Depressive Disorder*) stated that patients with major depression show a nervous system with sympathetic dominance when they are resting. However when stressed their nervous system undergoes a shift to parasympathetic dominance.

Obesity and Sympathetic Dominance
The November 2013 issue of the Oxford Journal of Clinical Endocrinology (*Obesity is Associated with an Altered Autonomic Nervous System Response to Nutrient Restriction*) stated in their report that

people who are obese show sympathetic dominance in their nervous systems.

Symptoms of Sympathetic Dominance
Appetite suppression is common in people with sympathetic dominance. This is because if the body is stressed for long periods of time, the body's vessels adapt to the constriction caused by the stress and end up becoming muscular, maintaining an extremely narrow diameter. This in turn requires larger amounts of blood to be pumped through the now narrow vessels increasing blood pressure. Those at high risk are people with low ranking jobs or stressful jobs or stay-at-home parents

Further Reading
The Parasympathetic Nervous System. Bruce Blaus. Wikiversity Journal of Medicine

Now that we have a better understanding how our nervous system(s) deliver information to our brain, let's take a look at one scientific study where HRV was used to predict roulette wheel behavior.

A Research Study Examining the Effects of Moon Phase on Intuition on Roulette
Researchers measured the body's electrical signals related to the pre-stimulus effect of a

roulette wheel during eight separate trials involving 13 volunteers using real cash.

Half of the sessions were conducted during full moons and the other half during the new moon and were designed to look for pre-stimulus responses (*a type of nonlocal intuition*). The study found significant differences between loss and win responses during both of the pre-stimulus segments. These segments provide valid information about nonlocal intuition.

The study found a significant pre-stimulus response existed approximately 18 seconds prior to the volunteers knowing the future outcome. However there was almost no relationship found in the pre-stimulus response and how much money the volunteers won or lost. The study also noted a significant difference in both pre-stimulus periods during full moons, but not during new moons. A chart summarizing the effect of the moon and its influence on intuition in this study is shown in the following graph.

Subject no.	Full Moon (52 sessions)		New Moon (52 sessions)	
	Win ratio	Amount won ratio	Win ratio	Amount won ratio
1	50%	−3%	50%	−10%
2	47%	−22%	42%	−36%
3	52%	9%	45%	−20%
4	54%	3%	39%	−29%
5	54%	18%	48%	10%
6	56%	30%	50%	0%
7	50%	−12%	41%	−48%
8	58%	18%	48%	−1%
9	48%	−13%	47%	−19%
10	56%	6%	49%	2%
11	43%	−37%	57%	36%
12	52%	4%	53%	15%
13	54%	14%	54%	11%
Average	51.8%	1.3%	48%	−6.9%

The study concluded that the methods used in the study exist as a reliable method of prompting physiological detection of pre-stimulus events and may be a valuable method for measuring aspects of nonlocal intuition. The study also found that if the volunteers became more attuned to their internal physiological responses (**HRV** or heart math), that their performance would have been much better on their betting choices.

We use heart math in our associative remote viewing sessions and have found it to considerably boost our remote viewing accuracy. We give details about how to do this in our second book **Remote Viewing. The Complete User's Manual on Experiencing Future Consciousness.**

Summary

Their overall win ratio was higher (Z=-2.2, P<.05) during full moons. The study also proves that the heart math win/loss response during full moons (but not new moons) contributed to the increased win ratio. This effect has also been shown to occur in research by Puharich and full moons and solar activity have been shown to affect psychokinesis experiments with one researcher suggesting the moon's interaction with earth's magnetosphere during the moon's passage through the magneto-tail (which takes place during full moons) may be responsible for the observed effects.

Key   Finding

The key lesson to be learned from this study is that it is possible to predict future events with associative remote viewing without using a blind target. The key element that makes this possible is to use HRV (heart math). Once resonant coherence has been achieved, the images begin flowing clearly during the session. It may be that common remote viewing works best using blind targets and associative remote viewing works best in association with HRV.

Reference - Electrophysiology of Intuition: Pre-stimulus Responses in Group and Individual Participants Using a Roulette Paradigm. Rollin Mc Craty and Mike Atkinson. March 2014.

This same effect also occurs as the sun's solar wind pushes upon Earth's magnetic bubble. As it does so, earth's magnetosphere stretches forming the magnetotail (*Interplanetary Shock Propagation Through the Magnetosphere to the Magnetotail. O. Goncharo et al.*). During new moons, solar wind speeds take longer to slow down and during full moons, solar wind speeds slow down faster.

In a research study titled How the solar wind may affect weather and climate, published in January 2015 by J. Wendel found that certain changes taking place in the IMF correlated with pressure changes above Earth's poles. Effects took place in the lower troposphere that were driven by a difference in electric potentials earth's ionosphere and the surface days sooner than the changes occurring in the mid-to-upper troposphere.

Pupil Dilation and Pre-Stimulus

Another study showed that a telling feature of

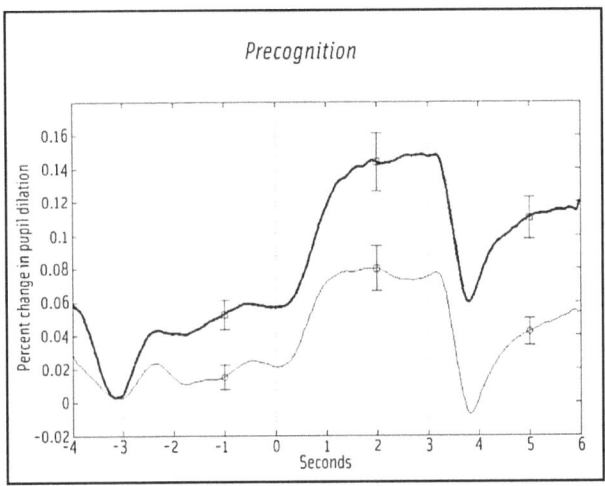

pre-sentient activity is a change in pupil dilation. This is because as light strikes our eyes, it registers changes in our nervous system. Shown in the following image is a change occurring in pupil dilation taking place a few seconds **BEFORE** an event takes place.

Reference

Predicting the Unpredictable: 75 Years of Experimental Evidence. Dean L. Radin. Consciousness Research Laboratory.

Further Reading
Precognition and real-time ESP performance in a computer task with an exceptional subject.
Journal of Parapsychology, 51, 291–320.
Honorton, C. (1987).

Now does solar weather affect the Parasympathetic Nervous System? Let's take a look at the data and find out.

The Effects of Solar Weather on Heart Rate Variability and the Body's Parasympathetic and Sympathetic Nervous Systems
Our research from performing over 70 associative remote viewing sessions on the future position of the Dow Jones industrial average found that our best results came when solar weather conditions are favorable. One of the main advantages of favorable solar weather conditions is the power of the Schumann resonance is stronger. In this book we shall clearly show the connection that exists between favorable solar weather conditions and a stronger Schumann resonance.

One of our major findings was that earth's geomagnetic activity influences the body's nervous system. This has now been verified by research studies which we are now going to show in greater detail.

A number of research studies have found an

association exists between magnetic storms and a decrease in HRV, most likely an occurrence of impaired cardiovascular system functioning [1] [2] [3] [4] [5] [6] [7] [8] [9] [10]. A ~25% reduction in the very low frequency (VLF) rhythm has been shown to occur [11] [12] [13] [14]. This is commonly associated with an increased risk of ill health [15]. Dimitrova et al. discovered that during geomagnetic storms that both the HF and the LF measures and the ratio between the low and high frequencies were reduced [16].

Studies now clearly show that the human nervous system exhibits "*anticipatory reactions*" (physiological) 2 to 3 days before a strong geomagnetic storm takes place. These anticipatory effects on the body include significant changes in blood pressure, heart rate, HRV, skin conductance and specific physiological complaints [17] [18] [19] [21] [22] [23] [24].

Studies by Chizhevsky during the 1920s suggested an unknown form of solar radiation was causing the anticipatory reaction [25]. The body's nervous system anticipatory activity is most likely coming from the solar wind. This is because after above average solar activity has taken place, the solar wind takes several days to reach Earth's magnetosphere, which in turn causes the above average geomagnetic activity to take place. The frequencies emitted during this time is dependent upon the magnetic field strength, magnetic field lines, density and speed of the sun's solar wind.

Research has also shown that an increase in the field-line resonances affect the body's cardiovascular system, possibly due to the frequencies overlapping with the rhythms that take place in the body's autonomic nervous and cardiovascular systems [27].

Studies conducted by Dimitrova and Khabarova found that ULF waves between 2 and 10mHZ caused increases in blood pressure (0.6) in comparison to geomagnetic measures (0.3) [28]. (ULF) waves may also be affecting health and/or physiological functioning. The most common are field-line resonances. These exhibit large amplitudes of magnetic waves taking place in earth's magnetosphere [26]. What is even more interesting is that Zenchenko et al. found in two-thirds of their experiments that a **synchronization between the body's heart rhythms** and the ultra-low frequency (0.5 to 3.0 mHz) of the geomagnetic field took place [29].

The Schumann resonance and its Influence on Human Brainwaves

During the late 1950s, Koenig and Schumann mapped out frequencies consistent with a mathematical that predicted an earth-ionospheric cavity resonance [30]. The fist Schumann resonance (SR) frequency is approximately 7.83 Hz. It varies (day/night) between + or - 0.5 Hz. Other SR frequencies include ~14, 20, 26, 33, 39, and 45 Hz respectively. These **frequencies closely overlap human brainwaves**, such as the **alpha** (8–12 Hz), the beta (12–30 Hz), and the **gamma** (30–100 Hz) brainwave frequencies. This close similarity between brainwave frequencies and the SRs and a tendency of the electroencephalogram rhythms becoming synchronous with the Schuman resonance was reported by Koenig [31].

Research by Pobachenko et al. [32] looked at the brainwaves of 15 individuals and the Schuman resonance frequencies for six weeks and discovered that variations in their brainwaves correlated with changes in earth's Schuman Resonance. The largest correlations in brainwave activity took place during periods of higher magnetic activity and also showed major changes in the Schuman resonance.

Research by Persinger et al. looked at brainwaves and the Schuman resonance in real-time and found that several Schuman resonance frequencies were identified in the spectral profiles of the brain [32] [33].

Research by Pobachenko et al. [32] looked at the brainwaves of 15 individuals and the

Schuman resonance frequencies for six weeks and discovered that **variations in their brainwaves correlated with changes in earth's Schuman Resonance**. The largest correlations in brainwave activity took place during periods of higher magnetic activity which also showed major changes in the Schuman resonance. Research by Persinger et al. looked at brainwaves and the Schuman resonance in real-time and found that several Schuman resonance frequencies were identified found in the spectral profiles of the brain [32] [33].

They discovered their brainwaves consisted of repeated periods of coherence occurring with the first three Schuman resonance frequencies (7–8 Hz, 13–14 Hz, and 19–20 Hz) (*in real-time*) [34].

Summary

Changes in earth's Schuman resonance are related to solar wind speed. Hence, solar radiation is affecting brainwave activity. This includes cognition, modulations and memory consolidation [35].

References. Introduction

1 - Cernouss S., Vinogradov A., Vlassova E. Geophysical hazard for human health in the circumpolar auroral belt: Evidence of a relationship between heart rate variation and electromagnetic disturbances. Nat. Hazards. 2001;23:121–135. doi: 10.1023/A:1011108723374. [Cross Ref]

2 - Cornélissen G., Halberg F., Breus T., Syutkina E.V., Baevsky R., Weydahl A., Watanabe Y., Otsuka K., Siegelova J., Fiser B. Non-photic solar associations of heart rate variability and myocardial infarction. J. Atmos. Sol. Terr. Phys. 2002;64:707–720. doi: 10.1016/S1364-6826(02)00032-9. [Cross Ref]

3 - Watanabe Y., Cornelissen G., Halberg F., Otsuka K., Ohkawa S.I. Associations by signatures and coherences between the human circulation and helio- and geomagnetic activity. Biomed. Pharmacother. 2001;55(Suppl. S1):76s–83s. doi: 10.1016/S0753-3322(01)90008-3. [PubMed] [Cross Ref]

4 - Dimitrova S., Angelov I., Petrova E. Solar and geomagnetic activity effects on heart rate variability. Nat. Hazards. 2013;69:25–37. doi: 10.1007/s11069-013-0686-y. [Cross Ref]

5 - Otsuka K., Cornelissen G., Weydahl A., Holmeslet B., Hansen T.L., Shinagawa M., Kubo Y., Nishimura Y., Omori K., Yano S., et al. Geomagnetic

disturbance associated with decrease in heart rate variability in a subarctic area. Biomed. Pharmacother. 2001;55(Suppl. S1):51s–56s. doi: 10.1016/S0753-3322(01)90005-8. [PubMed] [Cross Ref]

6 - Otsuka K., Ichimaru Y., Cornelissen G., Weydahl A., Holmeslet B., Schwartzkopff O., Halberg F. Dynamic Analysis of Heart Rate Variability from 7-Day Holter Recordings Associated with Geomagnetic Activity in Subarctic Area. IEEE; Piscataway, NJ, USA: 2000. pp. 453–456. Computers in Cardiology 2000.

7 - Otsuka K., Yamanaka T., Cornelissen G., Breus T., Chibisov S., Baevsky R., Halberg F., Siegelova J., Fiser B. Altered chronome of heart rate variability during span of high magnetic activity. Scr. Med. (Brno) 2000;73:111–116.

8 - Gmitrov J., Ohkubo C. Geomagnetic field decreases cardiovascular variability. Electromagn. Magnetobiol. 1999;18:291–303. doi: 10.3109/15368379909022585. [Cross Ref]

9 - Breus T.K., Baevskii R.M., Chernikova A.G. Effects of geomagnetic disturbances on humans functional state in space flight. J. Biomed. Sci. Eng. 2012;5:341–355. doi: 10.4236/jbise.2012.56044. [Cross Ref]

10 - Baevsky R., Petrov V., Cornelissen G., Halberg F., Orth-Gomer K., Akerstedt T., Otsuka K., Breus T.,

Siegelova J., Dusek J. Meta-analyzed heart rate variability, exposure to geomagnetic storms, and the risk of ischemic heart disease. Scr. Med. 1997;70:201–206. [PubMed]

11- Otsuka K., Cornelissen G., Weydahl A., Holmeslet B., Hansen T.L., Shinagawa M., Kubo Y., Nishimura Y., Omori K., Yano S., et al. Geomagnetic disturbance associated with decrease in heart rate variability in a subarctic area. Biomed. Pharmacother. 2001;55(Suppl. S1):51s–56s. doi: 10.1016/S0753-3322(01)90005-8. [PubMed] [Cross Ref]

12- Otsuka K., Ichimaru Y., Cornelissen G., Weydahl A., Holmeslet B., Schwartzkopff O., Halberg F. Dynamic Analysis of Heart Rate Variability from 7-Day Holter Recordings Associated with Geomagnetic Activity in Subarctic Area. IEEE; Piscataway, NJ, USA: 2000. pp. 453–456. Computers in Cardiology 2000.

13 - Otsuka K., Yamanaka T., Cornelissen G., Breus T., Chibisov S., Baevsky R., Halberg F., Siegelova J., Fiser B. Altered chronome of heart rate variability during span of high magnetic activity. Scr. Med. (Brno) 2000;73:111–116.

14 - Oinuma S., Kubo Y., Otsuka K., Yamanaka T., Murakami S., Matsuoka O., Ohkawa S., Cornelissen G., Weydahl A., Holmeslet B. Graded response of heart rate variability, associated with an alteration of geomagnetic activity in a subarctic area.

Biomed. Pharmacother. 2002;56:284–288. doi: 10.1016/S0753-3322(02)00303-7. [PubMed] [Cross Ref]

15- Tsuji H., Larson M.G., Venditti F.J., Jr., Manders E.S., Evans J.C., Feldman C.L., Levy D. Impact of reduced heart rate variability on risk for cardiac events. The Framingham heart study. Circulation. 1996;94:2850–2855. doi: 10.1161/01.CIR.94.11.2850. [PubMed] [Cross Ref]

16 - Dimitrova S., Angelov I., Petrova E. Solar and geomagnetic activity effects on heart rate variability. Nat. Hazards. 2013;69:25–37. doi: 10.1007/s11069-013-0686-y. [Cross Ref]

17 - Dimitrova S., Angelov I., Petrova E. Solar and geomagnetic activity effects on heart rate variability. Nat. Hazards. 2013;69:25–37. doi: 10.1007/s11069-013-0686-y. [Cross Ref]

18 - Khabarova O., Dimitrova S. On the nature of people's reaction to space weather and meteorological weather changes. Sun Geosph. 2009;4:60–71.

19 - Dmitreva I., Khabarova O., Obridko V., Ragulskaja M., Reznikov A. Experimental confirmations of bioeffective effect of magnetic storms. Astron. Astrophys. Trans. 2000;19:67–77. doi: 10.1080/10556790008241351. [Cross Ref]

20 - Khabarova O. Investigation of the tchizhevsky-velhover effect. Biophysics. 2004;49:S60.

21 -Dimitrova S., Stoilova I., Cholakov I. Influence of local geomagnetic storms on arterial blood pressure. Bioelectromagnetics. 2004;25:408–414. doi: 10.1002/bem.20009. [PubMed] [Cross Ref]

22 - Samsonov S., Sokolov V., Strekalovskaya A., Petrova P. On the Relationship of Cardiovascular Disease Exacerbation to Helio-Geophysical Disturbances; Proceedings of the XXVII Annual Seminar (Physics of Auroral Phenomena); Apatity, Russian. 2004; [(accessed on 3 July 2017)]. pp. 134–137.

23 - Azcárate T., Mendoza B., de la Peña S.S., Martínez J. Temporal variation of the arterial pressure in healthy young people and its relation to geomagnetic activity in Mexico. Adv. Space Res. 2012;50:1310–1315. doi: 10.1016/j.asr.2012.06.015. [Cross Ref]

24 - Dimitrova S., Mustafa F., Stoilova I., Babayev E., Kazimov E. Possible influence of solar extreme events and related geomagnetic disturbances on human cardio-vascular state: Results of collaborative Bulgarian–Azerbaijani studies. Adv. Space Res. 2009;43:641–648. doi: 10.1016/j.asr.2008.09.006. [Cross Ref]

25 - Khabarova O. Investigation of the tchizhevsky-velhover effect. Biophysics. 2004;49:S60.

26 - Southwood D. Some features of field line resonances in the magnetosphere. Planet. Space Sci. 1974;22:483–491. doi: 10.1016/0032-0633(74)90078-6. [Cross Ref]

27 - Kleimenova N., Kozyreva O. Daytime quasiperiodic geomagnetic pulsations during the recovery phase of the strong magnetic storm of 15 May 2005. Geomagn. Aeron. 2007;47:580–587. doi: 10.1134/S0016793207050064. [Cross Ref]

28 - Khabarova O., Dimitrova S. On the nature of people's reaction to space weather and meteorological weather changes. Sun Geosph. 2009;4:60–71.

29 - Zenchenko T., Medvedeva A., Khorseva N., Breus T. Synchronization of human heart-rate indicators and geomagnetic field variations in the frequency range of 0.5–3.0 mHz. Izv. Atmos. Ocean. Phys. 2014;50:736–744. doi: 10.1134/S0001433814040094. [Cross Ref]

30 -Schumann W., Konig H. Uber die beobachtung von "atmospherics" bei geringsten frequenzen. Die Naturwiss. 1954;41:183–184. doi: 10.1007/BF00638174. [Cross Ref]

31 - König H.L., Krueger A.P., Lang S., Sönning W. Biologic Effects of Environmental Electromagnetism. Springer; Berlin, Germany: 2012.

32 - Saroka K.S., Persinger M.A. Quantitative evidence for direct effects between earth-ionosphere Schumann resonances and human cerebral cortical activity. Int. Lett. Chem. Phys. Astron. 2014;20:166. doi: 10.18052/www.scipress.com/ILCPA.39.166. [Cross Ref]

33 - Persinger M.A., Saroka K.S. Human quantitative electroencephalographic and Schumann resonance exhibit real-time coherence of spectral power densities: Implications for interactive information processing. J. Signal Inf. Process. 2015;6:153. doi: 10.4236/jsip.2015.62015. [Cross Ref]

34 - Pobachenko S.V., Kolesnik A.G., Borodin A.S., Kalyuzhin V.V. The contigency of parameters of human encephalograms and Schumann resonance electromagnetic fields revealed in monitoring studies. Complex Syst. Biophys. 2006;51:480–483.

35 - Persinger M.A., Saroka K.S. Human quantitative electroencephalographic and Schumann resonance exhibit real-time coherence of spectral power densities: Implications for interactive information processing. J. Signal Inf. Process. 2015;6:153. doi: 10.4236/jsip.2015.62015. [Cross Ref]

Chapter 1. Solar Activity, HRV and the Nervous System

ow that we have a better picture of how earth's frequencies are interacting with our brain waves, we shall next move onto the 3 main factors that enhance the success of associative remote viewing sessions. 1 – Solar Weather. 2 – The Nervous System and 3 – Mind/Heart Coherence. These 3 are vital to successful ARV sessions (Associative Remote Viewing). Let's first take an in-depth look at how solar weather affects the nervous system.

A research study looked at the relationship between geomagnetic and solar activity and its impact on the body's human nervous system via changes taking place via HRV [1]. HRV is short for **Heart Rate Variability**. The study also looked at the intensity of the Schumann resonance taking place. The study involved 10 people over a month and found significant correlations occurred between the participant's HRV and the sun's solar wind speed, solar radio flux, Kp, Ap, cosmic ray counts, Schumann resonance power and total variations in earth's magnetic field. After looking at the data and removing the participant's circadian rhythms, it was discovered that the participants' HRV rhythms synchronized across the 31-day period with a period lasting about 2.5 days. This effect took place even though the participants were at separate locations[1].

Summary
Autonomic nervous system functioning in the body responds to changes in geomagnetic and solar weather conditions and is synchronized with time-varying magnetic fields connected with earth's Schumann resonance and geomagnetic field-line resonance.

Further Reading
Synchronization of Human Autonomic Nervous System Rhythms with Geomagnetic Activity in Human Subjects. Rollin Mc Craty et al. July 2917

HRV and Magnetic Storms
The study also showed that after a CME (*coronal mass ejection from the sun*) an immediate increase took place in the sun's 10.7cm radio flux. This took place at the very same time the X-class solar flare began. During this time a steep increase in the participant's HRV took place following the enhancement of the sun's 10.7cm radio flux. This then rapidly declined along with the sharp *'jump'* in the sun's solar wind speed and at the start of the severe magnetic storm.

The Sun's 10.7cm Solar Radio Flux Increases the body's Parasympathetic Nervous System
This next part of the study is a major finding, because it clearly shows why our **ARV sessions are more much more accurate** when the sun's 10.7cm radio flux levels are increasing or steady. We shall cover this important detail in greater depth later on. Now let's get back to the study.

The study found that a positive correlation existed between the sun's 10.7 cm radio flux (**F10.7**) and cosmic rays for the majority of the HRV variables. This also included negative correlations between the HF/LF ratios for the first two weeks (*unsettled period*) of the study.

This is a significant finding because a previous study also discovered that an increase in the sun's 10.7cm radio flux index was associated with increased mental clarity, lower fatigue and other positive effects. However increased solar wind speeds showed the opposite effects [2]. The study lasted 5 months and found that a time lag occurred in the body's autonomic nervous system in response to changes in magnetic and solar variables [3].

One interesting observation I noted from watching the science fiction movie "**Back to the Future**" is the mad scientist "Doc Brown" states that the "*Flux Capacitor*" is what makes time travel possible. It is interesting to note that our

ARV session accuracy always improves when the sun's 10.7cm radio flux is growing or steady, but never when it is declining or flat.

The KP "Sweet Spot" and the Sun's 10.7 cm Radio Flux

As mentioned throughout this book, the KP "**sweet spot**" takes place when the Middle Latitude Fredericksburg K-indices are between 11 and 7.

Date	A	Middle Latitude Fredericksburg - K-indices							
2017 03 23	9	4	3	2	2	1	2	1	1
2017 03 24	7	2	3	2	0	2	2	2	1
2017 03 25	3	0	2	0	0	2	2	1	1
2017 03 26	3	1	0	0	1	1	1	1	2
2017 03 27	34	2	3	5	5	4	5	4	
2017 03 28	-1	5	5	3	3	3	2	3	-1

This period takes place most often a few days later after geomagnetic activity has peaked. We shall show next why the sweet spot enhances ARV accuracy, most likely due to an interaction occurring between HRV and the sun's 10.7cm radio flux. Let's get back to the study [3].

The study found that HRV was positively correlated with the sun's 10.7 cm radio flux. After an increase in parasympathetic activity, the **main**

effects began approximately 20 hours after the sun's 10.7cm radio flux increased.

Summary
Increased Parasympathetic Nervous System Activity enhances the accuracy of Associative Remote Viewing. Favorable solar weather conditions are one period where the body's Parasympathetic Nervous System Activity becomes naturally stimulated. Hence, putting the body into a relaxed state of mind, especially utilizing the right essential oils that stimulate Parasympathetic Nervous System Activity during this period give a significant boost to ARV session accuracy.

The 10.7 cm radio flux and Anticipatory Reactions
The sun's 10.7 cm radio flux activity may be a mediator of anticipatory reactions discovered by Chizhevsky and additional radiation sources such as UV emissions, X–rays and cosmic rays that are emitted by the sun emitted during coronal mass ejections. These also very likely effect the body's autonomous nervous systems and its anticipatory reactions that take place prior to changes in the sun's solar wind speed and geomagnetic disturbances.

Another long-term study looked at activity taking place during magnetically quiet days.

Strong positive correlations were found to occur between HRV variables and cosmic rays. It suggested a beneficial response as cosmic rays increased [4].

Summary
This suggests that the body's parasympathetic nervous system becomes "enhanced" when the sun's 10.7cm radio flux is increased (and possibly cosmic rays). Also decreases in the HF/LF ratio show higher c activity that is relative to sympathetic activity as shown in ambulatory HRV recordings [1].

Inflammatory Responses and Cosmic Rays
One study found that serum C-reactive protein levels in a healthy population that was suspected of suffering from inflammatory-related problems found that a strong / inverse correlation existed between their C-reactive protein levels and the numbers of cosmic rays [5].

In the first 2-week period in this study, the cosmic ray counts were weakly and negatively correlated to HRV. The study also found strong positive correlations with their HF/LF ratios and that a sharp increase in the sun's solar wind speed along with a strong reduction in cosmic rays occurred during the early period in response to moderate geomagnetic storm activity and that it had an impact on their HRV measures.

Before the geomagnetic storm began, cosmic rays started to increase as the sun's solar wind decreased. When this occurred, a strong positive correlation with cosmic rays and negative solar wind speeds showed effects in all the participant's key HRV variables. As an added note, periods of higher Schuman resonance have also been found to be beneficial to health, such as lowering blood pressure [6] and ssignificant correlations have been found to exist between Schuman resonance power HRV measures (*Synchronization of Human Autonomic Nervous System Rhythms with Geomagnetic Activity in Human Subjects. Rollin McCraty, et al. July 2017*).

Summary

Cosmic rays may be showing a stronger influence on the body's autonomic nervous system than the sun's solar radio flux. From our own research at the Solar Institute, we have found that lower cosmic rays enhance remote viewing sessions. Lower cosmic ray counts are more likely to occur as solar radiation increases, which causes more air moisture.

Magnetic Fields Influence the Human Autonomic Nervous System

Another long-term study examined the time lags in the Autonomic Nervous System and its response over a 40 hour period following changes in Schumann resonance power (**SRP**). The study found a 9 hour lag time [7]. This study shows that a clear oscillatory pattern takes place due to environmental magnetic fields influencing the autonomic nervous system as shown by HRV activity. During quiet periods of magnetic activity, the Schuman resonance power plays **important roles in synchronizing the slow wave heart rhythm in people**.

Research has found that different people have varying degrees of sensitivity to Earth's magnetic fields [8].

Summary
Consistent studies now conclusively prove changes in geomagnetic and solar activity correlate with changes in the body's nervous system. These effects are synchronized with time-varying magnetic fields that are associated with geomagnetic field-line resonances and earth's Schumann resonances.

The most likely explanation for how geomagnetic and solar fields are influencing the human nervous system is via a resonant coupling effect that takes place between the nervous system and earth's geomagnetic frequencies. These are also known as **Alfvén waves** (or ultra low frequency standing waves) which occur in earth's ionospheric resonant cavity (Schumann resonances). These frequencies overlap with physiological rhythms.

Additional Reading
Synchronization of Human Autonomic Nervous System Rhythms with Geomagnetic Activity in Human Subjects. Rollin McCraty et al. Int J Environ Res Public Health. 2017 Jul; 14(7): 770. Published online 2017 Jul 13. doi: 10.3390/ijerph14070770. PMCID: PMC5551208

Geomagnetic Storms and Heart Rate Variability
This section shows information of key importance. This is because our ARV sessions have always been found to be most accurate a few days **after a major geomagnetic storm**, which is the time that earth's geomagnetic energy enters the "**sweet spot**" of between 11 and 7.

A study found that geomagnetic disturbances in earth's magnetic field affect the neural regulation of vascular tone, heart rate variability and the body's central nervous system. The study looked at the effect of geomagnetic fluctuations and its effect on the body in outer space. The study used an analysis of heart rate variability which allowed for a precise evaluation of the body's parasympathetic and sympathetic nervous systems. The study involved 30 cosmonauts on the spaceship Soyuz (32nd orbit).

The study found geomagnetic disturbances affected their autonomic nervous systems with increased nervous system activity taking place between 1 and 2 days after a geomagnetic storm.

Researchers at the Lithuanian University of Health Sciences developed a novel method based on HRV to determine a person's sensitivity Earth's magnetic field and its variations [9]. This would make a great tool for further studies involving solar weather and its effects on the body.

Reference
Regulation of autonomic nervous system in space and magnetic storms. Baevsky RM1, Petrov VM, Chernikova AG. Adv Space Res. 1998;22(2):227-34.

Now how do we "prime" our parasympathetic nervous system to get it into good shape before an ARV session. Let's explore this next.

References. Chapter 1

1 - McCraty R., Shaffer F. Heart rate variability: New perspectives on physiological mechanisms, assessment of self-regulatory capacity, and health risk. Glob. Adv. Health Med. 2015;4:46–61. doi: 10.7453/gahmj.2014.073. [PMC free article] [PubMed] [Cross Ref]

2 - The global coherence initiative: creating a coherent planetary standing wave. McCraty R, Deyhle A, Childre D. Glob Adv Health Med. 2012 Mar; 1(1):64-77.

3 - McCraty R., Deyhle A., Childre D. The global coherence initiative: Creating a coherent planetary standing wave. Glob. Adv. Health Med. 2012;1:64–77. doi: 10.7453/gahmj.2012.1.1.013. [PMC free article] [PubMed] [Cross Ref]

4 - Alabdulgader A., McCraty R., Atkinson M., Dobyns Y., Stolc V., Ragulskis M. Long-term study of heart rate variability responses to changes in the solar and geomagnetic environment. Nat. Commun. 2017 in review.

5 - Stoupel E., Abramson E., Israelevich P., Sulkes J., Harell D. Dynamics of serum c-reactive protein (CRP) level and cosmophysical activity. Eur. J. Int. Med. 2007;18:124–128. doi: 10.1016/j.ejim.2006.09.010. [PubMed] [Cross Ref]

6 - Mitsutake G., Otsuka K., Hayakawa M., Sekiguchi M., Cornélissen G., Halberg F. Does Schumann resonance affect our blood pressure? Biomed. Pharmacother. 2005;59:S10–S14. doi: 10.1016/S0753-3322(05)80003-4. [PMC free article] [PubMed] [Cross Ref]

7 - Alabdulgader A., McCraty R., Atkinson M., Dobyns Y., Stolc V., Ragulskis M. Long-term study of heart rate variability responses to changes in the solar and geomagnetic environment. Nat. Commun. 2017 in review.

8 - Khabarova O., Dimitrova S. On the nature of people's reaction to space weather and meteorological weather changes. Sun Geosph. 2009;4:60–71.

9 - Alabdulgader A., Maccraty R., Atkinson M., Vainoras A., Berškiene K., Mauriciene V., Navickas Z., Šmidtaite R., Landauskas M., Daunoraviciene A. Human heart rhythm sensitivity to earth local magnetic field fluctuations. J. Vibroeng. 2015;17:3271–3278.

Further Reading
Regulation of autonomic nervous system in space and magnetic storms. Baevsky RM1, Petrov VM, Chernikova AG. Adv Space Res. 1998;22(2):227-34.

Synchronization of Human Autonomic Nervous System Rhythms with Geomagnetic Activity in Human Subjects. Rollin McCraty et al. July 2017

Chapter 2. Essential Oils for a Healthy Parasympathetic Nervous System

This chapter will consist of a quick review of essential oils that specifically target the parasympathetic nervous system, strengthening it and enhancing it for successful ARV sessions.

Essential Oils and using the Heart Math EM meter are two great tools that greatly enhance the clarity of associative remote viewing sessions. Speaking from years of working with essential oils, most notably lavender, we have found using lavender in the late afternoon before an ARV session seems to help relax and calm the body before an ARV session.

The parameters and functioning of the Parasympathetic Nervous System strongly relate to that of successful associative remote viewing sessions, in that the mind and body are in a relaxed state of mind when the associative remote viewing session is being conducted. Hence in summary, a relaxed and alert Parasympathetic Nervous System adequately prepared before an ARV session greatly enhances the accuracy of Associative Remote Viewing Sessions. Now let's explore essential oils and their effects on the body's nervous system in greater detail.

Meniki and Hinoki Increase Parasympathetic Nervous System Activity

The essential oils of Hinoki (*C. obtusa*) and Meniki (*Chamecyparis formosensis*) are precious conifers that have wood properties and are used in furniture interiors in Taiwan. Both have a wood type aroma. A research study identified 36 compounds in Meniki essential oils, including linalyl acetate, cadinol, a-muurolene, cadinene, calamenene and myrtenol.

Twenty-nine compounds were found in Hinoki essential oil, including limonene, a-**terpineol**, a-pinene, cadinene, borneol and terpinolene.

The study next examined the effects of Hinoki and Meniki essential oils on the autonomic nervous system. The study found that after participant's inhaled the Meniki essential oil that their heart rate (HR) and systolic blood pressure decreased and their diastolic blood pressures increased. The study also found that their Sympathetic Nervous System Activity (SNS) was significantly decreased and that their Parasympathetic Nervous System Activity (PSNS) significantly increased.

Also after participants inhaled the Hinoki essential oil, their heart rate, systolic blood pressure and PSNS decreased. However their Sympathetic Nervous System Activity (SNA) increased. The study found that both the Hinoki and Meniki wood essential oils stimulated pleasant moods and strongly suggest these oils could be suitable agents for dysfunctions in the body's sympathetic nervous system.

This is a major finding because our **TXP formula**, which we shall cover in greater detail later on, utilizes limonene. Hence limonene may be increasing Parasympathetic Nervous System Activity.

Reference
Effect of Hinoki and Meniki Essential Oils on Human Autonomic Nervous System Activity and Mood States.Chen CJ, Kumar KJ, Chen YT, Tsao NW, Chien SC, Chang ST, Chu FH, Wang SY. Nat Prod Commun. 2015 Jul;10(7):1305-8.

Lavender's Effect on the Parasympathetic Nervous System
A research study with women volunteers examined if inhaling lavender essential oil could influence heart rate variability (HRV) and cause changes in their autonomic nervous systems. The women inhaled the lavender at 10, 20 and 30 minute intervals. The study found that increases in their parasympathetic nervous systems took place after the lavender was inhaled. Regions that were more active after breathing in the

lavender aroma included the **thalamus**, brainstem and cerebellum and reductions were seen in their frontal eye field (*a region located in the frontal cortex*).

Reference
Autonomic nervous function and localization of cerebral activity during lavender aromatic immersion. Duan X et al. 2007.

Research by Duan et al. found that after inhaling lavender a significant increase in both LF and HF of the Nervous System and that people watching horror movies showed reduced anxiety after inhaling lavender.

Reference
Duan, M. Tashiro, DI. Wu et al., "Autonomic nervous function and localization of cerebral activity during lavender aromatic immersion," Technology and Health Care, vol. 15, no. 2, pp. 69–78, 2007.

Further **Reading**
Autonomic nervous function and localization of cerebral activity during lavender aromatic immersion. Duan X1, Tashiro M, Wu D, Yambe T, Wang Q, Sasaki T, Kumagai K, Luo Y, Nitta S, Itoh M.

Juniper Essential Oil and the Parasympathetic Nervous System

A research study examining the effects of persons inhaling juniper essential oil and its effects upon the autonomic nervous system also looked at their blood pressure and heart rate variability (HRV).

The study found that their blood pressure decreased and that their parasympathetic nervous system activity increased while inhaling juniper essential oil as reflected via high frequency (HF). The study also showed that sympathetic nervous system activity was decreased by the juniper essential oil.

Reference
Effects of Juniper Essential Oil on the Activity of Autonomic Nervous System. Jong-Seong Park. Department of Physiology, Chonnam National University Medical School, Gwangju 61469, Korea.

Rose and Patchouli Essential Oils reduce Sympathetic Nervous System Activity

Rose or Patchouli essential oils have been shown to reduce sympathetic activity by up to 40%

Reference
Effects of fragrance inhalation on sympathetic activity in normal adults. Haze S, Sakai K, Gozu Y. Jpn J Pharmacol. 2002 Nov; 90(3):247-53.

Bergamot Stimulates the Parasympathetic Nervous System

Elementary school teachers that underwent two 10-minute aromatherapy sprays of Bergamot essential oil showed an enhanced parasympathetic nervous system.

Reference
Aromatherapy Benefits Autonomic Nervous System Regulation for Elementary School Faculty in Taiwan. Kang-Ming Chang and Chuh-Wei Shen. April 2011

Pepper, Estragon, Fennel and Grapefruit Increase Sympathetic Nervous System Activity

Nagai et al. looked at the effects of the essential oils of fennel, pepper, estragon or grapefruit and its effects upon the Sympathetic Nervous System. The study found that pepper, estragon, fennel or grapefruit essential oils increased Sympathetic Nervous System Activity compared to an odorless solvent (triethyl citrate) (*Pleasant odors attenuate the blood pressure increase during rhythmic handgrip in humans. Nagai M, Wada M, Usui N, Tanaka A, Hasebe Y. Neurosci Lett. 2000 Aug 11; 289(3):227-9*).

Further Reading
Influence of Fragrances on Human Psychophysiological Activity: With Special

Reference to Human Electroencephalographic Response. Kandhasamy Sowndhararajan and Songmun Kim. Helmut Viernstein. Technol Health Care. 2007;15(2):69-78. Sci Pharm. 2016; 84(4): 724–752. Published online 2016 Nov 29. doi: 10.3390/scipharm84040724. PMCID: PMC5198031

Summary
Essential oils delivered via aromatherapy shift the body more towards Parasympathetic Nervous System dominance via an increase in HF and HRV.

Final Conclusion / Summary
Favorable solar weather conditions are enhancing the strength of the heart and enhancing its ability to go into coherence most likely with the earth and the Schuman resonance.
   Favorable solar weather conditions also exhibit stronger Schuman resonance power. Specific essential oils affect heart rate variability (HRV) and the body's parasympathetic nervous system. Specific essential oils inhaled or placed upon the skin during these favorable solar weather conditions while performing HRV exercises such as heart math before or during an ARV session greatly enhances the accuracy and success of ARV sessions. Specific metals/minerals also enhance clarity during these times, as well as specific foods taken up to 24 hours before the

ARV session (*which we shall cover later on*). It may also be that these foods and essential oils target specific genes which enhance ARV sessions.

Anxiety During Remote Viewing
One of the major blocks to a successful remote viewing sessions is anxiety, which can simply be alleviated with lavender essential oil or other essential oils that relieve anxiety (*which we shall cover later on*). Also mindfulness exercises and meditations can also reduce anxiety.

Anxiety, St. John's Wort and Valerian
When St. John's Wort is combined with Valerian it greatly enhances the effects. A scientific research study found that this combination significantly reduced anxiety and that greater reductions in anxiety were seen with higher doses of valerian. The doses of St. John's Wort remained constant between the groups being treated, suggesting valerian has more of an effect on anxiety symptoms (*Treating depression comorbid with anxiety--results of an open, practice-oriented study with St John's wort WS 5572 and valerian extract in high doses. Müller D, Pfeil T, von den Driesch V. Phytomedicine. 2003; 10 Suppl 4():25-30*).

We shall cover additional essential oils and their effects on the nervous system in greater detail later on, but first let's take a look at the effects of the moon.

Chapter 3. Lunar Rhythms and Remote Viewing

A Summary of the Full Moon and Its Effects

Years of our associative remote viewing research found that our sessions were most accurate around the full moon. Further research by us found that the reason for this was due to enhanced water moisture occurring in earth's atmosphere which enhances the coherence occurring in the brain's microtubules. Let's look at a summarized picture of data involving moon phase and the absorption of water in various items.

Research conducted by Harold Burr at Yale recorded electrical activity in tree trunks for 9 years. There were no changes occurring in the tree's electrical activity when changes in humidity, atmospheric pressure or weather occurred. The only time a change occurred was when there was a full and new Moon (*Burr, H. S.. 'Diurnal Potentials in the Maple Tree, Yale Journal of Biology & Medicine, 1945,17,727-34*). The rhythm was also found to change during weaker sunspots (*Burr, H. S., The Fields of Life: Our Links with the Universe NY. 1973*).

Experiments by Lily Kolisko showed germination and first plant shoots were strongest in the days before the full moon with new moons showing the slowest responses (*Kolisko, L.. The*

Moon and the growth of Plants Anthroposophical Press, London, 1938, 1975) M. Maw of Canada's Department of Agriculture also came to the same conclusions during his studies (*Maw. M. G.. 'Periodicities in the Influences of Air Ions on the Growth of Garden Cress. Lepidium Sativum L.', Canadian Journal of Plant Science. 1967, 47, 499-505*).

Plant water absorption, germination metabolism and fertility have all been found to respond to the lunar/synodic cycle (*Graviou, E.. 'Analogies between Rhythm, in Plant Material in Atmospheric Pressure and Solar-Lunar Periodicities'. International Journal of Biometeorology, 1978. Vol. 22, No.2.*).

Researchers at North-western University in Illinois found that there was a 35% higher water absorption in beans just before the full Moon (*Brown, F., & Chow, C., 'Lunar-correlated Variations in water uptake by Bean seeds', Biological Bulletin, Oct., 1973, 145, 265-278*). This same pattern, including germination, was also found to take place in studies conducted by Dr. Jane Panzer of Tulane University. Her studies also found that if the pinto beans had been sterilized or pasteurized, that the effects were significantly reduced (*Panzer. J. J., 'Lunar Correlated Variations in Water Uptake and Germination in 3 Species of Seeds. U. of Tulane. 1976*).

One of the more interesting experiments involving over 1 million hours of potato oxygen-absorption was conducted by Professor Frank

Brown. Potatoes were kept in dark and their water moisture measured. The study found that water moisture peaked when the Moon was rising and at zenith (directly overhead) (*Panzer. J. J., 'Lunar Correlated Variations in Water Uptake and Germination in 3 Species of Seeds', PhD. U. of Tulane. 1976*).

Research by T.M Lai found that phosphorus and potassium absorption in the roots of corn seedlings **showed maximum absorption during full Moons** and minimal water absorption during new Moons. The potassium (flowers mainly and alkali) showed minimal absorption during full Moons and maximum absorption during new Moons (*Lai. T. M., 'Phosphorus and Potassium uptake by plants Relating to Moon Phases'. Biodynamics (US), Summer. 1976.*) As an interesting side not, Beehive traffic has been found to increase 100% during new moons, compared to a full moon (*Oehmke, M.G., 'Lunar Periodicity in Flight Activity of Honey Bees'. M Oehmke. Franfurt Germany. Journal of Interdisciplinary Cycles Research. 1973.4. 319-335*).

E. Graviou at Lyons University found dormant tomato seeds showed changes during new and full moons and tomato plants are one of the few plants that contain nicotine (*Graviou, E.. 'Analogies between Rhythm, in Plant Material in Atmospheric Pressure and Solar-Lunar Periodicities'. International Journal of Biometeorology, 1978. Vol. 22, No.2*).

Biodynamic gardening research by

Kollerstrom found that planting during full moons using unrotted organic manure or mineral fertilizers throughout the years brought higher yields. Moon phases have also been found to cause changes in Potato DNA according to research at the University of Paris. (*Rossignol M. et al.. 'Lunar Cycle and Nuclear DNA variations in Potato callus'. Geocosmic Relations (Ed Tomassen. Pudoc, Netherlands 1990). 116-126*).

Gout Attacks and the Full Moon
Gout attacks occur more often during the full and new moons (*Gout attacks and lunar cycle. Mikulecký M1, Rovenský J. July 2000*). It is interesting to note here that the microtubule-disabling drug known as colchicine treats gout due to its ability to immobilize neutrophil cells which are responsible for painful inflammation in joints. Colchicine happens to induce microtubule polymerization. Microtubules are nano-sized objects with a hexagon shape inside the neurons of our brain. We shall cover the relationship between microtubules and remote viewing in greater depth in a latter chapter.

Insect Flight and the Lunar Cycle
Using a light trap, Hartland-Row, Hora (1927), (1955) and Corbet (1958) discovered that some Trichoptera, Ephemeroptera and Diptera (Chironomidae) exhibited a cycle taking place at 2

to 5 or 23–26 days before or after a new moon (or within 5 days full moons (moon age 9–19 days) (*Lunar Periodicity of Insect Flight and Migration. W. Danthanarayana*).

Strokes and the Moon's First Quarter
Strokes have been shown to occur more often during the moon's first quarter (*Moonstroke": Lunar patterns of stroke occurrence combined with circadian and seasonal rhythmicity—A hospital based study. Yiting Mao et al. April 2015*). Full moons have been shown to shorten the length of a hospital stay up to 4 days (*The influence of seasons and lunar cycle on hospital outcomes following ascending aortic dissection repair. J. H. Shuhaiber et al. 2013*) Surgery done during a full moon or wanning moon shows better survival rates Impact of lunar cycle on heart rate variability (RR interval) (*Indian Journal of Basic and Applied Medical Research; December 2015. A. Ahamed Basha et al*).

Now that we have a better picture of how essential oils and lunar cycles affect us, let's take a look at brainwave rhythms and what state the brain is in **BEFORE** it makes a successful basketball throw or other sporting activity.

Chapter 4. Alpha Brain Waves and Performance

Research studies are beginning to reveal that alpha and theta brainwaves play a role in remote viewing and associative remote viewing. If this were indeed true, the brain would exhibit more of these brainwaves in sports involving precognition. For example, before making a successful basketball throw or golf shot. Let's take a look at the data.

Alpha Brainwave Activity during Air Pistol Shooting, Basketball free-throws and Golf Shots

A research study looking at alpha and theta brainwaves during air-pistol shots was conducted. The study involved 10 professional shooters that had extensive international experience. Brainwave monitoring equipment was set up to monitor the volunteer's brainwaves three seconds before making their air-pistol shots.

The brainwave measuring equipment looked for low alpha, high alpha and theta brainwave frequencies. Each volunteer executed 120 air-pistol shots. The study found that lower alpha band brainwaves (8–10 Hz) were associated optimal-automatic performance and that theta brainwaves were associated with optimal-controlled performance (more focus). Lower alpha band brainwaves include attentional processes and general task demands. Examples include vigilance and arousal. Upper alpha band (10–12 Hz) is associated with semantic performance (*Klimesch, 1999*) as well as task-related attention (*Klimesch, 2012*). Also theta brainwaves are thought to be linked to error monitoring *(Cavanagh, Cohen & Allen, 2009; Gevins & Smith, 2000; Luu, Tucker & Makeig, 2004; Trujillo & Allen, 2007; Yordanova et al., 2004)*. Studies on brainwave activity on sports performers making critical play shots have also found that anxiety was responsible for the majority of missed shots. Below is a picture of the volunteer's theta brainwave activity (4hz to 7hz) three seconds **before** the athletes made their shots.

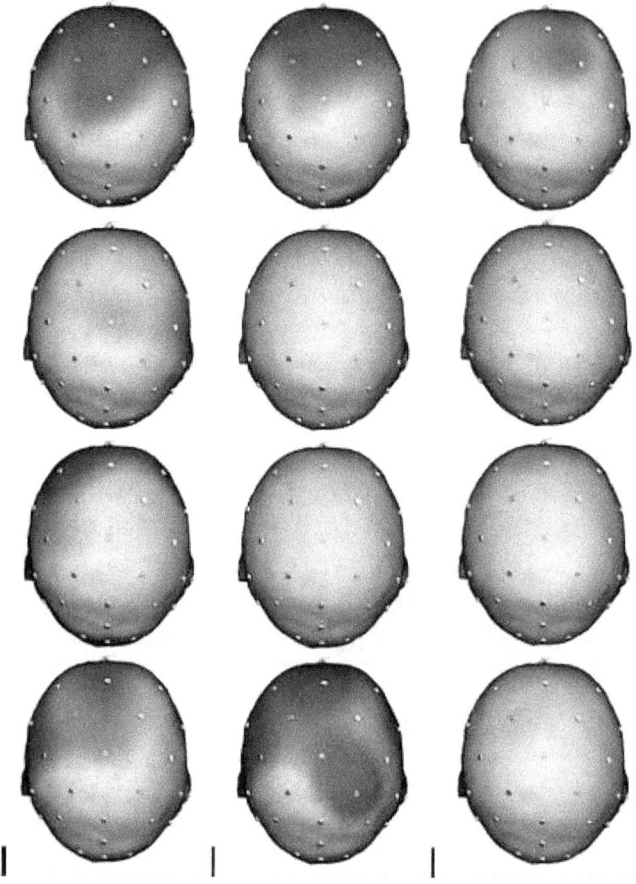

The next image shows low alpha brainwave activity (below 10hz) before the shots. Low alpha is more common in older adults.

Improve your Remote Viewing Accuracy Techniques
using Quantum Microtubules

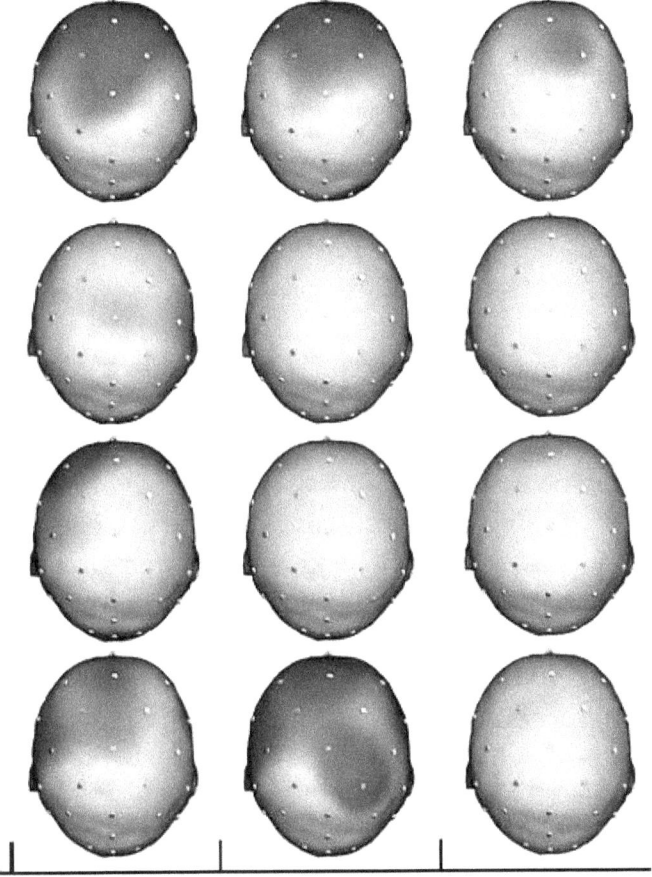

Reference
Proficient brain for optimal performance: the MAP model perspective. Maurizio Bertollo et al. May 2016.

Further Reading

The possible meaning of the upper and lower alpha frequency ranges for cognitive and creative tasks. Petsche H. et al. June 1997.

Another study looked at brainwave activity in athletes as they prepared to throw a basketball into a hoop.

Fifteen professional basketball players performed free throws while their brainwaves were measured 2 seconds **before** they threw the basketball. The study found the following: Theta (4 to 6Hz) and also upper theta (6 to 8Hz) brainwave activity was observed between successful and unsuccessful throws. Also the study found that brainwave patterns at the midline and right side frontal cortex fluctuated significantly when the athlete was going through his/her preparing process and then made an unsuccessful throw, compared to a successful throw.

Reference

The differences in frontal midline theta power between successful and unsuccessful basketball free throws of elite basketball players. L.Y. Chuang et al. Int J Psychophysiol. 2013 Dec;90(3):321-8.doi: 10.1016/j.ijpsycho.2013.10.002. Epub 2013 Oct 11.

Further Reading
Electroencephalogram Mapping of Brain States. 1stGerman Torres et al. Dec 2014

Professional Golfers and Alpha Brainwave States
Baumeister et al. (2008) compared novice and expert performance during golf putting. He found that the pro-golfers did better than novices overall and that they showed an increase in their upper alpha brainwaves which was accompanied with an increase in theta brainwaves.

The increase in theta power for the experts was stated as being in a more focused state of mind which was associated with enhanced attention. Alpha feelings were associated with upper alpha brainwaves and were interpreted as reflecting better inhibition of irrelevant sensory information.

The study also found an intra-hemispheric coherence occurring in low frequency alpha (8–10 Hz) between the brain's frontal and parietal central sites in both of the brain's hemispheres. This effect was enhanced during successful putts.

Alpha-Theta (A/T) Training
Alpha-Theta Training, also called A/T is used to reduce anxiety and encourage relaxation. It requires the person to raise their

levels of theta over levels of alpha (Peniston and Kulkosky, 1991; Raymond et al., 2005). It can also occur via the person raising their levels of both theta and alpha at the same time (Peniston and Kulkosky, 1989).

Sensory Motor Rhythm
Sensory Motor Rhythm training (SMR: see Kober et al., 2014) is used by individuals to reduce their motor interference and to enhance their cognitive performance. The participant is required to raise their SMR (12–15 Hz) levels while they control their beta levels.

Lavender essential oil and Brainwaves
Lavender essential oil has been shown to increase the power of both theta (4-8 Hz) and alpha (8-13 Hz) brain waves in the central area and bilateral temporal regions of the brain *(The effects of lavender oil inhalation on emotional states, autonomic nervous system, and brain electrical activity. W. Sayorwan et al. April 2012)*.

Further Reading
Further evidence of the possibility of exploiting anticipatory physiological signals to assist implicit intuition of random events. Journal of Scientific Exploration. Tressoldi PE, Martinelli M, Scartezzini L, Massaccesi S. 2010; 24(3):411

Implicit intuition: how heart rate can contribute to prediction of future events. Journal of the Society for Psychical research. Tressoldi PE, Massimiliano M, Zaccaria E, Massaccesi S. 2009;73:i–16

Physiological correlates of ESP: heart rate differences between targets and nontargets. J Parapsychol. Sartori L, Massaccessi S, Martinelli M, Tressoldi PE. 2004;68(2):351

Skin conductance prestimulus response: Analyses, artifacts and a pilot study. Journal of Scientific Exploration, in press; 2003. Spottiswood J, May E.

Anomalous anticipatory skin conductance response to acoustic stimuli. May EC, Paulinyi T, Vassy Z. Altern J Complement Med. 2005;11(4):587–8

Feeling the future: experimental evidence for anomalous retroactive influences on cognition and affect. Bem DJ. J Pers Soc Psychol. 2011;100(3):407–25

Predictive physiological anticipation preceding seemingly unpredictable stimuli: A meta-analysis. Front Psychol. Mossbridge J, Tressoldi P E, Utts J. 2012;3:390.

Unconscious perception of future emotions: An experiment in presentiment. Radin DI. Journal of Scientific Exploration. 1997;11(2):163–80

Electrophysiological evidence of intuition: Part 1. The surprising role of the heart. Journal of Alternative and Complementary Medicine. McCraty R, Atkinson M, Bradley RT. 2004;10(1):133–43 Myers DG.

Intuition: its powers and perils. New Haven, Connecticut: Yale University Press; 2002

Bradley RT, Murray G, McCraty R, Atkinson M. Nonlocal intuition in entrepreneurs and non-entrepreneurs: results of two experiments using electrophysiological measures. Int J Entrepreneurship Small Business. 2011;12(3):343–72

Nicotine and Precognition
We left off in book 2 of our Associative Remote Viewing series with the possibility that Nicotine may enhance intuition and / or precognition.
 For any of you who have watched the Science Fiction Television Series **'Time Trax'**, the very first episode shows the evil scientist who smoked a cigarette (nicotine) before stepping into the chamber that transported his body into

the past. The cigarette contained the substance "TXP", which allowed him to travel through time. It was after watching this episode that I wondered "*could nicotine be enhancing the brain's precognitive abilities?*". In this book we shall clearly show the following (*references will be shown later in this book*)-

- Nicotine enhances right brain functioning

- The left side of the brain, but not the right side emits photons and the left side of the brain rules logic and the right of the brain rules intuition.

- Low levels of nicotine activate the left side of the brain, whereas large doses create more activity on the right side of the brain.

- **Smoking tobacco produces dominant alpha brain waves**. Alpha waves enhance the brain's ability to handle emotion. Emotional control is a key part for successful associative remote viewing.

- Brief exposure to low levels of nicotine boosts the brain's 'reward' system and blocks the system that limits reward duration. Dopamine due to its reward inducing effects, plays a major role in intuition and precognitive behavior. Research shows that after taking L-DOPA, a substance that enhances dopamine, a significant increase alpha brainwaves occurs (*Alpha and beta EEG power reflects L-dopa acute administration in parkinsonian patients Jean-Marc Melgari, et al. Nov 2014*).

Another interesting fact is the science fiction movie **Millennium**, staring Kris Kristofferson and Cheryl Ladd, portrays Cheryl Ladd, who is a time traveler from the future. In the movie she is seen constantly smoking and is told by Bill Smith (Kris Kristofferson) that she smokes way too much and should consider quitting.

We shall show clearly in this book listing numerous researched peer reviewed published studies that nicotine may be enhancing remote viewing due to its effects on the brain's neurotransmitters. This is due to the fact that nicotine causes a release of glutamate which facilitates the release of dopamine *(Nicotine Addiction Neal L. Benowitz, M.D. August 2010)* and dopamine as we covered earlier enhances anticipatory effects.

The Hippocampus and Nicotine

The hippocampus region of the brain is related to emotion. Theta brainwaves have been found to be prominent in the hippocampus in mice during learning and memory retrieval (*Buszaki G (2006). Rhythms of the brain. Oxford University Press*) as well as humans (*Traveling Theta Waves in the Human Hippocampus Honghui Zhang and Joshua Jacobs. Sept 2015*).

Research has shown that people who are telepathic or that are distant healers possess the ability to activate on demand specific brain regions that are related to the brain's empathy circuit (the ability to infer and share the emotional experiences of others). People who have the ability to interpret the thoughts of others show cognitive empathy and that that the hippocampus region of the brain plays an important role in empathy *(The role of shared neural activations, mirror neurons, and morality in empathy – A critical comment. Claus Lamma,and Jasminka Majdandžića. Jan 2015), (Empathy and motivation for justice: Cognitive empathy and concern, but not emotional empathy, predict sensitivity to injustice for others. Jean Decety and Keith J. Yoder. April 2015).*

MRI tests conducted on psychic Mr. Sean Harribance showed that functional changes occurred in the hippocampus formation regions of his brain with large amounts of activity occurring in the right posterior cortical and hippocampus regions of the brain during experiments. We shall go into more detail about this later on in the book.

Scott Rauvers

Interesting Facts about the Hippocampus

- **Nicotine has been shown to enhance neuron activity** in the hippocampus regions in mice studies.

- Short-term administration of omega 3 fatty acids from fish oil increases transthyretin transcription in old rat hippocampus and that **gene expression of TTR** is similar to Ginkgo balboa extract.

- The hippocampus region of the brain is the part of the brain that is most **sensitive to light**, especially blue light and the brain's dopamine-mediated reward system is dependent upon the brain's hippocampus.

- A combination of vitamin B12 and Omega 3's increases neurotransmission in the hippocampus region in the brain's of mice.

- Linoleic acid enhances hippocampus synaptic transmission.

> Theta brainwaves in humans occur in the hippocampus while the brain is awake and also during REM sleep *(Moroni et al., 2007; Lega et al., 2012).*

> Plants use nicotine to produce NADP (which is short for nicotinic acid adenine dinucleotidephosphate). NADP breaks apart the water molecules during plant photosynthesis. This may mean that nicotine exhibits quantum effects due to photosynthesis exhibiting quantum effects.

Photosynthesis and Quantum Biology
The majority of plant life on earth takes in the energy it needs through one of two processes. Plants, some bacteria, and certain other organisms collect energy from sunlight through a process called photosynthesis. The light is converted to water and the carbon dioxide into more complex and energetic molecules called hydrocarbons, thus storing the energy so that it can be recovered later by breaking down the molecules through a process called oxidation.

Quantum Photosynthesis and the Human Heart

Part of the purpose of quantum effects that take place during photosynthesis is for it to **move the energy of sunlight into the plant's reaction center** in order to create energy. This same process may be taking place in our body during heart math with our heart acting as the reaction center and the brain's microtubules exhibiting the quantum effects.

Why photosynthesis in a remote viewing book?

We include a through explanation of photosynthesis in this edition due to the fact that photosynthesis exhibits quantum effects, which is key to associative remote viewing. Hence, if the process of photosynthesis is increased in the conditions of polarized light (especially right circularly polarized light) then the first quarter moon, which exhibits polarized light is a prime time for remote viewing (*The Effect of Circularly Polarized Light on the Photosynthesis and Chlorophyll a Synthesis of Certain Marine Algae. G. C. McLEOD. Oct 1957*). Hence an environment that is conductive to a clear communication link to the quantum realm may be responsible for enhancing the accuracy of associative remote viewing.

Further Reading
Focus: the quantum dimension of photosynthesis. February 13, 2009. Phys. rev. focus 23, 5.

Microtubules and Consciousness
A link exists between moisture, the brain's microtubules and the phase of the moon. Later in this edition, we shall clearly show that during the period from the moon's first quarter to just after the full moon earth's barometric air pressure is slightly higher and more water moisture exists in the air. This increased water moisture enhances the coherent resonation occurring in the brain's microtubules which exhibits quantum effects. Associative remote viewing conducted during this period, along with favorable solar weather conditions, intensifies earth's Schuman resonance. Specific gems and metals assembled in the correct order greatly enhance the clarity of associative remote viewing sessions.

Additional factors contributing to enhanced associative remote viewing accuracy is remote viewing at the LST peak hours (*Apparent Association Between Effect size In Free Response Anomalous cognition Experiments And Local sidereal Times. James P. Spottiswoode Cognitive Sciences Laboratory*) and being in heart math coherence (*Electrophysiology of Intuition: Pre-stimulus Responses in Group and Individual Participants Using* a Roulette Paradigm. Rollin Mc. Craty and Mike Atkinson. March 2014). It is also a fact that photosynthesis, besides its exhibiting quantum effects, that water moisture enhances Photosynthesis (*Responses of photosynthetic capacity to soil moisture gradient in perennial rhizome grass and perennial bunchgrass. Zhenzhu Xu and Guangsheng Zhou. January 2011*).

Further Reading
Electrophysiological Evidence of Intuition: Part 1. The Surprising Role of the Heart. Rollin Mc. Craty. Journal of Alternative and Complementary Medicine 2004

Summary
By utilizing exercises such as Heart Math in cohort with the right lunar phases, one can tap into their intuitive abilities with much greater ease and with far greater accuracy.

Water Moisture and Intuition
Just as the Schumann resonance varies with solar activity, the quantum realm may also do so, showing stronger intensity during favorable solar weather conditions.

Lithium and Moisture
Materials such as lithium absorb the new

moisture generated from first quarter moon to full via a capillary type action. Lithium has been found to stabilize microtubules and protect them against damage (*Stabilization of microtubules by lithium ion. Open overlay panel B. Bhattacharyya, J. Wolff. November 1976*). Because the quantum effect that occurs during photosynthesis seeks multiple pathways to find the most efficient route to the reaction center, this capillary action may be mirroring similar effects in materials that absorb water moisture.

It just so happens that the flow of sap in trees which works via capillary action, flows strongest through trees during spring time. Spring time happens to be the time of year 13:30 LST peaks in North America, showing a larger than normal spike in remote viewing accuracy (*Apparent Association Between Effect size In Free Response Anomalous cognition Experiments And Local sidereal Times. James P. Spottiswoode Cognitive Sciences Laboratory,*).

Sap Flow and Season
Sap has a capillary action. A study found that elevated sap flow during spring occurred in Norway in spruce trees when soil temperatures rose (*Bergh and Linder 1999*). The study found that temperature was the main driver of the flow of sap. Also the study stated that the onset of

photosynthesis increased in Korean pine trees as spring time soil temperatures began warming (*Wu et al. 2013*), and also that photosynthesis began increasing in Scots pine trees as the soil began thawing (*Strand et al. 2002*). The study concluded that the reason the flow of sap increased in early spring was due to warmer air temperatures as well as warmer soil which resulted in increased tree water uptake the exchange of canopy gas exchange. Our research on remote viewing over the years has shown that our best ARV sessions would always occur during early spring.

Reference
Ecosystem warming increases sap flow rates of northern red oak trees. Stephanie M. Juice. March 2016

Capillary Action type effects also take place in polymers and hydrogels. It just so happens that cytoskeletal proteins are polymers. Cytoskeletal proteins include tubulin which is the protein component of microtubules. Hence it is the cytoskeleton that is made up of long hollow cylinders which are microtubules and tubular cylinders are used to measure the Schuman resonance. It may be that hollow cavities resonate with the Schuman resonance. In Ireland there are large hollow tube-like towers known as "**Irish round towers**". In the region of these towers plant growth is stronger than that of the surrounding region. Hence, these towers may be amplifying the Schuman resonance, although further research is necessary to verify this. Perhaps the large cathedrals in Europe were designed to take advantage of the Schuman resonance. As we shall show later in this book, many of these churches have large stained glass windows that exhibit the properties of quantum dots. It is also interesting to note that pineapples, bananas and brown rice are tube shaped and as shown in this book bananas enhance dopamine levels.

Seasonal Variation of Photosynthesis

A research study looking at Taiwan spruce trees found that their rate of Photosynthesis exhibited a seasonal variation with a peak beginning from mid or late spring. This was accompanied by increased protein concentration with highest values observed during winter. The study also found that air temperature affected photosynthesis and that during early or mid spring as air temperatures increased, protein concentrations remained low. There was also a decrease in photosynthesis in winter, recovering with a peak in spring (*Seasonal variations in photosynthesis of Picea morrisonicola growing in the subalpine region of subtropical Taiwan. Weng JH et al*).

Further reading
The natural history of consciousness, and the question of whether plants are conscious, in relation to the Hameroff-Penrose quantum-physical 'Orch OR' theory of universal consciousness Peter W Barlow. July 2015

Quantum neurophysics: From non-living matter to quantum neurobiology and psychopathology. Tarlaci S, Pregnolato M. May 2016

How quantum brain biology can rescue conscious free will Stuart Hameroff. Oct 2012

Quantum physics in neuroscience and psychology: a neurophysical model of mind–brain interaction Jeffrey M. Schwartz, et al. June 2005

A New Spin on Neural Processing: Quantum Cognition Carol P. Weingarten, et al. Oct 2016

Chapter 5. Microtubules, Resonance and Precognition

Psychic abilities long ago were once considered to occur only in "gifted" individuals or during rare traumatic occasions. Psychic abilities are in fact, ongoing subconscious processes continuously influencing how all of us make everyday decisions. Our unconscious use of non-local information is associated with how we use our memory, subliminal perceptions and implicit physiological responses in response to emotional events.

The non-local manifestations of consciousness [1] [2], which are more common during remote viewing sessions, may be explained via quantum coherence as shown by Penrose [3].

Many other things all follow the same basic rules. William James (1898) proposed a transmission theory of consciousness [4] wherein the brain or mind was a passive system or, in principle, a consciousness receiver[5]. Like a radio receiver, the system would require an external signal, the identification of which would clarify the underlying mechanism. If indeed our bodies act as galactic antennas, perhaps the flow of information streams forth via the Schuman resonance which acts as a "step down transformer" where the information then resonates via the brain's microtubules then up into our waking conscious awareness.

This may explain therapeutic effects observed in several healing and religious meditative practices which use resonance in the form of prayers and chanting. Living systems have many resonant frequencies because of their unique degrees of freedom. This allows each system to vibrate as a harmonic oscillator that supports the accumulative progression of vibrations and/or waves which extend outward as a ripple within the entire system.

What is a Microtubule?
Microtubules were first discovered in the tips of the root of the plant Juniperus Chinesis[6]. It is interesting to note in our Solar Institute's second book on remote viewing titled: **Remote Viewing. The Complete User's Manual on Experiencing Future Consciousness**, that our device we use to enhance the clarity of remote viewing uses a ground rod to tap into the energy field exhibited by the soil.

In the human body, a microtubule is a nano-sized tubular structure contained in the brain's neurons which are made from two forms of tubulin. Communication occurring at the level of microtubules has been mathematically calculated

(Marcer and Schempp, 1997)[7]. Growing bodies of scientific evidence are beginning to show that microtubules possess three major properties related to intracellular and intercellular communications:

(1) they possess non-local information processing that is quantum in nature;

(2) they exhibit propagation of laser-like, coherent brief micro-pulses of light;

(3) they exhibit a collective, emergent and macroscopic property that arises from critical levels of coherence of quantum events[8].

Due to the small size of a microtubule, it is theoretically possible to consider microtubules behaving as quantum objects[9]. It may be that microtubules emit bio-photons (light pulses). Research suggests that micro pulses of light generate single-photon holograms, much like a laser beam, which is composed of numerous individual photons which is then able to create a hologram (Hirano and Hirai, 1986)[10]. If the trillions of microtubules existing in the human body create single-photon holograms, then the amount of holographically encoded information may be virtually unlimited.

Further Reading
Luminal particles within cellular microtubulesBoyan K. Garvalov. Sept 2011

Microtubules play a key role in non-local communication occurring across the human body. Research by Physicist Guenter Nimtz (1999) showed quantum photon tunneling through a barrier at a distance of approximately 5.5 inches with the time it took for the signal to travel through the barrier being instantaneous, regardless of the barrier length[11]. We explore the role microtubules play on consciousness in further detail later on in this book. Also it may be that the m-elements present in ormus enhance the energy flow occurring in the microtubules.

Quantum Effects observed at Room Temperature
The majority of quantum behavior takes place in significantly cooled, isolated environments. Research by physicists and chemists in Warsaw discovered quantum effects including tunneling occurring at room temperature and also at temperatures above boiling point and that it plays a major role in some chemical reactions in solutions *(Evidence for Dominant Role of Tunneling in Condensed Phases and at High Temperatures: Double Hydrogen Transfer in Porphycenes Piotr Ciąćka et al. Jan 2016)*.. The effect has also been seen to occur

in hydrogen nuclei which tunnels in particles floating in solution.

Summary
A basic chemical reaction occurs due to tunneling, and also in solutions and at room temperatures or higher. The same study (*shown above*) found that tunneling occurred in porphycene at room temperature. Professor Jack Waluk stated that quantum phenomenon may exist in the movements of the two protons occurring in porphycene and is further researching this *(Evidence for Dominant Role of Tunneling in Condensed Phases and at High Temperatures: Double Hydrogen Transfer in Porphycenes Piotr Ciąćka et al. Jan 2016).*

Further Reading
Direct Observation of Double Hydrogen Transfer via Quantum Tunneling in a Single Porphycene Molecule on a Ag (110) Surface. Matthias Koch et al. 2017.

What is Intuition or Psychic Awareness?
We are perpetually and unconsciously engaged in a universe of meaning extending far beyond our physical boundaries in time and space.

Intuition and the ability to tap into our psychic talents is not an ability. It is a universal characteristic of all living organisms. To put it simply, it exists as the standard feature of our existing-in-the-world. Intuition is not an ability that is stronger in some people and weaker in others. Intuition is always going on all the time for all of us. We can think of intuition as less like riding a bike and more like being engaged perpetually as physical bodies as we experience gravity.

At this point in time, our species has accumulated more knowledge than we might ever have thought possible on how intuition works. Radin (1997, 2004) proved that a person's subtle electrodermal responses to emotional pictures include within them an element preceding the exposure to the picture[12]. This has been confirmed in dozens of other studies. Carpenter (2002) showed that spontaneous social behavior is not a function of the unfolding stimulation that group members provide for each other, but that it also is reflecting the content of distant ESP targets that were being chosen at random by a computer in another city[13] [14].

Research by Bem (2005) showed people who are expressing emotions while observing pictures are showing not just the influence of the picture that is in front of them in the moment, but that they are also affected by whether or not they will soon be exposed to the picture in the near future[15].

How Fear Can sometimes be Mistaken for Intuition

Sometimes fear may cloud or even be misinterpreted as an intuitive concept, thought or feeling. To distinguish between fear and real intuition, know that true intuition flows from your higher being and fear cannot. A technique I have used for years is to perform a fear release exercise, which I show in greater detail on my website http://**ez3dbiz.com/red_tools.html**. Fear removal exercises allow you to more clearly connect with your true intuition and can help you see that some previous intuitive feelings were really fear. When fear is no longer the dominating force, intuition is enhanced because fear cannot exist on the higher planes which is the source from which all intuition comes from.

Summary

Intuition is not just an occasional ability. Instead it exists as an ongoing engagement that is linked to the unconscious where reality constantly expresses itself implicitly.

If indeed the information obtained via remote viewing occurs via intention and superposition, there would be studies showing this in further detail. One excellent method would be to explore the throwing of a dice and how intention contributed to the dice's outcome. Let's explore this next.

Using a Computer to Predict Dice Position

Dean Radin (1988) performed experiments where an operator attempted to predict the "tosses" of the computer which were equivalent of 6-sided dice. More than 60,000 trials were examined.

The data clearly showed that a person can achieve statistically significant results using this method. The study found the operator made 10,163 correct predictions with a probability of this event occurring of less than 1/20. The study also found that as soon as the computerized dice became biased, meaning one side of the dice had more chances to be selected compared to the other sides, and without the operator being aware of this bias, that the quality of the predictions increased[16]. With this approach the person guessed correctly 10,282 times, compared to the 10,000 random hits. The probability of this event was less than 1/1000.

Summary

It is easier for a person to intuitively see an event even if that person has no additional relevant information.

Can Computers Help Us Perfect our Psi Faculties?

The key point here is computers operate in binary fashion. This creates a distinct difference between messages and structure.

Experienced remote viewers are able to assemble mental images from subtle cues. This is because the information received during remote viewing appears as sounds, feelings, tastes, symbols, pictures and holistic impressions, instead of numbers and raw data. These sensory details are mostly emotion based. Hence computers cannot display emotions.

Certain electronic devices such as the EM Wave 2 can serve as tools that tell us when our body is in a state of coherence, which is key to successful ARV sessions.

References. Chapter 5

(1) PEAR Lab and Nonlocal Mind: Why They Matter. EXPLORE May/June 2007, Vol. 3, No. 3 191

(2) (2) Consciousness in the universe: A review of the 'Orch OR' theory. StuartHameroffa1RogerPenrose. August 2013

(3) Quantum computation in brain microtubules? The Penrose–Hameroff 'Orch OR' model of consciousness. Stuart Hameroff. March 2015

(4) William James, Gustav Fechner, and Early Psychophysics. Stephanie L. Hawkins. October 2011.

(5) William James and the "Theatre" of Consciousness

(6) Plant Cell Biology: From Astronomy to Zoology By Randy O. Wayne

(7) A Quantum Biomechanical Basis for Near-Death Life Reviews. Thomas E. BeckJanet E. Colli. March 2003

(8) A New Spin on Neural Processing: Quantum Cognition Carol P. Weingarten et al. Oct 2016.

(9) Quantum physics meets biology Markus Arndt et al. Nov 2009.

(10) Holography in the single-photon region. Hirano I, Hirai N.. June 1986
(11) The Superluminal Tunneling Story. Horst Aichmann, Günter Nimtz. April 2013
(12) Electrodermal Presentiments of Future Emotions. Dean L. radin. 2004
(13) Laboratory Psi Effects May Be Put to Practical Use: Two Pilot Studies. JAMES CARPENTER. 2010
(14) FIRST SIGHT: A MODEL AND A THEORY OF PSI. Jim Carpenter
(15) Feeling the future: A meta-analysis of 90 experiments on the anomalous anticipation of random future events. Daryl Bem, et al. Jan 2016.
(16) Effects of Consciousness on the Fall of Dice: A Meta-Analysis. Dean. L. Radin. 1991.

Further reading
Microtubules as Quantum Systems. Giulio D' Agostino. 2015

Meta-Analysis Of Esp Studies, 1987–2010: Assessing The Success Of The Forced choice Design In Parapsychology By Lance Storm, Patrizio . Re O D and Lorenzo Io.

A Proposed Process For Experiencing Visual Images Of Targets During An Esp Task By Yung-Jong Shiah

Additional References

BARGH, J. A. (1989). Conditional automaticity: varieties of automatic influence in social perception and cognition.

In James Uleman & John Bargh (Eds.), Unintended thought (pp. 3-51), New York: The Guilford Press.

BEM, D. J. (2005). Precognitive aversion. Proceedings of the Parapsychological Association, 48, 31-35.

CARPENTER, J. C. (2002). The intrusion of anomalous communication in group and individual psychotherapy: Clinical observations and a research project. Proceedings of the Symposium of the Bial Foundation, 4, 255-274.

CARPENTER, J.C. (2004). First Sight: Part one, A model of psi and the mind. Journal of Parapsychology, 68, 217-254.

CARPENTER, J.C. (2005). First Sight: Part two, Elaborations of a model of psi and the mind. Journal of Parapsychology, 69, 63-112.

CARPENTER, J.C. (2008). Relations between ESP and memory in terms of the First Sight model of psi.

Journal of Parapsychology, 72, 47-76.

CARPENTER, J.C. (2009). ESP contributes to the unconscious formation of a preference. Paper presented at the meeting of the Parapsychological Association, Seattle, August 2009.

CARPENTER, J.C. (book manuscript in preparation). First Sight: A Model and a Theory of Psi.

PALMER, J. (2006). Anomalous anticipation of target biases in computer guessing task. Proceedings of the Parapsychological Association, 49, 127-140.

RADIN, D. I. (1997). Unconscious perception of future emotions. Journal of Scientific Exploration, 11, 163-180.

RADIN, D. I. (2004). Electrodermal presentiments of future emotions .Journal of Scientific Exploration.18, 253-274

SCHACTER, D. L. (1997).Searching for Memory: The Brain, the Mind, and the Past. New York: Basic Books.

Chapter 6. Remote Viewing and Non-locality

When we look at human perception, research suggests humans perceive by means of both non-local and local processes. As an example, if a person looks at another nearby person, their eyes respond to the electromagnetic waves, however the mind also appears to respond to a non-local, instantaneous component (Mitchell, 1999)[1]. Remote viewing of persons located miles away from a target shows that the person experiences the non-local aspect. This experience is valid with experiences such as precognition, clairvoyance and telepathy.

The method that information travels (*as particles*) traverses the distance between a transmitter and a receiver in a measurable period of time that is finite in nature. It has been experimentally demonstrated that non-local interaction is occurring between photons (Aspect[2] [3], Dalibard[4], and Roger, 1982[5]). Hence, studies now conclusively confirm that non-local communication is possible across large distances. Bose-Einstein condensation can occur with photons, light particles or matter particles, which include protons, neutrons and electrons.

Due to the large amount of psychic research studies having been conducted over the course of the last 3 decades, researchers are now just beginning to connect anecdotal reports and search for the underlying mechanisms on the molecular level that govern psychic ability. Reports by people who have undergone near-death experiences report non-local perception such as watching their body being operated upon in an emergency room.

We outline in great depth in our second book on remote viewing Remote Viewing. The Complete User's Manual on Experiencing Future Consciousness, about how luminescent materials notably the beta emitter **tritium** enhances remote viewing. It is the Solar Institute's hypothesis that when earth's geomagnetic activity is at the right levels (*between 7 and 11*) that it significantly enhances the clarity of the information received possibly by enhancing the biophotons that occur in the brain.

Research has already shown that biophysical cells can act as EMF "sensors" (*Berzhanskaya et al., 1995, 1996; Potenza et al., 2004*) and that bacteria is particularly responsive to geomagnetic disturbances. Luminous bacteria has also been shown to exhibit increased photon emissions (*ultraweak, UPE*) **24 hours prior to a geomagnetic storm** outbreak and research by Berzhanskaya et al. (1995) discovered artificial magnetic fields affect the photon emission of luminous bacteria. They found that the specific frequencies of between 36 and 55 GHz (*adjacent to water absorption bands*) altered the photon emission of bacteria. Research by (*Persinger et al., 2012*) discovered two weeks before a magnitude 9 earthquake occurred that was miles away, that perturbations occurred in the background **photon emissions** from Sudbury Ontario. His research discovered that the perturbations occurred twice and that they produced persistent (~10 day) increases in the background photon emissions. Peak increases were noted to occur within 24 hours of the earthquake. After the earthquake, a consistent drop began taking place in the background photon emissions, which returned to normal levels in approximately 10 days. Hence we see a clear connection between seismic activity and photon activity. This is a perfect illustration of a quantum non-local effect taking place where two events miles apart from one another show a remarkable difference in their energy structure. In this case photons.

Reference
Electromagnetic fields as structure-function zeitgebers in biological systems: environmental orchestrations of morphogenesis and consciousness. Nicolas Rouleau and Blake T. Dotta. November 2014. Front Integr Neurosci. 2014; 8: 84. doi: 10.3389/fnint.2014.00084

The Schuman Resonance and Human Consciousness

The biorhythm of the human brain may be influenced by the Schumann resonance, due to its resonant oscillation which takes place in Earth's ionosphere (*Cherry, 2002*). Changes in blood pressure have been shown to take place as the Schumann resonance is impacted by changes in earth's geomagnetic field (*Mitsutake et al., 2005*). Blood pressure changes are linked as a factor in consciousness due to the rhythm of brain activity (*Persinger et al., 2013*).

How the Brain Receives Information via the Quantum Field During Remote Viewing

If one day we can mechanically harness the power of quantum computation, we may discover that consciousness exists as a type of quantum processor. Remote viewing utilizes non-local quantum states of superposition in order to achieve results. The effect occurs via the brain's microtubules that exist inside the neurons of the brain. These microtubules in turn are influenced by earth's Schuman resonance. Hence, when earth's Schuman resonance is at favorable levels, the ability to receive information via quantum effects is greater because the brain is able to process and hold larger amounts of information. Favorable Schuman resonance periods occur during favorable solar weather conditions, which we shall you how to find later on in this book.

What is a Quantum State?
A quantum state may exist as a superposition state. ie; two or more specific states co-exist independent of one another. This can be illustrated by having simultaneous thoughts of both ordering pasta and sushi before making the final decision. Once we order the pasta, it is akin to the wave function collapse / making the actual choice[6].

At the molecular scale, non-local communication exhibits its effects within the physical human body. Dick Bierman, a physicist at the University of Amsterdam, states that psi, or parapsychological effects, can only be explained by quantum non-locality[7].

A recent paper published in 2015 stated that unpublished studies showed a specific enhancement of power taking place in the brain coinciding with the Schumann Resonance [8]. Other research has demonstrated that an increase in spectral power densities in microtubules occurs within a frequency of between 7.7 and 7.8 Hz which just happens to be one of the frequencies of the Schumann resonance [8] [9].

Other research has shown that SPD's

(*spectral power densities occurring within Microtubules*) are enhanced at the frequency of 7.8Hz in the presence of weak magnetic fields[8]. Electroencephalograms of the human brain also show the existence of Schumann frequencies [10]. The 5 main Schumann frequencies are shown in the following picture.

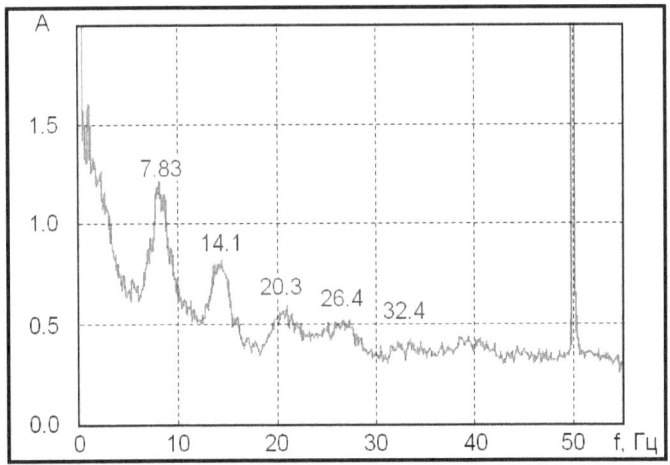

Chemical Reactions caused by Magnetic Fields

Magnetic fields also have been shown to affect chemical reactions. The reaction that takes place during this process is where two radicals are produced *(The effects of magnetic fields on chemical reactions. K. A. Mc. Lauchlan Science Progress (1933-)Vol. 67, No. 268 (Winter 1981), pp. 509-529).* Hence as earth's geomagnetic activity rises and falls, it may explain the effects caused on the body from geomagnetic activity due to its causing biochemical changes in the human body.

Further Reading
Spectral Power Densities of the Fundamental Schumann Resonance Are Enhanced in Microtubule Preparations Exposed to Temporally Patterned Weak Magnetic Fields: Implications for Entanglement. Blake Dotta et al. Jan 2015.

Testing Nonlocal Observation as a Source of Intuitive Knowledge. Dean Radin. Jan 2008

Incoherent spatial solitons in effectively instantaneous nonlinear media CARMEL ROTSCHILD et al. May 2008

Interpretation of "non-local" experiments using disentanglement. B. C. Sanctuary et al.

Non-local effects as a method of information received during remote viewing
Research by Dean Radin (1997) and his non-local experiments across a broad range of phenomena have shown that the viability of non-local perception is now a repeatable and valid concept[11].

What does Non-locality Mean?
Non-locality is a process whereby signals propagate across any distance instantly. These

signals may be telepathic, tachyonic or other yet undiscovered signals.

The signals that propagate in finite periods of time are known as local signals. An example is electromagnetic light to which our eyes respond to (*Goswami, 1993*)[12]. Incoherent electromagnetic waves such as common white light, radio or television transmissions, exhibit none of these three traits. Non-locality is sometimes easily confused with super-luminal signal propagation (faster-than-light which are signals faster than 186,000 miles per second) (*Nimtz, 1998*) [13] [14].

Summary
The human body possesses the necessary biomechanical mechanisms necessary to exhibit non-local communication.

Method of Information Transfer during ARV Sessions
With regard to explaining how time flows during a remote viewing session, the simple explanation is that the mind has the ability to hold and process larger amounts of information.

When projecting to the future date in time up to 4 days in advance during an associative remote viewing session, a panoramic review takes place in the mind in what seems like an instant of time. Everything is seen instantaneously, like a dense explosion, all information is there. Information is not displayed in a sequential fashion, such as going from point A and ending at point B, but instead exists as watching the information on an enormous TV screen, with the past being on the left and the future being on the right, with everything in-between, allowing one to see the whole event at the same instant.

In science supercomputers are used to create forecast models for future events. This is because it takes huge amounts of information that must be processed in order to generate a future forecast model. Perhaps the regions of our brain that are utilized used during remote viewing sessions are regions that allow the mind to comprehend larger than average amounts of data (the hippocampus?). This effect may only occur when earth's geomagnetic energy levels are between 7 and 11 (Middle Latitude Fredericksburg K-indices). Now this is just from my experience, and it may differ between remote viewers. This "sweet spot" is shown in the following illustration.

Improve your Remote Viewing Accuracy Techniques
using Quantum Microtubules

# Date	Middle Latitude - Fredericksburg - K-indices								
2017 03 23	9	4	3	2	2	1	2	1	1
2017 03 24	7	2	3	2	0	2	2	2	1
2017 03 25	3	0	2	0	0	2	2	1	1
2017 03 26	3	1	0	0	1	1	1	1	2
2017 03 27	34	2	3	5	5	5	4	5	4
2017 03 28	-1	5	5	3	3	3	2	3	-1

You can real time earth geomagnetic levels at the address below.

http://legacy-www.swpc.noaa.gov/ftpdir/indices/DGD.txt

A paper published in Nov 2008 titled: Quantum physics meets biology and published by Markus Arndt and colleagues stated that magnetic susceptibility *(Brukner et al., 2006)* or heat capacity may be good indicators for entanglement and further research is needed [15] [16]. Hence **magnetism exists as a quantum mechanical** occurrence. The paper also states that the process of vision, the sense of smell, the magnetic orientation of migrant birds and photosynthesis are hot topics and that these may exhibit quantum phenomena[15].

Ferromagnetism and Quantum Mechanical Effects
Ferromagnetism comes from two quantum mechanical effects.
1 – Spin

2 - The Fauli Exclusion Principle[17]

What is most interesting in our second book on Associative Remote Viewing titled: **Remote Viewing. The Complete User's Manual on Experiencing Future Consciousness** we discovered that certain ferromagnetic materials exhibited strong effects that enhanced our results. A research paper titled: Ferromagnetism induced by entangled charge and orbital orderings in ferroelectric titanate perovskites and published by N.C. Bristowe in March 2015 and colleagues, concluded that the Pauli exclusion principle causes anti-alignment between electron spins resulting in anti-ferromagnetic behavior [17].

Where does Consciousness Come From? While at this time (2017) we don't know the " why" of consciousness or how its sentience is produced in every cell, we can at least predict and hypothesize the " how " of consciousness is computed at the quantum scale, allowing us to understand and judge perceptions.
Romijn (2002) stated photons (which are

simply constituent units of electromagnetic fields) are carrier units of human consciousness[18]. Others have pointed out that hypothetical particles (*Eccles, 1992*), zero-energy tachyons (*Hari, 2008*) or even both of these are assumed to be responsible for experiencing consciousness[19] [20].

Consciousness exists as a series of discrete re-occurring events. Hence, this could account for why mantras or affirmations are so successful. The repeating of a single word over and over interacts with the wave of re-occurring events creating an accumulative effect via resonance/oscillation, much like pushing a child on a swing, taking less effort the more the word is repeated. Hence the discrete events in consciousness may also exhibit similar resonance like effects.

It is this series of events that gives rise to consciousness. Hence, conscious events begin at quantum levels and grow in size. Hence, what we are experiencing as a thought is actually the result of subtle actions that have taken place far downstream at the molecular level in our mind.

Intention and choices are cascading as quantum information in an upwards fashion forming a bio-directional feedback loop, similar to what's known as "Jacob's Ladder" shown in the bible. Superposition of quantum effects throughout the brain provide the "glue" essential to consciousness.

Quantum Behavior Powers the Sun
The immense power of the sun is released not just by particles, but also via waves. A proton exists as a quantum particle that contains a probability function describing its location. This enables the two wave functions of interacting particles to slightly overlap each other, even when the natural repulsive forces keep them entirely apart. When the particles undergo quantum tunneling and exist as a more stable bound state, such as deuterium, it causes a release of fusion energy which allows the chain reaction to begin.

This interaction/reaction is so small, it is no wonder that the weak force has been shown to also influence the release of energy from the sun. The quantum tunneling effect taking place via the sun is very small (somewhere in the order of 1-in-1028). This is the same as winning the powerball lottery three times in a row. However just that small number is enough to unlock the sun's power and shows that the power of quantum lies in its small size.

If all particles in our universe did not exhibit quantum behavior and that their positions existed as wave functions with quantum uncertainty due to their position which create the overlap enabling nuclear fusion, then nuclear fusion would never occur. Instead we would be living in a universe where almost all of today's stars in the universe would never have ignited. We would look up at a dark sky, desolate and frozen, with the night sky of full of dead stars and solar systems mildly lit by cold, rare, distant starlight.

Cordless Telephones affect the Brain's Microtubules

If indeed there is a quantum mechanical connection occurring within the brain's microtubules, then they may be affecting regions of the brain's microtubules[22]. Research has shown that cordless phones affect microtubules *(Effect of 1.8 GHz radio frequency electromagnetic fields on the expression of microtubule associated protein 2 in rat neurons. Zhao R et al. April 2006).* If this is so, could some cellular telephones be affecting consciousness? Further research is necessary to explore this hypothesis.

Further   Reading
Effect of 1.8 GHz radio frequency electromagnetic radiation on novel object associative recognition memory in mice. Wang K et al. March 2017

Electromagnetic fields (1.8 GHz) increase the permeability to sucrose of the blood-brain barrier in vitro. Schirmacher A et al. July 2000

An Experiment to "Stretch Time" to get More Done

Sometimes we may feel that time is slipping away, that we don't have enough time to complete the things we want to accomplish. Time is based on energy and how you mold that energy. This neat little exercise I have found can help give one much more time to catch up on little tasks, thus avoiding the "time shortage" effect.

If you have a set of duties or scheduled activities planned and you feel you may not have enough time to accomplish them, before you go about doing them, imagine another set of errands or activities with twice as much to do. Imagine yourself going twice the distance to fulfill them. For example if you needed to do shopping at your local supermarket and you feel you don't have the time, visualize yourself going on a shopping trip 50 miles away and buying twice as much stuff as you usually would. Next imagine yourself returning back to your present location. Visualize this clearly and as vividly as possible before you set out on your duties / errands. What you will find will happen is when you do your local shopping you will find that it will take much less time then you expect. This method works you are causing a "stretch" in the field of energy related to time. Even though you are not physically traveling there, the bio-field of time as it interacts with the present is making more time. This is because our perception of time is limited. Instead time, like consciousness, exists as units or "packets" of energy. By adding extra packets, you 'stretch' the rate at which time flows.

References. Chapter 6

(1) Mitchell, E. (1999). Nature's mind: The quantum hologram. Retrieved April 18, 2002 from the National Institute for Discovery Science Web site: http://www.nidsci.org/articles/naturesmind-qh.html

(2) Aspect, Alain (2007). "Quantum mechanics: To be or not to be local". Nature. 446 (7138): 866–867.

(3) Experimental Realization of Einstein-Podolsky-Rosen-BohmGedankenexperiment: A New Violation of Bell's Inequalities, A. Aspect, P. Grangier, and G. Roger, Physical Review Letters, Vol. 49, Iss. 2, pp. 91–94 (1982) doi:10.1103/PhysRevLett.49.91

(4) Quantum Mechanics. J. Dalibard and J.L. Basdevant. ISBN 1439-2674

(5) Experimental Test of Bell's Inequalities Using Time-Varying Analyzers. Alain Aspect, Jean Dalibard, and Gérard Roger. Dec 1982.

(6) Cohen-Tannoudji, C., Diu, B., Laloë, F. (1973/1977). Quantum Mechanics, translated from the French by S. R. Hemley, N. Ostrowsky, D. Ostrowsky, second edition, volume 1, Wiley, New York, ISBN 0471164321.

(7) CONSCIOUSNESS INDUCED RESTORATION OF TIME SYMMETRY (CIRTS): A

PSYCHOPHYSICAL THEORETICAL PERSPECTIVE. DICK J. BIERMAN

(8) Spectral Power Densities of the Fundamental Schumann Resonance Are Enhanced in Microtubule Preparations Exposed to Temporally Patterned Weak Magnetic Fields: Implications for Entanglement. T. Blake et al. Sept 2015.

(9) A. Nickolaenko and M Hayakawa, Schumann Resonance for Tyros: Essentials of Global Electromagnetic Resonance in the Earth-Ionospheric Cavity, Springer, 2014.

(10) Schumann Resonance Frequencies Found Within Quantitative Electroencephalographic Activity: Implications for Earth-Brain Interactions. Michael A. Persinger. International Letters of Chemistry, Physics and Astronomy

(11) Distant Healing Intention Therapies: An Overview of the Scientific Evidence. Dean Radin, et al. Nov 2015

(12) Quantum Nonlocality. AmitGoswami.

(13) Superluminal signal velocity Authors G. Nimtz. Dec 1998.

(14) Basics of superluminal signals Authors G. Nimtz, A. Haibel. Jan 2002

(15) Quantum physics meets biology. Markus Arndt et al. Nov 2009

(16) Quantum Entanglement in Nitrosyl Iron Complexes. S. M. Aldoshin, E. B. Feldman, and M. A. Yurishchev

(17) Ferromagnetism induced by entangled charge and orbital orderings in ferroelectric titanateperovskites. N. C. Bristowe et al. Mar 2015.
(18) Are virtual photons the elementary carriers of consciousness? H. Romijn. Jan 2002
(19) Quantum aspects of brain activity and the role of consciousness. F Beck and J C Eccles. Dec 1992
(20) Eccles'sPsychons Could be Zero-Energy Tachyons. SyamalaHari. June 2008
(21) Impact of Electromagnetic Waves Generated by Cellular Phones on Male Fertility: A Review ShilpaKhullar

Chapter 7. The Hypothalamus and its Sensitivity to Light

The hippocampus region of the brain is the part of the brain that is most sensitive to light[1].The dopamine-mediated reward system in the brain facilitates memory formation, which is dependent upon the brain's hippocampus[2] [3] [4]. If indeed photons are responsible for the effects of superposition which convey information received during remote viewing, the region of the brain known as the hypothalamus, which is the region of the brain most sensitive to light may be acting as the primary receiver for the information.

Dopamine is also sensitive to light [5].Hence because dopamine exhibits a seasonal variation with lower amounts produced in the body during the spring and summer[6] [7], it may be that excess light depletes dopamine levels, much like excess light depletes melatonin levels (less in summer, more in winter [8]) . Our ARV sessions have always been harder to do during the summer and summer is also the time that the LST time related to remote viewing is weaker, as we shall show later on.

Further **Reading**
Melatonin and seasonal rhythms. Wehr TA J Biol Rhythms. 1997 Dec; 12(6):518-27

What is Glutamate?

Glutamate is an amino acid that occurs in cyanobacteria[9]. Glutamate is a memory neurotransmitter that allows nerve cells to send their signals to other nerve cells and is one of the most abundant neurotransmitters in the nervous system. It accounts for over 90% of synaptic activity occurring in the human brain. Research has also shown an increase in glutamate concentrations in the body before an epileptic seizure[10]. Cyanobacteria found in freshwater ponds contain glutamate. Cyanobacteria is also being looked into as an electrical power source.

Nicotine and Glutamate

Nicotine protects against glutamate neurotoxicity when administered to cultures of cerebellar neurons (*Nicotine prevents glutamate-induced proteolysis of the microtubule-associated protein MAP-2 and glutamate neurotoxicity in primary cultures of cerebellar neurons. Miñana MD et al. Neuropharmacology. 1998 Jul;37(7):847-57*).

Further Reading

Thermal release of nicotine and its salts adsorbed on silica gel. Qing Hua et al.

Linoleic Acid found in Cyanobacteria

A research study found that Cyanobacteria contains linoleic acid, as well as other omega 3's

(*Lipid and fatty acid composition of freshwater cyanobacteriaA. K. SALLALN, et al. 1990*). A

Another study found that applying linoleic acid to the skin caused the skin to turn lighter after UV application on studies with brownish guinea pigs. The study concluded this effect was due to delayed production of melanin due to the linoleic acid. The strongest effects came from the alpha-linoleic acid, followed by the linoleic acid. Linoleic acid, or omega-6 fatty acid. Hence this could be why Cyanobacteria contains linoleic acid perhaps as a way for the Cyanobacteria to absorb more light (*Linoleic acid and alpha-linolenic acid lightens ultraviolet-induced hyperpigmentation of the skin. H. Ando et al. July 1998*).

An interesting fact about UV light and nicotine is when ultraviolet light is shone onto nicotine it undergoes a catalyst reaction, turning it yellow (*The Effect of Ultraviolet Radiation on Nicotine. C. H. Rayburn et al. January 1941. J. Am. Chem. Soc.,1941*).

Digoxin and the Hypothalamus
The hypothalamus produces the substance Digoxin which is also found in the Foxglove plant[12].

Digoxin has been found to be at elevated levels in people who show the following behaviors[12].

- Creativity and high IQ

- Hypersexual behavior
- Reduced appetite
- Insomnia
- An increased tendency for spirituality
- An increased tendency for addiction
- Less bonding and affectionate behavior
- Left handedness/right hemispheric dominance (*intuitive*)

If indeed photons are responsible for conveying information received during remote viewing, the hypothalamus may be acting as the primary receiver for the information due its sensitivity to light.

Blue light and the Thalamus
When light enters the eyes, certain nerves become stimulated which sends small signals to the brain. The brain then unscrambles the signals giving the objects we see form and shape.

Research by Dr. W.R.A Muntz at MIT looked at activity in the thalamus in a frog's brain as it was exposed to different colors of light. His experiment discovered that the brain's responses to blue light were 10 times as strong compared to when the brain was exposed to yellow, red or green light.

His study also found that frogs can only see small insects if they are moving, but not standing still, which is why frogs can starve to death if around large numbers of dead flies. The study concluded that a frog can't see water but instead sees water as an open space covered in blue, as well as the sky's reflection on the water, especially if it is cloudy. This study may also show that in humans that blue light also stimulates the thalamus (*The Photopositive* Response Of The Frog (Rana Pipiens) Under Photopic And Scotopic Conditions. By W. R. A. Muntz University of Sussex. 11 February 1966). J. Exp. Biol. (1966), 45, 101-111*).

Blue enhances the strength of white. This is why some detergents add a touch of blue to their white laundry powder to enhance the whiteness of fabrics. The same type of effect may be occurring when small amounts of blue light are added to white light.

The first quarter moon rises at midday, when the sky's color is deep blue (*blue strengthens white*) and sets at midnight. Hence the photons in earth's atmosphere may have a capacitor/retaining effect being charged via the white light of the moon (reflected light from the sun), and are intensified from this combination of white and blue light. The star Sirius is composed of a strong bluish white light.

When the ancient Egyptians observed Sirius rising a little before the Sun rose, they knew it was a time the moisture along the Nile River would increase, leading to flooding.

The Dogon Tribe and Sirius

From 1931 to 1956 French anthropologist Marcel Griaule studied the Dogon Tribe in West Africa (*Ciarcia, Gaetano "Dogons et Dogon. Retours au 'pays du reel'", L'Homme 157 (janvier/mars): 217–229), (Imperato, Pascal James, Historical Dictionary of Mali Scarecrow Press, 1977 ISBN 978-0-8108-1005-1 p.53*). His studies found the Dogon believe Sirius has two companion stars (the first and second companions of Sirius A).

During 1976 Robert K. G. Temple showed in his book, **The Sirius Mystery**, that the Dogon's incredible accurate knowledge of cosmological facts that Sirius, which is part of a binary star system, and that its second star (Sirius B, a white dwarf) is astounding due to the fact that Sirius B is completely invisible to the human eye and that it takes about 50 years to make a complete orbit. Existence of Sirius B had been verfied by Friedrich Bessel during the year 1844. Temple stated the Dogon's information may have come from ancient Egyptian sources and myth and a possible extraterrestrial transmission of knowledge of these stars (*Robert K. G. Temple, The Sirius Mystery, 1975*). Also reports state that the Dogon know of yet another star that exists in the Sirius star system called, *Ęmmę Ya*, or "larger than Sirius B yet lighter and dimmer in magnitude." In the year 1995, gravitational studies did show the possibilty that a brown dwarf star may indeed orbit Sirius (a Sirius-C). It' orbital period is estimated to be 6 years (*Benest, D., & Duvent, J. L. (1995) "Is Sirius a triple star?". Astronomy and Astrophysics 299: 621–628*).

Studies conducted by French anthropologists Griaule and Germaine Dieterlen during the 1930's and 40's found the Dogon had extensive anatomical and physiological knowledge including their own systems of astronomy and calendrical measurements as well as methods of calculation and a pharmacopoeia (*Griaule (1970), p. xiv*). The key spiritual figures in their religion were the **Nummo**/Nommo twins who were referred to as the Serpent which were amphibians that were compared to lizards, chameleons, and sloths due to their slow movements and shapeless necks. They were described as fish that were capable of walking on land and stood upright on their tails. Their skin was green like a chameleon having all the colours of the rainbow (*Shannon Dorey, The Master of Speech, Elemental Expressions Ltd., Elora, Canada, 2002, reprint 2013. p.13*). The Nummo were also referred to as "**Water Spirits**" (*Griaule (1970), p. 97*) or "Dieu d'eau" (Gods of Water). They (Nummo) were identified as hermaphrodites and appeared on the female side of the Dogon sanctuary (*Griaule (1970), p. 105*) and they were symbolized by the sun, and the color red, female symbols in their religion.

Guidelines that Enhance Associative Remote Viewing
A more comprehensive step by step instruction method on how to perform an Associative Remote Viewing session can be found in our second book **Associative Remote Viewing Remote Viewing. The Complete User's Manual on Experiencing Future Consciousness**.

The Star Arcturus and Remote Viewing
The star Arcturus rises in the east at 5:30 p.m. at my location in Honolulu, Hawaii during late spring (April 28th). During this time, the second peak period of remote viewing (**8:45 LST**) takes place at approximately 6:30 p.m. During this time Sirius rises at 11 a.m. in the morning and is zenith (directly overhead) at 4:30 p.m.

Improve your Remote Viewing Accuracy Techniques
using Quantum Microtubules

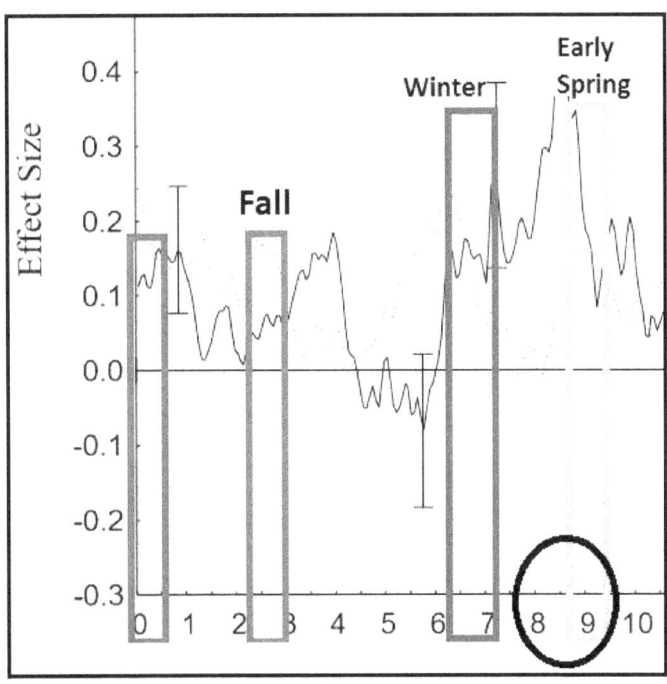

Reference
Apparent Association Between Effect Size In Free Response Anomalous Cognition Experiments And Local Sidereal Time. *Seasonal Emphasis by author.*

Arcturus is a golden yellow orange colour. A little bit below the star Arcturus is the star Spica in Virgo. Spica happens to be a bright whiteish bluish star which is the brightest star in the Virgo constellation.

Light, Alertness and the Hypothalamus
Numerous studies published since 2004 show

that light affects the brain's cognitive functions and alertness [13]. Additional studies found that when **blue light** was present during cognitive tasks, that it **enhanced** the activity in the brain's brainstem (LC), **hypothalamus, thalamus** and limbic regions first (hippocampus and amygdala) then spread to the brain's cortical regions[14] [15] [16] [17] [18] [19] [20] [21] [22] [23]. These regions govern our emotions. It's interesting to note that police emergency vehicles utilize blue light for their sirens and the science fiction television show Dr. Who utilizes a police call box to travel through time.

Blue Light Enhances Photosynthesis
Plants exposed to blue light exhibit faster photosynthesis rates compared to plants grown under green or red light.

A research study conducted at the Chinese Academy of Sciences in Beijing showed blue luminescence from luminol initiated photosynthesis in geranium leaves and that carbohydrate production was related to the strength of the bioluminescence. The bioluminescence was controlled by varying luminol levels. Hence, this research study shows that bioluminescence can be used as a light source for photosynthesis. Also extreme conditions can reduce photosynthesis. Hence there exists a "sweet spot' (*Bioluminescence powers photosynthesis. Amy Middleton-Gear14 October 2013*), (*Chemical Communications. Bioluminescence as a light source for photosynthesis. Huanxiang Yuan et al.*).

Could blue light be enhancing quantum effects? It is interesting to note that the science fiction time travel movie back to the future used a lot of bluish white light when the Dolorean car travelled through time.

Further **Reading**
Light-sensitive brain pathways and aging. V. Daneault et al. March 2016

Because dopamine exhibits a seasonal variation with lower dopamine occurring during the spring and summer (*Seasonal variation in human central dopamine activity. Karson CN et al. Feb 1984*), it may be that light influences dopamine levels. This could explain why on a seasonal basis, our most inaccurate remote viewing sessions would always occur during summer (in North America). Research has shown that excess light affects dopamine functioning [24]. Bright light has been shown to be beneficial to patients with Parkinson's and in some cases reduced the amount of therapy (*Bright light therapy in Parkinson's disease: a pilot study. Paus S et al. July 2007*).

Light and Tubulin
After baby rats are born, when they first open their eyes, the neurons in their visual cortex start producing large quantities of tubulin. Tubulin is one of the components of the brain's microtubules[26]. Production of tubulin and microtubule health have been correlated with memory, peak learning and experience in baby chick brains (*Mileusnic et al, 1980*)[27].

Hypothalamus Activity During Midnight
A region inside the brain's hypothalamus known as the suprachiasmatic nucleus (SCN) has its own circadian rhythm. At midnight the body's cortisol levels (which are controlled by the hypothalamus) are lowest. After midnight they gradually begin building up, peaking in the morning. The body's cortisol levels then start to decline slowly throughout the day (*Debono et al. 2009; Krieger et al. 1971; Weitzman et al. 1971*)[28].

Further Reading
Replication of cortisol circadian rhythm: new advances in hydrocortisone replacement therapy. Sharon Chan and Miguel Debono. June 2010

Melatonin and Microtubules
Melatonin may play a role in the activity taking place in Microtubules. Research has found that there exists an interaction between melatonin

and microtubules[29]. I mentioned heavily in our second book on associative remote viewing **Remote Viewing. The Complete User's Manual on Experiencing Future Consciousness** that the very durable metal Tungsten was discovered by Dr. Kozyrev to exhibit strong effects on space/time[30]. Also our research using Tungsten metal or minerals that contained Tungsten greatly enhanced the accuracy of our associative remote viewing sessions of the Dow Jones. Let's find out why this is.

Why the metal Tungsten Enhances Associative Remote Viewing Sessions
Time consists of a high frequency. Physics Professor and co-founder of modern string theory, Michio Kaku states that the radio in your living room is made up of numerous frequencies. However your radio is usually tuned to only one frequency. This means you are decohered from all other frequencies. While listening to your favourite radio station, you are coherent or in alignment either exactly with or in a whole number with that single one frequency. Our universe is vibrating and there are vibrations of other universes right where you sit reading this book. Hence frequency is the dividing factor between them all.

It may be that the frequency of time is of a very high frequency associated with the microwave/far-infrared region. Far Infrared sources include silicon carbide, tungsten lamps and black body radiation. Metals such as Tungsten are suited for high frequency applications. Hence Tungsten wires were used as radio frequency detection devices in early crystal radio sets (*Uncle Tungsten: Memories of a Chemical Boyhood. Oliver Sacks*). During 1902 G.W. Pickard found that a microphone could be used to detect radio waves and found that combining well conducting materials with a highly resistive material worked well. After years of searching for suitable materials, he found that silicon contacted by a tungsten wire delivered the very best results. Carborundum can also be used as a substitute for silicon. This was eventually replaced by vacuum tubes due to the cumbersome ability to press the tungsten wire to the silicon. What is also interesting is the mineral quartz crystal is made of silicon powder.

Tungsten is also used to make electron emitters, being mixed with barium and strontium (*Microwave Electronic Devices. T.G. Roer*). Tungsten wire is used in radiation detectors and frequency multipliers (harmonic generators with silicon crystals (*Nuclear Moments. H. Kopferman, E. E.

Schneider). Tungsten wire is used to detect microwaves and microwaves are emitted by sunspots. The spectrum may include the millimeter wave (MMW) band of frequencies. These extend from 30GHz to 300 GHz. The terahertz (THz) band ranges from 200 GHz to approximately 30 THz. These bands are known as the far-infrared, submillimeter and near-millimeter frequencies. Tungsten is also used in near-infrared single-photon detection devices (*Transition-edge sensor. Wikipedia*). When tungsten is heated it emits light in the 320nm range (*Physical Chemistry. Peter Atkins, Julio de Paula*). A heated tungsten wire is also used to detect microwaves. (*Microwave Spectroscopy. C. H. Townes, A.L. Schawlow*).

The Schottky Diode
The Schottky diode is a semiconductor diode that is formed by a junction of a semiconductor with metal material. Its forward voltage drop is very low and it has a very fast switching action.

Cat's-whisker wire detectors used early in the days of wireless and the metal rectifiers used in early power are forms of primitive Schottky diodes. Some of these used tungsten wires. Tungsten is used in point-contact diodes which depends on the pressure of its contact on a semiconductor crystal. One section of a diode is made up of a small rectangular crystal of n-type silicon. Next to this is a fine bronze-phosphor, fine berylium-copper or tungsten wire known as the CATWHISKER which is pressed up against the crystal, forming the other part of the diode. Point contact diodes are known as the oldest microwave semiconductor devices. They were commonly used in world war 2 microwave receivers and even today are used in detectors and receiver mixers (*Basic Electr-Msbte. Malvino/patil*). The lead bearing mineral, Galena was also used in early crystal radio sets as a "cats whisker" to receive radio frequencies. Galena is a mineral with an extremely high lead content and when tungsten is mixed with lead sulfide you get Tungsten disulfide which is used as a dry lubricant and sold commercially under the brand **Dicronite**. Dicronite can also be found as a coating on Bolts and Screws. It is also used as a catalyst to hydrotreat crude oil and is used to create tubular shaped inorganic nano-tubes. When Manganese is combined with Lead Sulfide it results in enhanced sensitivity in quantum dot solar cells (*Exploring the effect of manganese in lead sulfide quantum dot sensitized solar cell to enhance the photovoltaic performance Dinah Punnoose, et al*).

Tungsten Disulphide also has been shown to exhibit chirality (*Chiral superconductivity experimentally demonstrated for the first time February 27, 2017 by Lisa Zyga. February 27, 2017),*

(Tech ref - Superconductivity in a chiral nanotube F. Qin et al. Jan 2017).

Further Reading
Characteristics of Tungsten-Nickel Point Contact Diodes Used as Laser Harmonic-Generator. Mixerseiichi Sakuma And Kenneth M. Evenson/ August 1974

Further Research
Internet Search Term - tungsten wire + microwave detector

Wulfenite
The mineral wulfenite (*also called lead molybdate*), which we use in our ARV sessions, happens to contain large amounts of Tungsten.

Lead Molybdate
Minerals associated with Lead Molybdate include vanadinite, limonite, galena, smithsonite and mimetite.

Wulfinite is composed of mostly of Lead Molybdate arranged in the structure of plate like tetragonal crystals (pictured is the tetragonal shape). It has a hexagonal crystalline structure which is similar in shape to molybdenum disulfide. It commonly occurs with lead ore

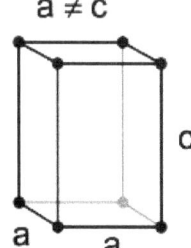

and is also a rich source of molybdenum. Wulfenite has a crystal lattice with a tetrahedral structure containing four oxygen ions that occupy the four tetrahedral corners around a molybdenum ion. Research has shown that TMD's can emit quantum light via single photons of light. Pictured is wulfenite.

What does TMD stand for? TMD is short for Transition Metal Dichalcogenide, which is a semiconductor that can be layered a few atoms thick[31]. Tungsten disulfide and its closely related companion metal, Molybdenum, happen to be TMD's. These materials can easily be purchased online. Tungsten diselenide and Tungsten disulphide have been used to build quantum emitters[32] [33].

What is Tungsten Disulfide?

Tungsten disulfide is a hexagonal structure similar in appearance to molybdenum disulfide and occurs naturally in the mineral Tungstenite.

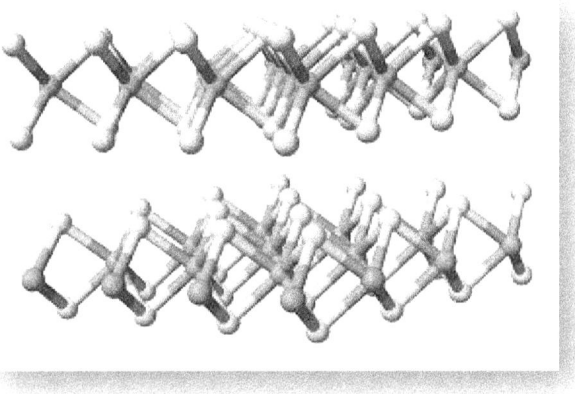

The atoms are bonded to six selenium ligands with a pyramidal geometry. Tungsten Disulfide is used to create a catalyst reaction when hydrotreating crude oil. Because these materials are able to readily absorb large amounts of light, they are used in electrochemical solar cells, giving them the ability to pass 95% of incident light.

Tungsten Disulfide's unique structure also allows it to be used to manufacture LED's of any color. Weak van der Walls coupling has also been observed to take place in Graphene-Tungsten Disulfide[34] [35].

In the second book of our remote viewing series The Complete User's Manual on Experiencing Future Consciousness magnetic susceptibility seemed to enhance the receptivity of the images received during remote viewing.

Hence there may exist a connection between magnetic susceptibility and entanglement as entanglement effects may be the mechanism for which the information is received during remote viewing.

Research by Brukner et al., 2006 found that magnetic susceptibility or heat capacity might provide good indicators for entanglement (*Wiesniak et al., 2008*). There may also exist entanglement properties occurring in biological systems and more research is needed to explore this further *('Brukner C, Vedral V, and Zeilinger A (2006). Crucial role of quantum entanglement in bulk properties of solids.) (Wiesniak M, Vedral V, and Brukner C (2008). Heat capacity as an indicator of entanglement).*

Chemical reactions also occur in nature as catalyst reactions. Kozyrev found that chemical reactions (a form of catalyst) also had effects on time. Some common catalysts include platinum, iron, vanadium, and molybdenum.

Kozyrev also studied the chemical reaction taking place in Potassium permanganate (*On The Way To Understanding The Time Phenomenon: The Constructions Of Time In Natural Science, Part 2.*

Levich A P. May 1996). Potassium permanganate is a very strong oxidizer that does not generate toxins. Potassium permanganate is used to heal fungal infections and dermatitis. It also removes iron and hydrogen sulfilde from water. It also contains Manganese heptoxide. Tungsten Disulfide also promotes catalyst reactions in the artificial leaf energy generator we shall explore later. Also certain radioactive minerals decay rates vary according to solar activity and it may be that solar activity has an effect on catalyst reactions involving beta decay. Pictured is the molecular composition of Potassium permanganate.

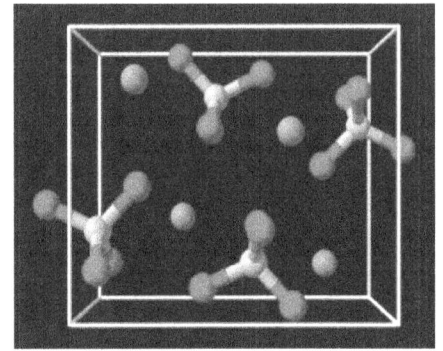

Quantum Dots

Many materials that readily absorb light are also materials used to make quantum dots. Materials include: Tungsten Diselenide (WSe2), Titanium dioxide (TiO2), Zinc Oxide, Zircon and Selenium. Lead Sulfide is also used to make quantum dots

(*Photoluminescence of Lead Sulfide Quantum Dots of Different Sizes in a Nanoporous Silicate. Glass Matrix. Aleksandr P. Litvin et al. March 27, 2017*).

Mixing Tungsten diselenide (WSe2) with Gold causes a 20,000-fold Increase in Photoluminescence

Research has shown that metal dichalcogenides can increase photoluminescence up to 100-fold. When Tungsten Diselenide flakes are placed into sub-20-nm-wide trenches in gold substrate, a photoluminescence enhancement of ~20,000-fold takes place. This occurs due to the coherent light being able to absorb more of the light (*Giant photoluminescence enhancement in tungsten-diselenide-gold plasmonic hybrid structures. Z. Wang et al. May 2016. Nat Commun. 2016 May 6;7:11283. doi: 10.1038/ncomms11283*).

Anglesite

A mineral known as Anglesite contains lead sulfate and occurs as an oxidation product of galena, primary lead sulfide ore.

Metaphysical properties of Anglesite state that it causes intensification of psychic communication in various ways. These include enhancing contact with the angelic realm and helping one find their spirit guide. It also helps clarify spiritual based communication. As an interesting side note, galena which contains high levels of lead, is used to make zero point energy batteries and to pick up radio waves. You can find numerous videos of how to make zero point energy batteries using galena. Also the mineral spodumene contains lithium / aluminium and was used by Dr Thomas Henry Moray of Salt Lake City to build a zero point energy device that extracted energy from the environment. He called his device the Swedish Stone.

Dr. Moray placed the spodumene inside a crystal radio set which caused it to generate extremely strong signals that destroyed headphones. It also made large speakers crackle and roar. Moray stated the crystal drew energy from cosmic rays emitted by the sun and nearby stars. He named it the **Cosray Receiver** (cosmic ray receiver). The desktop sized instrument was strong enough to run a small factory in 1929 (*A radiant sea of energy. Mark Pilkington. June 2003. Theguardian*).

Further Reading
The Sea of Energy in Which the Earth Floats. 1978 Edition: Cosray Research Institute, Salt Lake City, UT. Thomas Henry Moray.

Quantum Dots and Stained Glass
The magnificent colored stained glass windows displayed in some churches in Europe are made with leaded glass.

Artisans during the middle ages discovered that adding gold chloride to hot molten glass caused a strong red tint. Also they found that the addition of silver nitrate turned the glass into a vibrant yellow. Recently researchers looked at the micro-chemical-composition of these stained glass windows and discovered signs of nanotechnology. Their analysis found that the gold and silver nano-particles contained within the stained glass windows acted as quantum dots; reflecting yellow and red light respectively.

Carbon Nano-materials
In 2006 researchers discovered traces of nano-wires and carbon nano-tubes in Damascus steel blades. They theorized that these nano-wires, encapsulated by the carbon nano-tubes, were responsible for Damascus steel's legendary sharpness and durability. Damascus steel was produced from the 12th to 18th century by Damascene metal smiths that used steel ingots imported from India. The Damascene blades were more durable and sharper than traditional western blades.

During the late 19th century industrialists used carbon black for making special materials. Today we know carbon black is a nano-material and that carbon exhibits quantum effects. At the turn of the century, researchers discovered carbon black reinforced rubber and also increased the hardness of vulcanized natural rubber. It wasn't until 1910 that BF Goodrich started using carbon black filler to extend the lifespan of tires.

Today scientists know the reason adding carbon black to rubber enhances it strength is due to the interaction between the nano-sized carbon particles and the rubber. Rubber is a tightly compacted material which causes it to bend Nanosized carbon particles are small enough to get in between these tight spaces. Hence we can see that throughout history there are specific cases where scientific theory was thoroughly understood well before the application was developed. Surfaces of nanosized carbon materials also show collective polarization (*Nanocarbon-Inorganic Hybrids: Next Generation Composites. Dominik Eder*).

Carbon **Materials**
Soil, silicon, graphene, silicon carbide, diamond carbon, titanium, tar, diamond, coal, carbon fiber and carbon nanotubes.

Another example of nano-particles at work is water and oil. Many people know that only a small amount of oil added to turbulent water will calm the turbulence of the water. This effect occurs due to a mutual repulsion that exists between the water and the oil. This effect is so strong it causes oil to push itself away from the water creating a widespread, almost invisible film on top of the water.

Experiments by Lord Rayleigh, Irving Langmuir, as well as other researchers, discovered that oil floating above dense water caused a monolayer film that was just a few nanometers thick. This simple observation has made possible thin film coating technologies that exist in our everyday lives.

During the 1920s Katherine Blodgett and Langmuir immersed a substrate into a solution and coated the substrate with a barium separate film that measured exactly one molecule thick. Today this discovery is known as the Langmuir–Blodgett Film and it allows researchers to create extremely precise thin films for use in materials science and electrical engineering.

Additional experiments by Blodgett and Langmuir discovered that a thin film of 44 or 46 molecule layers will cancel out the reflection of light on glass. Hence today almost all lenses that need minimal reflection, such as telescopes or cameras are coated with no reflective thin films.

Quantum Dot Size = Frequency
One of the beautiful things about quantum dot technology is that the frequency can be changed

by changing the size. This is akin to a series of glass tubes filled with various levels of water. As one strikes the tube with a small hammer, a different tone is emitted, depending upon how much water is in the tube.

Quantum dots emit light of varying frequencies when light or electricity is applied. The frequencies emitted can be tuned to precise levels by changing the dots' size. Precision exists in this medium due to the very small nano-particle size. Quantum dots have potential use in solar cells, transistors, LEDs, quantum computing, diode lasers and medical imaging. When suspended in solution, they can also be used in spin-coating and inkjet printing.

The Selenium Cadmium Photoresistor
When selenium is combined with cadmium it greatly enhances its sensitivity. This allows it to pick up very weak sources of light.

A research study looking at a photo detector made with selenium and cadmium found that it was able to detect volume pulses from the ear in humans and animals. The element of the photo detector exists as a film-type selenium-cadmium photoresistor. The detector uses a tungsten glow filament and detector frame made from organic glass (*Use of highly sensitive cadmium selenide photoresistors to study the peripheral circulation in man and animals. V. B. ZakharzhevskiiA. O. Olesk. 23 November 1964*).

What is Cadmium Selenide?
Cadmium Selenide (CdSe) is a black to red-black solid that has n-type semiconductor properties. Much of today's research on cadmium selenide is on its potential use in nanoparticles. Three types of Cadmium Selenide exist in nature. They are sphalerite (cubic), wurtzite (hexagonal) and rock-salt (cubic). Sphalerite converts to wurtzite upon moderate heating. The transition begins at around 130 °C, and after reaching 700 °C for a day, the transition is final. Rock salt structures occur under high pressures. Himalayan rock salt lamps consist of rock salt.

Associative Remote Viewing and Geographical Location / Region
Over the last 3 years of doing ARV sessions in different regions, I have found that the regions that get the best results are regions where the **Gravity Bouguer Anomaly shows little to no anomalies**. It is interesting to note that the Climax Mine region in Colorado happens to be a region with a low Gravity Bouguer Anomaly. We shall explore this in detail later on.
Wulfenite is also called lead molybdate and

is commonly found with lead-rich deposits. Associated minerals that are also found with Wulfenite include powellite, molybdenum and molybdenite. Mineral deposits occur Colorado, Alaska, California, Oregon, Utah, Washington, South America, China and elsewhere. Powellite occurs in hydrothermal deposits of molybdenum and with the minerals basalt, pegmatite and tactite. Other minerals found with powellite include ferrimolybdite, stilbite, molybdenite, laumontite and apophyllite.

Wulfenite has a flattened tetrahedral structure containing four oxygen ions that occupy the four tetrahedral corners around a molybdenum ion. Lead ions bind with wulfenite. Wulfenite is also often associated with barite [barium sulfate, $BaSO4$], sphalerite [zinc sulfide, ZnS] and galena.

The world's largest molybdenite mine was once at Climax, Colorado called the Climax Mine. Additional mines include shenandoah lead-zine mine in Good Spring Mining District (Osgood Mountains), Grethcell Mine in Nevada.

Hexagon Shaped Minerals
I state in our second edition of remote viewing, Remote Viewing. The Complete User's Manual on Experiencing Future Consciousness, that Hexagon Shaped structures have a lot to do with devices

that enhance remote viewing. This may be due to the fact that microtubules are hexagon shaped.

Here are some Hexagon shaped Minerals - Amethyst, Graphite, Hexagonal Boron Nitride, Selenium, Carbon Graphite and Telluriuim[36].

During our remote viewing sessions we had used the minerals amethyst, selenium and hexagon boron nitrate with success. We also state that Tellurium is one of the "rare elements" and that it also may be a remote viewing enhancer. Hence, materials with hexagonal shapes may be exhibiting quantum coherence (*Wolfram Sander et al. 2017. Experimental Evidence for Heavy Atom Tunneling. Angewandte Chemie International Edition, 2017. The Cope Rearrangement of 1,5 Dimethylsemibullvalene*).

Research has shown that the atoms of Carbon (a heavy atom) behave like particles and also like waves. This quantum-mechanical effect often occurs in light particles such as hydrogen atoms or electrons. Research scientists have also found that carbon atoms display tunneling effects. A study stated that solvents enhance the ability for certain materials to tunnel (*The Cope Rearrangement of 1,5 Dimethylsemibullvalene-2(4)-d1: Experimental Evidence for Heavy Atom Tunneling. Angewandte. Tim Schleif, Joel Mierez-Perez, Stefan Henkel, Melanie Ertelt, Weston Thatcher Borden, Wolfram Sander. Chemie International Edition, 2017; DOI: 10.1002/anie.201704787*).

Does Carbon Dioxide Increase Photosynthesis?

A research study looked at whether enrichment of carbon dioxide increases photosynthesis in plants. The study found that concentrations of carbon dioxide will stimulate photosynthesis by 20 to 40% by the year 2050. The study also found that photosynthesis was increased under water stress conditions. (*Hourly and seasonal variation in photosynthesis and stomatal conductance of soybean grown at future CO(2) and ozone concentrations for 3 years under fully open-air field conditions. Bernacchi CJ et al. Plant Cell Environ. 2006 Nov;29(11):2077-90*).

Further   Reading
Projected land photosynthesis constrained by changes in the seasonal cycle of atmospheric CO_2.
Sabrina Wenzel et al. 27 October 2016.

How Tungsten Diselenide is used to Make an Artificial Leaf Solar Cell

The process of photosynthesis is basically a catalyst reaction. Nanoflake tungsten diselenide happens to be a perfect catalyst for artificial photosynthesis. It may be that nano particles/quantum dot type substances enhance the effect / rate for energy to travel multiple pathways

Researchers from the University of Illinois at Chicago have developed a new solar cell which converts hydrocarbons into fuel, using only sunlight for energy. Standard solar cells convert sunlight into electricity. The problem with this is that the energy must be stored in bulky and heavy batteries. The new solar cell does the work of plants. It converts atmospheric carbon dioxide into fuel. Hence in the future, we may witness solar farms composed of artificial leaves, removing carbon from the atmosphere while at the same time producing energy-dense fuel efficiently.

One of the researchers, Amin Salehi-Khojin, states that this new solar cell is not photovoltaic. Instead it is photosynthetic. This means instead of creating unsustainable energy (from fossil fuels to greenhouse gas) it generates power in reverse (by absorbing carbon dioxide from earth's atmosphere
and converting it into fuel using sunlight). The final byproduct of plant photosynthesis is a form of sugar. The new CO2 cell delivers syngas. This consists of a mixture of hydrogen gas and carbon monoxide. Syngas is able to be burned directly and it can also be converted into diesel or other types of fuels that exist as hydrocarbons. This in turn will render fossil fuels obsolete because of its ability to turn CO2 into fuel at a cost far less than a gallon of gasoline.
The majority of chemical reactions that allow

for a conversion of CO_2 into burnable forms of carbon rely on expensive metals such as silver. However the researchers in this study focused on nano-structured compounds known as transition metal dichalcogenides (also called TMDC's) which are catalysts.

TMDC Catalysts

TMDC catalysts are commonly used to produce hydrogen. The best catalyst they discovered was nanoflake tungsten diselenide. Tungsten diselenide is able to more completely break carbon dioxide's chemical bonds. The study found that the catalyst reaction of tungsten diselenide occurs 1,000 times faster than standard noble-metal catalysts.

The researchers used an ionic fluid known as ethyl-methyl-imidazolium tetrafluoroborate mixed at 50-50 with water to generate the tungsten diselenide reaction. The water and ionic liquid preserved the catalyst reaction allowing it to reduce its ability to consume itself in the reaction process. (*Breakthrough solar cell captures CO2 and sunlight, produces burnable fuel July 28, 2016 UC Today*), (*Technical reference - Nanostructured transition metal dichalcogenide electrocatalysts for CO2 reduction in ionic liquid. Mohammad Asadi et al. Jul 2016*).

References. Chapter 7

(1) Light exposure before learning improves memory consolidation at night Li-Li Shan et al. Oct 2015

(2) Hippocampus, amygdala and stress: Interacting systems that affect susceptibility to addiction. Pauline Belujon and Anthony A. Grace. Jan 2011

(3) Reward, dopamine and the control of food intake: implications for obesity Nora D. Volkow et al. Nov 2010

(4) The role of the basal ganglia in learning and memory: Insight from Parkinson's disease. Karin Foerde and Daphna Shohamy. Nov 2011

(5) Dopamine and light: dissecting effects on mood and motivational states in women with subsyndromal seasonal affective disorder Elizabeth I. Cawley, et al. Nov 2013

(6) Seasonal Effects on Human Striatal Presynaptic Dopamine Synthesis Daniel P Eisenberg, et al. Nov 2010

(7) Circadian rhythms of dopamine, glutamate and GABA in the striatum and nucleus accumbens of the awake rat: modulation by light. T.R. Castañeda et al. Apr 2004

(8) Melatonin and seasonal rhythms. T.A. Wehr. Dec 1997

(9) Extraction and Quantification of GABA and Glutamate from CyanobacteriumSynechocystis sp. PCC

6803. Simabkanwal and AranIncharoensakdi. September 20, 2016
(10) The role of glutamate in epilepsy and other CNS disorders. B.S. Meldrum. Nov 1994
(11) Trash to Treasure: From Harmful Algal Blooms to High-Performance Electrodes for Sodium-Ion Batteries. XinghuaMeng et al. Sept 2015.
(12) Central role of hypothalamic digoxin in conscious perception, neuroimmunoendocrine integration, and coordination of cellular function: relation to hemispheric dominance. Kurup RK1, KurupPA.. June 2002
(13) Light-sensitive brain pathways and aging V. Daneault et al. March 2016
(14) Light as a modulator of cognitive brain function. G. Vandewalle et al. Oc 2009
(15) Vandewalle G, Maquet P, DijkDJ.Light as a modulator of cognitive brain function. Trends Cogn Sci. 2009;13(10):429–438. doi: 10.1016/j.tics.2009.07.004
(16) Vandewalle G, Schwartz S, Grandjean D, Wuillaume C, Balteau E, Degueldre C, Schabus M, Phillips C, Luxe A, Dijk DJ, Maquet P, et al. Spectral quality of light modulates emotional brain responses in humans. ProcNatlAcadSci U S A. 2010;107(45):19549–54. doi: 10.1073/pnas.1010180107

(17) Vandewalle G, Archer SN, Wuillaume C, Balteau E, Degueldre C, Luxen A, Dijk DJ, Maquet P, et al. Effects of light on cognitive brain responses depend on circadian phase and sleep homeostasis. J Biol Rhythm. 2011;26(3):249–259. doi: 10.1177/0748730411401736

(18) Vandewalle G, Schmidt C, Albouy G, Sterpenich V, Darsaud A, Rauchs G, Berken PY, Balteau E, Degueldre C, Luxen A, Maquet P, Dijk DJ, et al. Brain responses to violet, blue, and green monochromatic light exposures in humans: prominent role of blue light and the brainstem. PLoS One. 2007;2(11) doi: 10.1371/journal.pone.0001247. [PMC free article] [PubMed] [Cross Ref]

(19) Vandewalle G, Gais S, Schabus M, Balteau E, Carrier J, Darsaud A, Sterpenich V, Albouy G, Dijk DJ, Maquet P, et al. Wavelength-dependent modulation of brain responses to a working memory task by daytime light exposure. Cereb Cortex. 2007;17(12):2788–2795. doi: 10.1093/cercor/bhm007. [PubMed] [Cross Ref]

(20) Vandewalle G, Collignon O, Hull JT, Daneault V, Albouy G, Lepore F, Phillips C, Doyon J, Czeisler CA, Dumont M, Lockley SW, Carrier J, et al. Blue light stimulates cognitive brain activity in visually blind individuals. J CognNeurosci.

2013;25(12):2072–2085. doi: 10.1162/jocn_a_00450. [PMC free article] [PubMed] [Cross Ref]

(21) Vandewalle G, Hebert M, Beaulieu C, Richard L, Daneault V, Garon ML, Leblanc J, Grandjean D, Maquet P, Schwartz S, Dumont M, Doyon J, Carrier J, et al. Abnormal hypothalamic response to light in seasonal affective disorder. Biol Psychiatry. 2011;70(10):954–961. doi: 10.1016/j.biopsych.2011.06.022. [PMC free article] [PubMed] [Cross Ref]

(22) Vandewalle G, Balteau E, Phillips C, Degueldre C, Moreau V, Sterpenich V, Albouy G, Darsaud A, Desseilles M, Dang-Vu TT, Peigneux P, Luxen A, Dijk DJ, et al. Daytime light exposure dynamically enhances brain responses. Curr Biol. 2006;16(16):1616–1621. doi: 10.1016/j.cub.2006.06.031. [PubMed] [Cross Ref]

(23) Perrin F, Peigneux P, Fuchs S, Verhaeghe S, Laureys S, Middleton B, Degueldre C, Del Fiore G, Vandewalle G, Balteau E, Poirrier R, Moreau V, Luxen A, et al. Nonvisual responses to light exposure in the human brain during the circadian night. Curr Biol. 2004;14(20):1842–1846. doi: 10.1016/j.cub.2004.09.082.

(24) Dopamine and light: dissecting effects on mood and motivational states in women with subsyndromal seasonal affective

disorder Elizabeth I. Cawley, et al. Nov 2013.

(25) Primary and secondary features of Parkinson's disease improve with strategic exposure to bright light: a case series study. Willis GL, Turner EJ Chronobiol Int. 2007;

(26) Specificity of Regenerating Optic Fibres for Left and Right Optic Tecta in Goldfish. R Cronly-Dillon et al. Nature 251 (5475), 505-507. 1974 Oct 11

(27) Passive Avoidance Learning Results in Region-Specific Changes in Concentration of and Incorporation into Colchicine-Binding Proteins in the Chick Forebrain. Radmila Mileusnic et al. May 1980

(28) Replication of cortisol circadian rhythm: new advances in hydrocortisone replacement therapy. Sharon Chan and Miguel Debono. June 2010

(29) Effects of melatonin on microtubule assembly depend on hormone concentration: role of melatonin as a calmodulin antagonist. Huerto-Delgadillo L1 et al. Sep 1994.

(30) Experimental detection of the torsion field.. Yu.V.Nachalov, E.A.Parkhomov

(31) Cascaded exciton energy transfer in a monolayer semiconductor lateral

heterostructure assisted by surface plasmon polariton. J. Shi et al. June 2017
(32) Large-scale quantum-emitter arrays in atomically thin semiconductors. Carmen Palacios-Berraquero et al. Sept 2016
(33) Atomically thin quantum light-emitting diodes. Carmen Palacios-Berraquero et al. September 2016.
(34) Electron transfer and coupling in graphene–tungsten disulfide van der Waals heterostructures. Jiaqi He et al. Nov 2014.
(35) New ultrathin semiconductor materials exceed some of silicon's 'secret' powers, Stanford engineers find. Stanford University Article published August 11, 2017.
(36) Materials Handbook. Francois Cardarelli.

Chapter 8. Quantum Transitioning from Photons as the carrier of Information during Remote Viewing

The non-local manifestations of consciousness [1], which are more common during remote viewing sessions, may be explained via quantum coherence as shown by Penrose [2].

What is Quantum Coherence? Herbert Fröhlich was one of the first to predict quantum coherence in living cells. He was also a major contributor to explaining how superconductivity works. His research on quantum coherence in living cells was built upon from earlier research by Oliver Penrose and Lars Onsager, 1956[3] [4].

Quantum coherence occurs as a result of two sources of waves that become coherent as long as their phase difference, wave form and frequency are the same. To put it simply, quantum coherence exists as form of waves allowing stationary interference and is the result of all the properties between single waves, packets of waves or several waves.

Further Reading
Bose-Einstein Condensation and Liquid Helium. Oliver Penrose and Lars Onsager. Nov 1956

The information received during remote viewing sessions is most likely quantum in nature and quantum entanglement may exist at the very heart of remote viewing. Since 2001 it has been science fact that quantum transitions occur during the process of electron transport in plants during the light properties of photosynthesis[5].

Frohlich suggested macroscopic quantum phenomena occurring in matter existed as a form of energy transport without loss based on collective coherent oscillations, also called the Frohlich frequency[6].

Proof that the receiving of information during remote viewing sessions is due to quantum effects may be due to the fact that the effects of remote viewing or psychic ability cannot be affected by any type of shielding or any distance [7], of which electromagnetic energy is susceptible to shielding. Umezawa suggested that the brain's cells exist as a spatially distributed system capable of quantum mechanical degrees of freedom exhibiting quantum phenomena[8]. Davydov also studied Frohlich's model and found thin layers near cell membranes exhibit "solitonic excitation states" causing dissipation-free energy waves in proteins[9]. Jibu et al. states that microtubule quantum functioning links with other neurons via quantum photons which are coherent and tunnel through membranes[10] [11]

(12).

The theories of Hameroff and Penrose suggested quantum effects in brain functioning, including the emergence of consciousness, occurred via microtubules (MT) in the brain's neurons. This allows the arrangement of disordered oscillations into coherent photon modes inside the brain's microtubules[13].

Further Reading
Holographic View of the Brain Memory Mechanism Based on Evanescent Superluminal Photons. Takaaki Musha.

Remote viewing and Time
Remote viewing is not constrained by time. Gerard Hooft stated the multi-verse we exist in is made up of four-dimensional space-time that is holographed onto a two-dimensional membrane, much like the membranes that make up the cells in our body [14]. We can physically move freely about in three dimensions, however this is really an illusion that is caused by the architecture of quantum interference.

It may just be that information received during remote viewing comes from interference and that holography is somehow involved.

Further Reading
The Cellular Automaton Interpretation of Quantum Mechanics. Gerard 't Hooft. Dec 2015.

What is The Universal Wave Function?
The universal wave function is a term introduced by Hugh Everett[15]. It forms a core concept in the many-worlds interpretation of quantum mechanics[16]. It received recent investigation from Stephen Hawking and James Hartle[17].

During remote viewing our consciousness is able to access the master hologram and decode the information contained within it. As the hologram interacts with the Universal Wave Function it creates PSI effects. PSI exhibits weak effects, and it is hard to access on demand because we are victims of the interaction of the effects of interference and don't have a direct link to the Universal Wave Function. It only appears to us to exhibit faster than light properties because of the interaction between the two.

Entanglement effects occurring during remote viewing may be influencing perceptions experienced. Entanglement could also explain why remote viewing is so effective, yet utilizes such little energy. Hence no black hole in the center of a galaxy has to be physically harnessed to pierce the fabric of space/time.

British Physicist Sir Roger Penrose states that human consciousness emerges from quantum vibrations occurring in protein polymers called microtubules in the brain's neurons (which we shall discuss in greater detail later on)[18]. These vibrations exhibit interference type properties which are responsible for the "collapse of the wave function" in quantum mechanics and their resonation controls the process of which the brain's neuron's fire. This in turn generates consciousness which is connected to a "deeper order" via ripples in the space/time geometry. Hence consciousness behaves more like music or a quantum computer.

References. Chapter 8

(1) PEAR Lab and Nonlocal Mind: Why They Matter. EXPLORE May/June 2007, Vol. 3, No. 3. Page 191
(2) Quantum computation in brain microtubules? The Penrose–Hameroff 'Orch OR' model of consciousness Stuart Hameroff. March 2015
(3) Long-range coherence and energy storage in biological systems Authors H. Fröhlich. 1968
(4) Orchestrated Objective Reduction of Quantum Coherence in Brain Microtubules: The "Orch OR" Model for Consciousness Stuart Hameroff and Roger Penrose
(5) Bacterial photosynthesis begins with quantum-mechanical coherence. H. Sumi. 2001
(6) Trajectory of frequency stability in typical development. J. Frohlich et al. Mar 2015
(7) Book. A guide to remote viewing and transformation of consciousness by Russell Targ
(8) Advanced Field Theory Micro, Macro, and Thermal Physics Authors: Umezawa, Hiroom. ISBN 978-1-56396-456-5
(9) The Problem of Bioenergetics. DAVYDOV SOLITON
(10) Jibu, M.; Pribram, K.H.; Yasue, K. From conscious experience to memory storage and retrieval: The role of quantum brain

dynamics and boson condensation of evanescent photons. Int.J. Mod. Phys. B 1996, 10, 1753–1754. [Google Scholar]

(11) Jibu, M.; Yasue, K. What is mind?— Quantum field theory of evanescent photons in brain as quantum theory of consciousness. Informatica 1997, 21, 471–490. [Google Scholar]

(12) Jibu, M.; Yasue, K.; Hagan, S. Evanescent (tunneling) photon and cellular vision. BioSystems 1997, 42, 65–73. [Google Scholar] [CrossRef

(13) Consciousness in the universe: A review of the 'Orch OR' theory Author links open overlay panel. Stuart Hameroff & Roger Penrose. March 2014

(14) The Cellular Automaton Interpretation of Quantum Mechanics Gerard 't Hooft. Dec 2015.

(15) Hugh Everett, Relative State Formulation of Quantum Mechanics, Reviews of Modern Physics vol 29, (1957) pp 454–462. An abridged summary of The Theory of the Universal Wavefunction

(16) Many-Worlds Interpretation of Quantum Mechanics First published Sun Mar 24, 2002. Stanford Encyclopedia of Philosophy

(17) Stephen W Hawking, James B Hartle "The Wave Function of the Universe," Physical Review D, vol 28, (1983) pp 2960–2975

(18) Consciousness in the Universe: Neuroscience, Quantum Space-Time

Geometry and Orch OR Theory. Roger Penrose and Stuart Hameroff. Journal of Cosmology, 2011, Vol. 14.

Further Reading

The Many-Worlds Interpretation of Quantum Mechanics. Princeton Series in Physics. Bryce Seligman DeWitt, R. Neill Graham, eds. Princeton University Press (1973), ISBN 0-691-08131-X Contains Everett's thesis: The Theory of the Universal Wave Function, pp 3–140.

"The Wave Function of the Universe," Physical Review D, vol 28, Stephen W Hawking, James B Hartle (1983) pp 2960–2975

Relative State Formulation of Quantum Mechanics, Reviews of Modern Physics vol 29, (1957) pp 454–462. Hugh Everett, An abridged summary of The Theory of the Universal Wavefunction

Assessment of Everett's "Relative State Formulation of Quantum Theory", John Archibald Wheeler, Reviews of Modern Physics, vol 29, (1957) pp 463–465

Quantum Mechanics and Reality. Bryce Seligman DeWitt,, Physics Today,23(9) pp 30–40 (1970) also April 1971 letters follow up.

Chapter 9. The Hippocampus, Empathy and Psychic Ability

One of the interesting functions of the hippocampus region of the brain is that it is related to emotion. The ability to control emotion plays a key role in accurate remote viewing sessions.

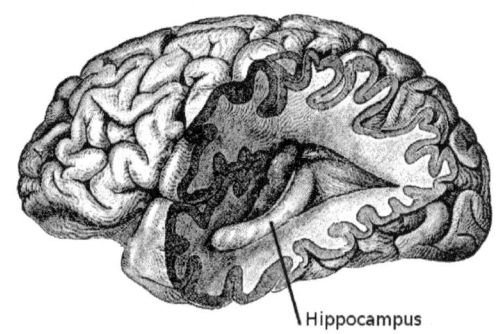
Hippocampus

In a research study involving black-capped chickadees conducted in upstate New York, the study showed that chickadees had a tendency to exhibit food-hoarding during the month of October each year. In the study, researchers went out and captured chickadees six times of the year and then measured the volume of several of their brain structures. They found that the hippocampus region in the brain had a larger volume that was relative to the rest of the brain during October compared to any other time of the year. The study concluded an association exists between the intensity of food hoarding and the size of hippocampus formation, suggesting the enhanced size may be caused by the increased use of spatial memory. Hence, October may be a time of year we are more prone to emotion, which may explain why all the horror movies are designed to be released during October each year (*Seasonal variation in hippocampal volume in a food-storing bird, the black-capped chickadee.Smulders TV et al. May 1995*). Pictured is a chickadee.

Where is the Brain's Hippocampus? The hippocampus exists as a small organ in the brain's medial temporal lobe. It forms a major part of the limbic system, which is the region of the brain that regulates emotions. The hippocampus is also primarily associated with memory, and in particular long-term memory.

This explains why traumatic events have such a long term impact on a person's personality due to the hippocampus which governs long term memory.

Extrasensory Perception and Hippocampus
A research study examined the brainwave activity occurring in the psychic Mr. Sean Harribance. The "extrasensory" processes performed by Mr. Harribance during the study showed functional changes occurring in his right parietotemporal cortices (or its thalamic inputs) and the hippocampus regions of his brain[2] [3].

Summary
One of perhaps multiple regions of the main, the brain's right cerebral hemisphere may be associated with enhanced psychic intuition. Especially in the right posterior cortical and hippocampus regions of the brain. The parahippocampal region of the brain is closely linked with the brain's hippocampus[3].

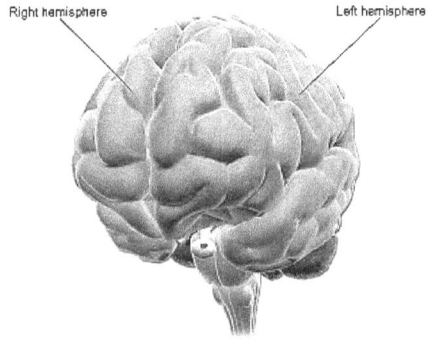

Cerebral Hemispheres
Right hemisphere   Left hemisphere

Hippocampus Empathy and Psychic Ability

Research is beginning to show that people who are telepathic or that are distant healers may possess the ability to activate on demand specific brain regions that are related to the brain's empathy circuit.

Empathy is the ability to infer and share the emotional experiences of others. A study showed that people who had the ability to interpret the thoughts of others had greater cognitive empathy compared to individuals without these abilities[5] [6]. Studies have also shown that the hippocampus region of the brain is important for empathy [7]. Additional research has shown that psychic mind readers have greater cognitive empathy than those without this ability [8]. Superior empathizing abilities have also been seen as important traits for telepathy and distant intentionality[9] [10]. Also a study showed that charisma plays an important role in telepathy [11].

The cuneus (the brain region that is associated with empathy) is reported to be linked with distant intentionality [12] and the hippocampus region (which is associated with empathy) plays an important function in telepathy[13].

What is Telepathy?
Telepathy is the communication of impressions from one mind to another independently of the common channels of sense [14].

The Results of an MRI Study on Telepathy
An MRI study looked at the brains of people who were performing telepathy. The study found that the region of brain that was most active during telepathy sessions was the right parahippocampal gyrus[15].

Summary
An association exists between the right parahippocampal gyrus and telepathy.

Where is the Parahippocampal Gyrus?
The parahippocampal gyrus is an isolated region of grey matter in the cortical region of the brain. It is part of the limbic system and surrounds the hippocampus. This region of the brain plays an

important role in memory retrieval and encoding[16].

One study titled Neuroprotective effect of olive oil in the hippocampus, published in April 2013 by M. Zamani and colleagues, found that when mice were fed olive oil, that it significantly reduced the cell death and decreased memory loss by **affecting activity in the hippocampus** region of the brain.

Oleic Acid is a pale yellow oily liquid that is found widely in nature. It is found in large quantities in olive oil, avocados and canola oil.

The Substance Bergamot and the Hippocampus

Essential oil of Bergamot has been shown to release neurotransmitters in the hippocampus regions of the brains of mice. Another study found that limonene acted very similar to Bergamot by enhancing GABA, dopamine and the hypothalamic-pituitary-adrenal regions of the brain [16].

The Brainwave Frequency of 7hz. The Key Frequency to Remote Viewing?

The parahippocampal gyrus in the human brain acts as a multimodal integrator of sensory information coming from regions of the neocortices. It is intimately involved with the processing of visuo-spatial information. When Ingo Swann, one of the early Masters of Remote Viewing, had his brain examined while performing remote viewing, anomalous MRI signals were noted in these same regions. The research study found that when Ingo Swann remotely viewed objects, that it showed a highly stereotyped 7Hz spike-like paroxysmal electroencephalographic activity located over the right occipital portion *(Remote viewing with the artist Ingo Swann: neuropsychological profile, electroencephalographic correlates, magnetic resonance imaging (MRI), and possible mechanisms. M.A Persinger. MA, Roll WG).*

References. Chapter 9

(1) Seasonal variation in hippocampus volume in a food-storing bird, the black-capped chickadee. T. V. Smulders et al. may 1995
(2) Protracted parahippocampal activity associated with Sean Harribance Michael A Persinger and Kevin S Saroka. Dec 2012
(3) Investigating paranormal phenomena: Functional brain imaging of telepathy Ganesan Venkatasubramanian et al. Dec 2008
(4) The role of the parahippocampal cortex in cognition Elissa M. Aminoff, et al. Jul 2013
(5) The role of shared neural activations, mirror neurons, and morality in empathy – A critical comment. Claus Lamma,and Jasminka Majdandžića. Jan 2015
(6) Empathy and motivation for justice: Cognitive empathy and concern, but not emotional empathy, predict sensitivity to injustice for others. Jean Decetyand Keith J. Yoder. April 2015
(7) HIPPOCAMPAL CONTRIBUTIONS TO THE PROCESSING OF SOCIAL EMOTIONS. Mary Helen ImmordinoYangand Vanessa Singh. Oct 2011.
(8) In search of "master mindreaders": are psychics superior in reading the language of the eyes? Dziobek I, Rogers K, Fleck S, Hassenstab J, Gold S, Wolf OT, Convit A Brain Cogn. 2005 Jul; 58(2):240-4.

(9) In search of "master mindreaders": are psychics superior in reading the language of the eyes? Dziobek I, Rogers K, Fleck S, Hassenstab J, Gold S, Wolf OT, Convit A Brain Cogn. 2005 Jul; 58(2):240-4

(10) Evidence for correlations between distant intentionality and brain function in recipients: a functional magnetic resonance imaging analysis. Achterberg J, Cooke K, Richards T, Standish LJ, Kozak L, Lake J J Altern Complement Med. 2005 Dec; 11(6):965-71

(11) Reinterpreting telepathy as unusual experiences of empathy and charisma. J.M. Donovan. Aug 1998.

(12) Evidence for correlations between distant intentionality and brain function in recipients: a functional magnetic resonance imaging analysis. Achterberg J, Cooke K, Richards T, Standish LJ, Kozak L, Lake J J Altern Complement Med. 2005 Dec; 11(6):965-71

(13) Cognitive and behavioral profile in a case of right anterior temporal lobe neurodegeneration. Gorno-Tempini ML, Rankin KP, Woolley JD, Rosen HJ, Phengrasamy L, Miller BL Cortex. 2004 Sep-Dec; 40(4-5):631-44

(14) Myers FW. Human personality and its survival of bodily death. London: Longmans; 1903.

(15) **Investigating paranormal phenomena: Functional brain imaging of telepathy. Ganesan Venkatasubramanian et al. Dec 2008.**
(16) **Investigating paranormal phenomena: Functional brain imaging of telepathy. G. Venkatasubramanian et al. July 2009**

Chapter 10. Substances that Enhance Remote Viewing

While our second book **Remote Viewing. The Complete User's Manual on Experiencing Future Consciousness** focused more on the hardware and devices you can build yourself to enhance your remote viewing sessions, this book is focused on not only the substances that enhance associative remote viewing, but why they do, how they work and where they come from.

Now let's examine substances that enhance associative remote viewing. This chapter will delve deeply into how dopamine is one of the best substances to enhance remote viewing. Below is the element symbol for Dopamine.

[Chemical structure diagram of dopamine showing a benzene ring with HO and OH groups, connected to a chain ending in NH_2]

First, let's look at the quantum effect and the substance known as linoleic acid. A research study concluded that linoleic acid may act as a control parameter inducing ordered states (which varies depending upon the linoleic acid's concentration). Hence the fatty acid couples to the ion channels which operate at quantum levels [1]. Hence, linoleic acid may strengthen the brain's microtubules due to its ability to induce ordered states. This in turn may enhance coherence, allowing for a longer lasting / stronger resonance enhancing the quantum wave effect. Below is the symbol for Linoleic Acid.

18:2n-6

Linoleic Acid

Resveratrol Synergizes with Linoleic Acid (*The combination of resveratrol and conjugated linoleic acid. N. Arias. July 2013*)

Is ATP Exhibiting Quantum Effects?
ATP is produced in our cells' mitochondria. It occurs as electrons move through a chain of intermediate molecules. When scientists calculate how fast this process occurs, the math shows that the process is occurring much faster than it should be.

Vlatko Vedral, a quantum physicist at the University of Oxford, has worked on this problem and thinks he has found a solution. Vedral proposes the excess speed is due to "superposition". Hence allowing a sort of quantum-mechanical wave effect to take place, allowing the electrons to be in two places at the same time. He has proposed that "quantum omnipresence" is what might be responsible for speeding the electrons' passage through the reaction chain. Vedral has done the math and his calculations support the idea[2] [3] . However he states we yet have no experimental studies to prove this. However one example has already been proven, and that is that some bacteria and plants receiver their energy via superposition.

During the days we perform our remote viewing sessions, we take extra Omega 3's, which may be acting as a fuel for ATP.

Linoleic Acid as Quantum Fuel
Because low dopamine levels have been associated with mood disorders[4], linoleic acid may be restoring dopamine levels. People with very low concentrations of linoleic acid in their bodies have been shown to be susceptible to mood disorders, especially in the brains of depressed people who completed suicide[5].

The School of Agriculture and Veterinary Medicine at the University of Bologna recently discovered that the fatty acid compositions of the brains of very obese rats contained a statistically significant reduction of linoleic acid [6] [7]. Lalovic et al. showed that depressed or bipolar suicidal subjects showed a decrease of linoleic acid in the ventral prefrontal cortex (*the decision making region of the brain in humans*) of their brains[8].

Research has also shown that very low levels of linoleic acid exist in platelets in people who contemplated suicide[9] [10] [11]. And another study found that piglets fed a diet with no linolenic acid from birth showed lower neurotransmitter activity taking place in the frontal cortex of their brains[12].

It may be that linoleic acid plays a major role in the connections that the membrane makes to the cytoskeleton. Hence low linoleic acid levels may block or reduce the flow of information processing.

Further Reading
Quantum human & animal consciousness: a concept embracing philosophy, quantitative molecular biology and mathematics. M. Cocchi et al. 2011 .

What is the Cytoskeleton?
The cytoskeleton contains microtubules, which are hollow crystalline self-assembling cylinders which exist as hexagonal lattices that are made of subunit proteins known as tubulin. The term

Hamiltonians, which is a quantum mechanical term for an operator that corresponds to the total energy of the system, shows a strong link between quantum effects in the membrane and in the cytoskeleton.

Cytoskeletal branches move about, they're living organisms which grow extensions such as dendrites or axons. The cytoskeleton decides which genes to turn on in the human brain. Quantum coherence most likely is taking place in the cytoskeletal microtubules which are hollow. The cytoskeleton also builds polarized cells and microtubules build on the asymmetry of the cytoskeleton and maintain the polarized environment[13].

What is Linoleic Acid?
Linoleic acid is a polyunsaturated omega-6 fatty acid. It is required during the molecular breakdown occurring in prostaglandins and cell membranes. Its structure is composed of 32 hydrogen atoms, 18 carbon atoms, and two oxygen atoms. Linoleic acid is also required during the molecular breakdown occurring in prostaglandins and cell membranes.

Linoleic acid is found in cooking oils and cannot be produced by the human body. Linoleic Acid is present in moderate amounts in Olive oil (3.5% to 21%) and high levels of linoleic acid are found in hemp (cannabis sativa) seed oil (*hemp seed oil is better absorbed*), flaxseed, walnut (juglans regia) seed oil (59% la), evening primrose (o. biennis) seed oil from canada (64% la),, borage seed oil (35–38%) and the fungi zygomycetes. Linoleic acid is also found in black raspberry seed oil (rubus occidentalis l., cv. jewel) [14].

Linoleic Acid Synergy
Linoleic Acid synergizes with Paclitaxel (taxol)[15]. Linoleic Acid also synergizes with Safflower Oil, Oleic Acid and Conjugated Acid[16].

A research study found that oral administration of evening primrose oil to mice exposed to radiation reduced the sensitivity of the radiated skin and prevented the radiation-associated increase in the blood flow of tissue. The study also found that evening primrose oil caused changes in the plasma levels of linoleic acid whether the mice had received radiation or not. Mice that took the evening primrose oil and received radiation, showed increased linoleic acid levels after 20 days. The study concluded that evening primrose oil reduced skin sensitivity of radiation-induced mice *(The effect of evening primrose oil on the radiation response and blood flow of mouse normal and tumour tissue. F. Rahbeeni et al).*

This is a major finding in our research because in our second book on remote viewing, **Remote Viewing. The Complete User's Manual**

on Experiencing Future Consciousness, we show that substances that protect against radiation, such as Gingko extract enhance ARV sessions and when solar activity is lower, which is a form of radiation, our ARV sessions are more accurate. This makes perfect sense because when solar activity is stronger, solar radiation levels are higher and earth's geomagnetic activity is also stronger and above average geomagnetic activity has been shown to reduce the accuracy of remote viewing sessions.

Linolecic Acid and Taxane Synergy
A research study found that conjugated linoleic acid exhibits anti-tumor effects when combined with taxane and is researching this combination as a powerful cancer treatment substance[16]. Taxane is found in the pacific yew tree. The seeds of the Pacific Yew are commonly eaten by wildlife, but are extremely toxic if humans eat it. It is interesting to note that one species of Yew Tree known as Himalayan Yew happens to contain abundant oleic and linoleic acids and that the substance Taxol found in Pacific Yew happens to synergize with Linoleic Acids[17].

Linoleic Acid Amounts in Some Oils[18]

- Safflower Oil 78%
- Grape Seed Oil 73%

- Poppy Seed Oil 70%
- Sunflower Oil 68%
- Flax seed oil also contains adequate amounts of Linoleic Acid

Gems and Minerals that Enhance Remote Viewing

In our second book we go into one of the better minerals that enhance remote viewing, **Wulfenite**. Here in this next edition we shall now cover some additional minerals that enhance remote viewing.

Cavansite

Cavansite was originally discovered in 1967 in Malheur County, Oregon Cavansite is composed of calcium vanadium oxide. In metaphysics it is said to aid intuition, remote viewing, psychometry and mediumship. *(The Pocket Book of Stones. Robert Simmons)*. Cavansite also contains the mineral vanadium. Associated minerals include calcite quartz and zeolite. Its colour is rich blue (*Staples, L.W., Evans, H.T. Jr., and Lindsay, J.R., "Cavansite and pentagonite, new dimorphous calcium vanadium silicate minerals from Oregon", American Mineralogist, Vol. 58, pg 405-411, 1973*).

Additional minerals that enhance remote viewing include:

- **Amazez**
- **Benitoite**
- **Ulexite**. High in Boron. The fiber-optic effect is the result of the *polarization of light* into slow and fast rays within each fiber, the internal reflection of the slow ray and the refraction of the fast ray into the slow ray of an adjacent fiber
- **Clear Apophyllite**
- **Blue Hemimorphite.** Hemimorphite contains an abundance of zinc and is a component of calamine. It has been historically mined from the upper regions of zinc and lead ores and is associated with smithsonite. Hemimorphite is also found in association with sphalerite.

Reference
The Book of Stones, Revised Edition: Who They Are and What They Teach. By Robert Simmons, Naisha Ahsian.

References. Chapter 10

(1) Linoleic acid: Is this the key that unlocks the quantum brain? Insights linking broken symmetries in molecular biology, mood disorders and personalistic emergentism . Massimo Cocchi et al. April 2017
(2) Inflationary Cosmology as a Probe of Primordial Quantum Mechanics. Antony Valentini. May 2008
(3) Evidence of Primordial Non-Gaussianity (f NL) in the Wilkinson Microwave Anisotropy Probe 3-Year Data at 2.8 σ. Amit P. S. Yadav and Benjamin D. Wandelt. May2008
(4) The role of dopamine in mood disorders. D.J. Diehl and S. Gershon. March 1992
(5) Lalovic A, Levy É, Canetti L, Sequeira A, Montoudis A, Turecki G. Fatty acid composition in postmortem brains of people who completed suicide. J Psychiatry Neurosci. 2007;32:363–370.
(6) Cardenia V, Vivarelli F, Canistro D, Estrada MTR. Linoleic acid in brain rat and obesity disease: a new starting point for chronic disease? Commun Ital Soc Exp Biol (Sect Bologna). 2015
(7) Preference for linoleic acid in obesity-prone and obesity-resistant rats is attenuated by the reduction of CD36 on the tongue. Christina S.-Y. Chen et al. Dec 2013

(8) Lalovic, A., Levy, E., Canetti, L., Sequeira, A., Montoudis, A., & Turecki, G. (2007). Fatty Acid Composition in Post-Mortem Brains of People Who Completed Suicide. Journal of Psychiatry & Neuroscience, 32, 363-370.

(9) [Lipids, depression and suicide. A. Colin et al. Feb 2003

(10) Omega-3 Fatty Acids and Depression: Scientific Evidence and Biological Mechanisms Giuseppe Grosso, et al. March 2014

(11) Omega-6 fatty acids and greater likelihood of suicide risk and major depression in early pregnancy Juliana S. Vaz, et al. May 2013.

(12) S. de la Presa Owens, S.M. Innis Docosahexaenoic and arachidonic acid prevent a decrease in dopaminergic and serotoninergic neurotransmitters in frontal cortex caused by a linoleic and α-linolenic acid deficient diet in formula-fed piglets J. Nutr., 129 (1999), pp. 2088–2093

(13) Beyond polymer polarity: how the cytoskeleton builds a polarized cell Rong Li1 & Gregg G. Gundersen. Nov 2008

(14) Long-chain polyunsaturated fatty acid sources and evaluation of their nutritional and functional properties Elahe Abedi and Mohammad Ali Sahari. Sept 2014.

(15) The therapeutic efficacy of conjugated linoleic acid - paclitaxel on glioma in the rat. Ke XY et al. Aug 2010

(16) In vitro synergistic efficacy of conjugated linoleic acid, oleic acid, safflower oil and taxol cytotoxicity on PC3 cells Sadi Kızılşahin et al. May 2014
(17) Medicinal Plants. Alice Kurian, M. Asha Sankar
(18) Comparison of dietary conjugated linoleic acid with safflower oil on body composition in obese postmenopausal women with type 2 diabetes mellitus1,2,3,4 Leigh E Norris et al. Sept 2009

Further Reading
Understanding nutrition, depression and mental illnesses T. S. Sathyanarayana Rao et al. April 2008

Chapter 11. Polarized Light

Polarized Light and Plant Growth
In my second book of remote viewing Remote Viewing. The Complete User's Manual on Experiencing Future Consciousness, I mention the importance of circularly polarized light, which is present in the light of some stars. Stars happen to emit coherent light [1] and our most successful associative remote viewing sessions always occurred during nighttime, especially around midnight.

Left-handed Circularly Polarized Light and its effects on Lentil and Pea plant Growth
A study looking at the growth characteristics of lentil and peas using left-handed and right-handed circularly polarized light found the shoots of the plants grew faster when they were placed under left-handed polarized light. The study found that in both plants (*the lentil and pea*) that the circular polarization of light did not significantly change when the light penetrated the outer layer of leaf or stem, resulting in a small property of birefringence. Hence the accelerated growth of the shoots exposed to left handed circular polarized light came from the absorption in the interior of the stems or leaves. This also happens to be the region of the plant where photosynthesis takes place, as well as the region

where linoleic acid is found in highest concentrations[2]. This could mean that there exists a connection between quantum coherence and polarized light, specifically circularly polarized light, due to the fact that the photosynthesis in plants utilizes quantum effects. Also it is interesting to note that dopamine has a beautiful glow under polarized light.

In another research study, circular polarization was detected in photosynthetic microbes. The study concluded that circular polarization spectroscopy may act as a powerful remote sensing technique for searches of generic life[3].

Substances that turn the plane of polarization (rotate) include turpentine and lemon essential oil. M. Wiedemann (1851) stated that when submitting these substances to magnetism, that the rotation would-be proportional to one another through all colors of the spectrum (*Report of the 13th meeting of the British Association for the Advancement of Science held at Oxford University. Page 55. June / July 1860. London. John Murray. Albemarle Street. 1861*). Also it is interesting to note that mung beans which are high in molybdenum, are sensitive to magnetic fields (*The effects of inverter magnetic fields on early seed germination of mung beans. Huang HH and Wang SR. Dec 2008*).

Further Reading

Biological Effect of Audible Sound Control on Mung Bean (Vigna radiate) SproutW. Cai et al. Aug 2014

Also not only does water create coherence, as shown to occur in microtubules[4] [5], but water can also polarize light either via reflection or by being viewed from while underwater. This is explained in further scientific detail in our second remote viewing book **Remote Viewing. The Complete User's Manual on Experiencing Future Consciousness**.

References. Chapter 11

(1) Optical analogues of the Newton–Schrödinger equation and boson star evolution Thomas Roger et al. Nov 2016
(2) The Effect of Circularly Polarized Light on the Growth of Plants. Pavel P Shibayev and Robert Pergolizzi. Jan 2011.
(3) Detection of circular polarization in light scattered from photosynthetic microbes. W.B. Sparks et al. May 2008.
(4) Quantum optical coherence in cytoskeletal microtubules: implications for brain function. M. Jibu et al. 1994
(5) Quantum-mechanical coherence in cell microtubules: a realistic possibility? N.E. Mavromatos. May 1999

Chapter 12. The Mid-Brain Dopamine System

The Mid-Brain Dopamine System (MDS) is an ancient evolutionary reward system which provides signals to the brain when the unexpected occurs[1].

The dopaminergic region of the brain is the part of the brain that regulates feedback processing (*Keitz 2008*) [2]. Bananas contain very high levels of dopamine, which is a reward activator in the brain. It is kind of interesting to note that bananas are used as the main reward food for chimpanzees when they have successfully performed a task.

What is Dopamine?
Dopamine is known as a caltecholamine neurotransmitter which is important for the

regulation of the body's movements. Hence, healthy amount of dopamine in the body translate to the ability for your brain to be more alert.

What are the Effects of Dopamine?
Dopamine enhances feelings of satisfaction and well-being, especially when an especially challenging task has been completed and as we shall show later on in this book, theta brainwaves manifest themselves after a series of tasks have been completed. Dopamine also reduces compulsive behavior. Anti-psychotic medications block dopamine levels in the brain. This is because schizophrenics have been shown to have excess levels of dopamine in their brain[3].

The Zacks Functional MRI Experiment
In a research study known as the Zacks Functional MRI Experiment, subjects were shown movies. The movies were stopped part-way through the movie while the subjects were texted what was going to happen next in the movie. MRI scans of the brains of the subject's during the study found that significant activity occurred in the substatianigra, which happens to be the part of the brain impacted by people with Parkinson's disease. This substatianigra region also controls adaptive decision making and movement. The experiment concluded that the mid-brain reacted

strongly when the subjects found that they had gotten the answer wrong[4].

What is Parkinson's disease?

Parkinson's disease (PD) is the second most common neurodegenerative disorder caused by massive losses of dopamine[5]. As we stated earlier in this book, Michael J. Fox, who played the role of Marty in Back to the Future has been diagnosed with Parkinson's and dopamine plays a major role in associative remote viewing. Was this a coincidence by the casting crew?

The Reward Effect, Dopamine and Enhanced Precognition

Precognition acts as a system which alerts our attention to threats. The response to the threat acts as a form of reward. It feels good to be right and to reward ourselves after accomplishing something. Achieving success using our psychic gifts is one example.

Stronger emotion of reward = enhanced sensitivity to future events.

Dopamine and Feelings of Satisfaction

Precognition is simply explained as experiencing enjoyment and reward. It is closely related to trauma and pain in that one receives enjoyment knowing that they heeded their intuition to avoid a serious encounter of future pain. This explains why positive signals received via reward are "*perceived*".

Modern psychiatry was founded by studying traumatic experiences *(Vander Kolk and van der Hart, 1991)*[6]. Frederic Myers stated that trauma was what powered strong psychic phenomena[7]. Perhaps traumatic events cause shockwaves through space and time via a type of psychic ether.

Further   Reading
Brain, Mind, and Medicine: Charles Richet and the Origins of Physiological Psychology. New Brunswick, NJ: Transaction. Wolf S. (1993)

Relationship between trauma and dissociation. Dorahy, M. J. and Van der Hart, O. (2007)

The Case of Eusapia Palladino: Gifted Subjects' Contributions to Parapsychology'. Journal of the Society for Psychical Research. Alvarado C. S. (1993).

Two of the most powerful reward electors in nature are food and sex. A study involving college students who were looking at random pictures on a screen, moments **BEFORE** pornographic or violent images were displayed showed there was a change in their nervous system[8].

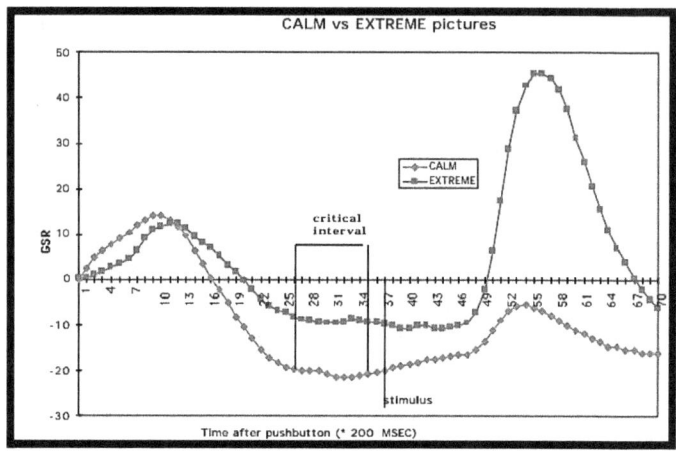

As the study shows, EXTREME emotions increased the presentiment effect. Shown below is the response lapse time.

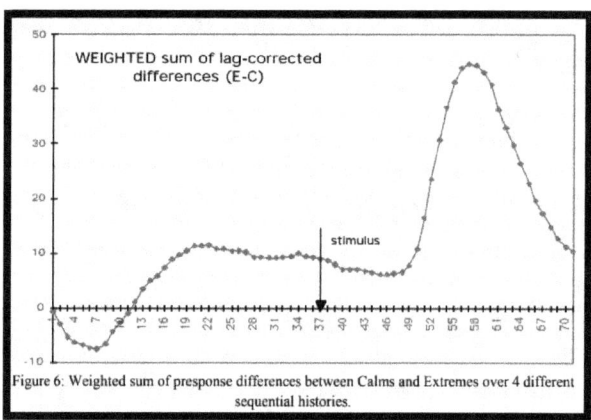

Figure 6: Weighted sum of presponse differences between Calms and Extremes over 4 different sequential histories.

Image courtesy of Dick J. Bierman. Emotion & Intuition. University of Utrecht & Amsterdam.

This study has since been repeated numerous times and exhibiting the same results. This shows that dopamine is closely linked to the nervous system.

Earth's geomagnetic activity has been shown to respond to events before they occur, as witnessed in the 911 attacks which we outline in great detail in our second remote viewing book.

Summary

Presentiment research shows a reaction occurs in the body's nervous system a few seconds before an emotional / extreme event occurs. By using methods to go Into coherence such as heart math, one's nervous system connects with the earth, extending the length of presentiment action substantially, due to the heart's ability to resonate with the energies of earth while it is in a state of coherence.

Further   Reading
Deciding Advantageously before knowing. The Advantageous Strategy. Science, feb. 28 1997, 275-5304, pp. 1293-1295.Bechara, A., Damasio, H., Tranel, D. & Damasio, A.R. (1997).

Anomalous Anticipatory Response on Randomized Future Conditions, Perceptual and Motor Skills, 84, pp. 689-690. Bierman, D.J. & Radin, D.I. (1997).

An Alternative Method for Significance testing of Waveform difference potentials. Psychophysiology, 30, pp.518-524. Blair, R.C. & Karniski, W. (1993)

CNV as an index of precognitive information. European Journal of Parapsychology, Hartwell, J. (1978).

Affective judgment and psychophysiological response: Dimensional covariation in the evaluation of pictorial stimuli. Journal of Psychophysiology, 3-1, pp. 51-64. Greenwald, M.K., Cook, E.W. & Lang, P.J. (1989).

Measuring the relative magnitude of unconscious influences. Consciousness & Cognition, 4, pp. 422-439. Bierman Emotion & Intuition. Merikle, P.M., Joordens, S. & Stolz, J.A. (1995).

Affect, Cognition, and Awareness: Affective priming with Optimal and Suboptimal Stimulus Exposure. Journal of Personality and Social Psychology, 64, 5, pp. 723-739. Murphy, T.S. & Zajonc, R.B. (1993).

Unconscious Perception of Future Emotions. An experiment in Presentiment.Procs.of the 39th PA Convention, San Diego, 1996, pp. 171-185. Radin, D.I. (1997).

Dopamine, Reward and Pre-sentiment

Early research with mice showed dopamine release occurred in the brain after being rewarded in a behavior known as "food rewarded learning" [9]. More recent findings have discovered that it is **the anticipation of reward**, rather than the actual reward itself that triggers the release of dopamine[10] [11]. This effect has also been shown to occur in people who gamble [12].

In a Swiss study, research on the brain's of monkeys found that the rate of the firing of their neurons increased significantly right **before** a highly probable reward was provided[13]. It could be that remote viewing may be addictive for some people due to the anticipation of reward. Reward is a major player in remote viewing, producing dopamine. Our research shows that performing ARV sessions no more than once per week on average, with sessions separated apart from one another achieves the best results, in order to avoid 'burn out'.

Further Reading

Reward-related signals carried by dopamine neurons. Wolfram Schultz et al.

Predictive Reward Signal of Dopamine Neurons. Wolfram Schultz. July 1998

Principles of Pleasure Prediction: Specifying the Neural Dynamics of Human Reward Learning. Todd S. Braver and Joshua W. Brown. April 2003

The Immune System and Dopamine
When the immune system becomes activated, the cells release cytokines which activate signaling pathways. These in turn regulate the secretion and synthesis of the neurotransmitters acetylcholine, dopamine, norepinephrine and encephalin [14]. The neurotransmitter norepinephrine is made from dopamine. Dopamine production in the body comes from the amino acids phenylalanine and tyrosine. Phenylalanine is the precursor to tyrosine. Phenylalaninecan is found in Aspartame.

Tyrosine is made in the body from eating foods rich in Phenylalanine. Foods that contain lots of Phenylalanine include: Basil, cottonseed and sunflower seed oils. Also milk, cheese, nuts, fish and eggs also contain high levels of phenylalanine[15]. If taking a Phenylalanine supplement, it works best taken on an empty stomach. Basil essential oil contains large amounts of Linalool [16].

Foods highest in Tyrosine
Egg Whites and Spirulina.

Glutamine, which is responsive to light as we showed earlier, is another substance that is a precursor to dopamine.

Dopamine extends Lifespan in Worms
Compounds that target dopamine or serotonin in the body were found to extend lifespan in worms [17].

Linoleic acid Protects against loss of Dopamine
Linoleic Acid has been shown to protect dopaminergic neurodegeneration in studies conducted on worms [18]. The research study exposed Caenorhabditis elegans worms to a neurotoxin and then determined how well linoleic acid protected the worms. The worms that did not receive linoleic acid showed a 48% significant reduction in their "locomotion" rate. However the worms that received linoleic acid showed increased "locomotion" rates and exhibited significant recovery. The control worms survived on average 26 days and the worms exposed to the neurotoxin lived on average 21 days. Worms that were co-exposed to the neurotoxin and received linoleic acid supplementation showed a significant reduction in their neuronal degeneration. The study concluded that linoleic acid significantly suppresses movement disorder and dopaminergic neurodegeneration in C.

elegans worms [18].
Was the reason the worms survived longer with the linoleic acid was because of the quantum effects of linoleic acid we covered earlier? The study [18] stated that the reason the worms' lifespan was extended was because Alpha-linolenic acid suppressed the dopaminergic neurodegeneration in the brain regions. Because microtubules exist in the brain, it may be that the Alpha-linolenic acid is targeting regions in the brain responsible for quantum effects thus extending lifespan. Further research is necessary to explore this further.

Excess Linoleic Acid Accelerates Aging
ARV sessions are not recommended to be performed for excessive periods of time as they can contribute to the aging process through the excessive loss of dopamine. This is why our ARV sessions are performed between 1 to 2 weeks apart.

Studies have shown that excess linoleic acid shortens telomeres. This has been shown in a study where women who consumed large amounts of linoleic acid displayed shorter telomeres [19].

Also from experience over the years, I have found that ARV sessions do indeed put a strain on the body, possibly by enhanced oxidative stress caused by the mental stress. There is a "sweet spot" of linoleic acid that extends lifespan.

Summary
Dopamine is closely related to the success of precognition because our minds tend to be attracted to pleasure occurring in the future. Hence, experiences of pleasure in the future send signals back in time to the present where the connection becomes 'closed' due to the receiving of the information.
   Now that we have a clearer understanding of Dopamine and its effects, let's next examine some substances, methods and techniques used to enhance Dopamine levels in the body.

References. Chapter 12

(1) Midbrain dopamine neurons encode a quantitative reward prediction error signal. H.M. Bayer and P.W. Glimcher. July 2005
(2) Prefrontal cortex and striatal activation by feedback in Parkinson's disease. M. Keitz et al. Oc 2008
(3) The Role of Dopamine in Schizophrenia. Ralf Brisch, et al. May 2014
(4) Event Perception: A Mind/Brain Perspective Jeffrey M. Zacks et al. March 2007
(5) Multiple hit hypotheses for dopamine neuron loss in Parkinson's disease. D. Sulzer. May 2007
(6) VanderKolk BA., vanderHart O. The intrusive past: the flexibility of memory and the engraving of trauma. Am imago. 1991;48:425–454
(7) Trevor Hamilton, Immortal Longings: F.W.H. Myers and the Victorian Search for Life after Death (Exeter: Imprint Academic, 2009), pp. 359, ISBN: 9-781845-401238.
(8) Emotion and Intuition I, II, III, IV & V Unravelling variables contributing to the presentiment effect. Dick J. Bierman. Emotion & Intuition. University of Utrecht & Amsterdam
(9) Role of brain dopamine in food reward and reinforcement. Roy A Wise. June 2006

(10) Feedback that confirms reward expectation triggers auditory cortex activity. T. Weis et al. Oct 2013
(11) "Liking" and "Wanting" Linked to Reward Deficiency Syndrome (RDS): Hypothesizing Differential Responsivity in Brain Reward Circuitry. Kenneth Blum et al. May 2013.
(12) Neurobiological underpinnings of reward anticipation and outcome evaluation in gambling disorder. Jakob Linnet. Mar 2014
(13) Responses of Monkey Dopamine Neurons to Reward and Conditioned Stimuli during Successive Steps of Learning a Delayed Response Task. The Journal of Neuroscience, March 1993, 13(3): 900-913
(14) Molecular Biology of the Cell. 4th edition.
(15) Treating phenylketonuria by a phenylalanine-free diet. K. Start. 1988
(16) Linalool - a marker compound of forged/synthetic sweet basil (Ocimum basilicum L.) essential oils. N.S. Radulović et al. Oct 2013.
(17) A pharmacological network for lifespan extension in Caenorhabditis elegans. Ye X1 et al. April 2014
(18) Alpha-linolenic acid suppresses dopaminergic neurodegeneration induced by 6-OHDA in C. elegans. S. Shashikumar et al. Nov 2015
(19) Associations between diet, lifestyle factors, and telomere length in women. Aedín Cassidy. et al. May 2010.

Chapter 13. Methods that Enhance Dopamine

Dopamine levels in the body exhibit a seasonal variation with the lowest levels occurring during the spring and summer. One study measured the rate of blinking in people with schizophrenia and found during spring and summer dopamine levels were significantly enhanced [1], although this seasonal variation can vary with healthy adults [2]. This could account for the above average levels of suicide that occurs from late spring into early summer.

Our remote viewing research over the years also showed that the accuracy of our remote viewing sessions would always drop during summer [3]. In studies conducted on rats, stress has been shown to increase dopamine levels in young and old rats, but not in very old rats [4]. Most protein type foods will increase dopamine levels [5], which is why protein bars can be so addictive. However, as I show in almost all my anti-aging books, excess proteins can contribute to a build up ammonia in the body, one of the main mechanisms of aging.

Tyrosine

Tyrosine foods trigger a chemical reaction that releases dopamine [5]. Foods that contain Tyrosine include - Almonds, Bananas, Avocados, Chocolate, Coffee and Eggs.

The Herb White Peony and Dopamine

White peony is commonly used to heal neurodegenerative disorders such as Parkinson's Disease. This is due to the fact that Parkinson's Disease depletes dopamine levels. A study found that white peony was responsible for keeping dopamine levels at healthy levels in the body[6].

Gingko

Gingko may help keep dopamine in the brain circulating longer (increasing bandwidth) as has been proven in mice studies [7]. From our years of researching remote viewing, we have found taking 2 to 5 drops of gingko extract in lemon juice before our remote viewing sessions is extremely effective at contributing towards the success of our ARV sessions.

Geraniol and Dopamine

A research study found that Geraniol, which is found in the essential oils of rose, lemon and some spices, restored depleted dopamine levels in studies involving mice[8]. Another study found that Geraniol showed significant protection against Parkinson's disease in experiments done on fruit flies [9]. And in a final study, Curcumin (*found in the spice Turmeric*) was found to not only provide neuroprotective effects, but significantly reduced dopaminergic neuronal

oxidative damage in mice[10].
Geraniol happens to be a prime component of our TXP formula, which we use to enhance remote viewing by rubbing into our hands and breathing it in just before practicing the Heart Math Coherence Exercise. The TXP formula is shown later in this book.

Cacao Essential Oil
Cacao Essential Oil, also known as chocolate essential oil, has been shown to stimulate both dopamine and serotonin production[11].

Pistachio
Pistachio happens to be one of the few, if only nuts, that contains anthocyanins. As I have written in many of my anti-aging books, anthocyains play a key anti-aging role due to their anti-inflammatory and antioxidative effects. Research has found that Pistachio has been effective for the release of dopamine [12].

Another study conducted on mice showed that oral administration of anthocyanins suppressed dopamine abnormalities[13]. And in a final research study, extracts of blackcurrant and boysenberry showed a significant protective effect and also restored calcium buffering ability of cells that were subjected to oxidative stress induced by dopamine. The study found that blackcurrant polyphenolics showed the strongest protective effect[14].

Protecting Dopamine Flow
A study found that naringin, which is found in abundant levels in grapefruit and citrus fruits, **enhances the flow of dopamine** to the striatum via protective like effects [15] and studies have shown that resveratrol protects dopamine neurons against lipopolysaccharide-induced neurotoxicity due to its anti-inflammatory Actions[16].

Lower Lipopolysaccharide Levels
Because above average levels of lipopolysaccharide have been found to induce degeneration of dopaminergic neurons[17] eating yogurt [18] and/or bromelain [19] have been shown to lower lipopolysaccharide levels.

Spinach, Blueberry and Strawberry
Research has shown that dried aqueous extracts of strawberry, blueberry and spinach were able to enhance the release of dopamine in studies done on mice[20]. This could also be why chocolate covered strawberries are so popular, as both chocolate and strawberries enhance dopamine in the body.

Brilliant Blue

The food dye brilliant blue has been shown to restore dopamine levels [22].

Selenium and Dopamine

Our research has shown that the mineral selenium is a strong absorber of light. Hence this may be why selenium enhances dopamine levels in the body when taken as a supplement.

Excess selenium may show inhibitory effects. This is due to the fact that research on mice that were fed low amounts of selenium in their diets showed increased dopamine turnover[22]. Selenium found in blue corn is better absorbed by the body. This is similar to the effects of consuming calcium. Taking a lot of calcium in one meal in a day does not allow the body to absorb as much calcium as if you had taken small amounts of calcium 2 to 3 hours apart from each other through the course of a day. Riboflavin and selenium may be synergistic and provide neuro-cognitive protection. Plants and Fungi synthesize riboflavin. One study showed that rats that were fed a diet low in selenium or riboflavin had reduced glutathione levels[23]. This is because glutathione increases selenium.

Garlic and Dopamine Interaction

Garlic has been shown to interact with dopamine[24].

Dopamine and Onion Powder
Because onions have an antidepressant-like effect it would make sense that they enhance dopamine levels. Onions are also closely related to garlic. However, a study involving rats who took onion powder did not show an increase in their dopamine levels. Instead their dopamine levels were raised by the physical exercise given in the experiment[25].

Clary Sage
Clary sage essential oil has been shown to exhibit antidepressant like effects (*Lee et al 2014, Seol et al 2010), (Antidepressant-like effect of Salvia sclarea is explained by modulation of dopamine activities in rats. Seol GH et al. July 2010*).

Creatine Boosts Dopamine Levels
A research study examining the effect of creatine on the brain's neurotransmitters after exhaustive exercise, found that after participant's took the supplement creatine and then performed an exhaustive aerobic exercise, that they had higher dopamine levels during both the period they were exercising and also during their recovery. This was in comparison to the group that did not take creatine before the aerobic workout[26].

I mention in many of my anti-aging books that creatine is a powerful short term energy booster that helps keep the energy in the body burning for longer periods. Excess levels of creatine are detrimental. However just the right amount works very well.

Also creatine may play a role as intensive exercise such as in marathons has been shown to increase creatine levels [27] and from my experience, mild exercise the day an ARV session is about to take place seems to enhance remote viewing accuracy.

The Thyroid and Dopamine Function
In our second remote viewing book **Remote Viewing. The Complete User's Manual on Experiencing Future Consciousness**, I show that tachyon devices are related to the thyroid. It is interesting to note here that the thyroid happens to be an important part of dopamine production in the body.

A healthy functioning thyroid is the key to healthy dopamine levels[28] and the thyroid requires tyrosine in order to properly function [29]. Low tyrosine levels in the body cause low dopamine levels.

The three main essential elements for proper thyroid functioning include: Iodine (*available from Kelp*), Selenium and Tyrosine[30] [31]. Take Vitamin B6 with Tyrosine to enhance dopamine absorption. For example a banana with a vitamin B6 supplement. Tyrosine may also synergize with

Vitamin B6 as both are neurotransmitter enhancers although further research is necessary to confirm if this synergy exists.

Selenium
Selenium provides a long lasting effect against a loss of dopamine in the body[32].

L-DOPA
L-Dopa is also called Levodopa and is an amino acid made in the bodies of humans and some animals. It is also made in plants. L-Dopa is usually made in the body by the amino acid tyrosine and is a precursor to the neurotransmitter dopamine. It is also used in the treatment of Parkinson's disease. Research studies involving Levodopa found that Levodopa decreased the sensitivity in cognitive decisions in skeptics by making skeptics slightly more conservative[33].

Further Reading
Blackmore, S., & Moore, R. (1994). Seeing things: Visual recognition and belief in the paranormal. European Journal of Parapsychology, 10, 91–103.

A neural substrate of prediction and reward. Science, 275, 1593–1599. Schultz, W., Dayan, P., & Montague, P. R. (1997).

Are creativity and schizotypy products of a right hemisphere bias? Brain and Cognition, 49, 138–151. Weinstein, S., & Graves, R. E. (2002).

Mucuna Pruiens and Dopamine
L-Dopa readily crosses the blood brain barrier, compared to dopamine which does not. This gives it a significant advantage in treating Parkinson's disease.

When L-Dopa first enters the nervous system it becomes converted to dopamine. Because of its ease passing the blood brain barrier, it may cause adverse side effects, hence the substance known as Carbidopa is often given to offset the side effects[34].

L-Dopa can be purchased online often in the form of a Mucuna Pruiens extract. Pictured are the seeds.

Mucuna Pruiens seeds are also called Velvet Bean and have the following properties:

> ➤ A front line treatment for Parkinson's Disease[35]

- Shown in mice studies to be active against Snake Venom[35]
- Shown in mice studies to increase the levels of oxidized glutathione[35]
- Protects against glutathione loss[35]
- Reduce depression[35]

In India the seeds of Mucuna Pruiens have often been used as a nerve tonic and for male virility. The powdered seeds also reduce glutathione depletion. Research has also shown that L-Dopa reduces depression [35]. Other studies have shown that Mucuna Pruiens seeds show strong neuro-protective properties and that just a small amount produces strong effects. A double blind clinical and pharmacological study concluded that Mucuna Pruiens could be a very effective substance in treating Parkinson's Disease[36]. Another study titled: Short-term administration of omega 3 fatty acids from fish oil results in increased transthyretin transcription in old rat hippocampus, published in February 2003 and conducted by Laszlo G. Puskas and colleagues found that fish oil fed to elderly mice showed that 23 of their genes were altered and **effects were observed in the hippocampus regions** of the brain. The study found that gene expression of TTR was **similar to the effects caused by**

nicotine and Ginkgo extract. Pictured below is the transthyretin gene.

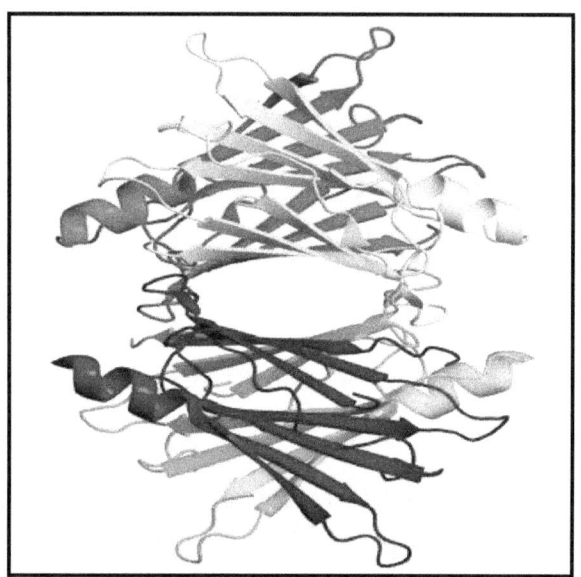

Transthyretin

Is Transthyretin (TTR) the Psychic Gene?
TTR is short for transthyretin. Transthyretin is a transport protein existing in the body's serum and cerebrospinal fluid. Its purpose is to carry the thyroid hormone thyroxine (**T4**). It also carries theretinol-binding protein bound to retinol (Vitamin A). Hence the name, transportation of retinol andthyroxine. The liver

releases transthyretin into the bloodstream and the choroid plexus releases transthyretin into the body's cerebrospinal fluid. The pineal gland consists of specialized secretory cells called pinealocytes which synthesize melatonin and secrete it directly to the body's cerebrospinal fluid, which then moves it into the blood[37].Nicotine has been shown to enhance the secretion of transthyretin from the choroid plexus in mice studies[38].

Other small molecules that bind to thyroxine binding sites include resveratrol and the drugs Tafamidis (Vyndaqel), flufenamic acid, diflunisal and toxicants (PCB)[39]. Transthyretin cerebrospinal fluid levels have been found to be lower in people with schizophrenia and other neurobiological disorders[40]. Glutamic Acid has been shown to enhance TTR stability (*Role of the Glutamic Acid 54 Residue in Transthyretin Stability and Thyroxine Binding. Masanori Miyata et al. 2010)*.

Further **Reading**
New Approaches to Assess the Transthyretin Binding Capacity of Bioactivated Thyroid Hormone Disruptors Mauricio Montaño et al. Aug 2012

Nicotine Protects Against Alzheimer's
Epidemiological studies have shown that cigarette smoking protects against several neurodegenerative disorders among them Alzheimer's disease.

A research study found that nicotine increased transthyretin levels and concluded the mechanism responsible for this was due to the increased secretion and biosynthesis of transthyretin from the choroid plexus. The study concluded that nicotine administration may offer therapeutic effects for patients with Parkinson's[39]. Also research has shown that Nicotinic receptor agonists enhance cognitive functioning (*Nicotinic and muscarinic agonists and acetylcholinesterase inhibitors stimulate a common pathway to enhance GluN2B-NMDAR responses. Masaru Ishibashia et al. July 2014*). Nicotine also Synergizes with Vitamin D3 (*Exp Biol Med (Maywood). 2009 Aug;234(8):908-17. doi: 10.3181/0811-RM-346. Epub 2009 Jun 22*).

Additional Substances that increase Transthyretin in the body
Omega 3 fatty acids (*Puskas et al 2003)*, Ginkgo biloba *(Watanabe et al 2001*), nicotine (*Li et al 2000*) and the female hormone estradiol (*also called oestradiol*) (*Tang et al 2004*) increase Transthyretin levels.

Could the reason women are more intuitive

then men be due to this specific hormone? and would men become more intuitive if given this hormone for short periods of time? Research is needed to verify this hypothesis unless the men start turning into women, it probably would not be a research avenue to consider. It is however interesting to note that the movie Predestination involved a male time traveler from the future who went back in time and had a sex change. When he became a women, he then had sex with the male from the future time, with a baby being born that is the result of the origin of him/herself. It sounds very confusing, but watching the movie makes it so much more clearer.

The movie Predestination was a perfect example of the predestination paradox, which is also known as causality loop, causal loop or closed time loop.

Fish Oil and Transthyretin
A study found that feeding fish oil to rats for a short period of time enhanced transthyretin in the hippocampus of aged rats[42]. Another research study found that stress and glucocorticoids *(steroid hormones)* increase transthyretin expression in rat choroid plexus[43].

Another study involving 3 month old rats that were fed fish oil for a month showed over expressed genes of Transthyretin (*Effects of dietary* omega-3 polyunsaturated fatty acids on brain gene expression Klára Kitajka, et al. July 2004).

Transthyretin Synergy
As we showed earlier small molecules bind to thyroxine. Hence other molecules that bind to thyroxine binding sites such as resveratrol should exhibit synergy with transthyretin. Omega 3's may also offer synergy. Let's take a look at the data and find out.

A research study titled: Effect of Resveratrol and Nicotine on PON1 Gene Expression that was published in February 2013 and conducted by Gupta, et al. found that combining resveratrol with nicotine had a *significant effect* on **PON1 gene** activity and concluded that it may offer significant protection against cardiovascular disease. Transthyretin also synergizes with IGF-I (*Evidence for synergistic action of transthyretin and IGF-I over the IGF-I receptor. Marta Vieira et al. 2016*).

What is IGF-I?
IGF-I stands for insulin-like growth factor which is found in animal proteins and zinc. Fish oil has been shown to enhance IGF-I levels in humans (*The effects of n-3 long-chain polyunsaturated fatty acids on bone formation and growth factors in adolescent boys. Damsgaard CT et al. June 2012*).

The PON1 Gene

PON1 is short for Serum paraoxonase/arylesterase and is also called A esterase or homocysteinethiolactonase. PON1 is a human enzyme encoded by the PON1 gene. PON1 exists in all mammals but is absent in the serum of reptiles, birds, fish and insects. PON1 is also used to hydrolyse pesticides and nerve gasses. The effects of PON1 have been found to exhibit anti-aging effects, however the sole mechanism is unknown[44].

Resveratrol and Fish Oil reduces Catecholamine levels

As we showed earlier in this book, Catecholamines enhance neurotransmitter function by strengthening the signals flowing between the brain's neurons. A research study found that combining reservatol with omega 3's reduced the decline of catecholamine in obese rats[45]. Omega 3's also increase dopamine and Omega 3's synergize with Vitamin B12. One study found that feeding rats omega 3 PUFA gave them a 40% increase in their dopamine levels in the frontal cortex region and also increased the binding to the dopamine D2 receptor [46]. A research study found that a combination of B12 and Omega 3's increased neurotransmission in the hippocampus region of mice[47].

Aspirin and Salicylate Protect Dopamine

Aspirin and Salicylate have been shown to protect against the depletion of dopamine in studies conducted on mice. A study concluded that the protective effects were possibly due to hydroxyl radical scavenging[48]. We have not found good results taking asprin before an ARV session. This may be due to its sedative/blood thinning effects.

Methods that Enhance the Release of Dopamine

Nicotine

The structure of Nicotine is similar in molecular structure to sodium. The pure forms of nicotine (freebase nicotine) react strongly with oxygen, water, and living tissues. Hence nicotine may have potential as a battery electrolyte.

Freebase nicotine is extremely poisonous and is used as an insecticide, lasting a short term in the environment (up to 30 minutes). Plants use nicotine to produce NADP (which is short for nicotinic acid adenine dinucleotide phosphate). NADP breaks apart the water molecules during plant photosynthesis. Plants make NADP from the nitrates and phosphates in soil (fertilizer).

If Plants contain nicotine would that not kill them?

Plants mix their nicotine with the substances

called nicotine sulfate, nicotine oxide (*cotinine*), nicotine citrate, nicotine malate and nicotinic acid (*vitamin B3, niacin*). These types of nicotine are non-toxic and highly stable and are essential to good plant health. Tobacco plants also contain a mix of nicotine malate and nicotine citrate. The reason tobacco plants have more nicotine then other plants is because they grow so fast, which is why hemp and corn also store and produce nicotine due to their rapid growth.

Nicotinamide / NAD
The purpose of NAD is the transfer of electrons from one molecule to another. Nicotinamide is a coenzyme that exists in all living cells. The coenzyme NAD+ was discovered by biochemists William John Young and Arthur Harden during 1906 while observing adding boiled and filtered yeast extract to uncoiled yeast extracts accelerated the alcoholic fermentation process.

Coenzymated B-3 Nicotinamide Adenine Dinucleotide is available as a supplement or can be bought as a pure chemical extract. Nicotine dinucleotide phosphate is also available in chemical form from suppliers. While you could take standard niacin, Coenzymated B-3 works better because it enters the bloodstream in its active form. This avoids the loss that occurs during digestion, and the liver's conversion process.

A research study found that brief exposure to low levels of nicotine boosted the brain's 'reward' system and blocked the system that limits reward duration (*Short- and Long-Term Consequences of Nicotine Exposure during Adolescence for Prefrontal Cortex Neuronal Network FunctionNatalia A. Goriounova and Huibert D. Mansvelder. Dec 2012*).

Another study found that significantly enhanced perceptual functioning and attention to detail (*also called eagle-eyed visual acuity*) occurs in some persons with autism spectrum conditions. The study concludes that the effect may be due to high numbers of dopamine receptors at the retinal or neural level due to perhaps increased levels of dopamine in these regions[50]. Nicotine also enhances neuron activity in the hippocampus region in mice brains (*The linoleic acid derivative FR236924 facilitates hippocampus synaptic transmission by enhancing activity of presynaptic alpha7 acetylcholine receptors on the glutamatergic terminals. Yamamoto S. etal. 2005*).

Further   Reading
Nicotine extends duration of pleasant effects of dopamine. John Easton. Medical Center Public Affiars. University of Chicago Chronicle

Rethinking Extrasensory Perception: Toward a Multiphasic Model of Precognition

L-DOPA

A study involved giving forty subjects L-dopa which increases dopamine in the brain[51]. They then showed the participants a slide show with scrambled or real faces and words. The increased dopamine caused both believers and skeptics to readily identify jumbled words, scrambled faces and words as being completely normal. The study suggests patternicity may be associated with above average levels of dopamine in the brain. The study also found that the effects of L-dopa was stronger on skeptics than on believers[51]. Hence, more dopamine makes skeptics less skeptical than in making believers more believing. A similar effect has also been shown to occur in newborn babies [52]. Additional research is beginning to show that people who believe in the paranormal have a greater tendency to perceive "patterns in noise," and are better inclined to attribute meaning to random connections[53]. This means these people have a fine eye for detail and can better spot small details in large amounts of random data.

Research scientists found dopamine agonists don't just enhance learning, but in higher doses trigger symptoms of psychosis, such as hallucinations[54]. This may be closely related to the fine line between madness (indiscriminate patternicity) and creativity (discriminate patternicity). The key is knowing the proper dose (*sweet spot*). If one takes too much, one is likely to make lots of Type I errors = false positives. Too little, Type II errors = false negatives.

The release of dopamine in the brain occurs as a stream of information; a message saying "*Do that again.*". It produces sensations of pleasure that accompanies mastering a task or reaching a goal, making the person wanting to repeat the behavior. **Behavior—Reinforcement—Behavior. Repeat...**

The gene that programs the production of dopamine is called **DRD4** (*dopamine receptor D4*)[55].

When dopamine is released by specific certain neurons in the brain, it becomes picked up by other neurons which are receptive to its chemical structure. This establishes dopamine pathways where certain regions become more active which encourages reward repeating behaviors. Studies show if you remove dopamine production from a human or rat they become catatonic. If dopamine production is over stimulated, frenetic behavior occurs in rats and schizophrenic behavior occurs in humans.

Do Nicotine Patches Increase Endurance?

A research study looking at 12 men that were 22 years of age who wore nicotine patches and exercised on a bicycle exercise machine, showed that ten out of the 12 men were able to cycle for longer wearing the nicotine patches and that they showed a significant 17 +/- 7% improvement in their performance ($P < 0.05$). No negative side effects were observed in their heart rate, perceived exertion or ventilation. Also no differences existed in their circulating fatty acids, plasma glucose or lactate. The study concluded that nicotine prolongs endurance.

Reference
Effect of transdermal nicotine administration on exercise endurance in men. Mündel T and Jones DA. Exp Physiol. 2006 Jul;91(4):705-13. Epub 2006 Apr 20.

Lithium Enhances Nicotine Sensitivity

Also lithium enhances sensitivity to nicotine. This may be due to its capillary action, enhancing the rate at which moisture is absorbed and concentrated *(Chronic treatment with lithium produces super sensitivity to nicotine. Dilsaver SC and Hariharan M. Biol. Psychiatry. 1989 Mar 15;25(6):795-9).*

Nicotine as a Plant Defense Mechanism and Nicotine in Food

Nicotine is used as an insecticide and the tobacco plant makes nicotine as a natural insecticide to stop bugs from attacking it. As a matter of fact, you can steep a cup of tobacco in one gallon of water overnight, then strain and spray plants that have, aphids, caterpillars and other insects to protect the plants. (*Nicotine Keeps Leaf-Loving Herbivores at Bay. 2004 Public Library of Science. August 2004*). Nectar coated with nicotine has been shown to accelerate the rate at which bumblebees learn flower colors (*Nicotine in floral nectar pharmacologically influences bumblebee learning of floral features. D. Baracch et al. May 2017*).

Jasmonic acid

As tobacco plants perform chemical reactions to defend themselves against insect attack, they begin making jasmonic acid. The jasmonic acid is used to make nicotine in the plant.

Research with Nicotiana sylvestrisplants (*wild tobacco*) showed jasmonic acid increased in about 5 minutes after insect attack. In 2 hours jasmonic acid was present in the roots, and after 5 hours the nicotine reached the leaves (*Pests leave lasting impression on plants, Stephanie Pain. March 1995. New Scientist*). The tobacco plant Nicotiana attenuate, found in Utah, doubles its nicotine production after a caterpillar attacks it

(*Nicotiana attenuata lectin receptor kinase Suppresses the Insect-Mediated Inhibition of Induced Defense Responses during Manduca sexta Herbivory. Paola A. Gilardoni et al. Sept 2011*).

Another research study found that plants previously exposed to jasmonic acid will produce nicotine faster compared to those not having been previously exposed to jasmonic acid (*Medicinal Natural Products: A Biosynthetic Approach. Dewick, Paul (2009). United Kingdom: John Wiley & Sons, Ltd. pp. 42–53. ISBN 978-0-470-74168-9*), (*Success for plants' pest control". BBC News. 2008-10-07*).

Jasmone is found in the oil of jasmine flowers and exists as a pale yellow liquid that is sometimes colorless. Pictured is the symbol for Jasmonic Acid.

Jasmine may synergize with Bergamot. Jasmine also protects against E. Coli (*Antibacterial Potential Assessment of Jasmine Essential Oil Against E. Coli. C. C. Rath et al. March 2008*) and Bergamot has been shown to increase the skin's sensitivity to light (*Essential Oils Loaded in Nanosystems: A Developing Strategy for a Successful Therapeutic Approach. Anna Rita Bilia et al. May 2014*) and as an interesting side note, thyme essential oil has been shown to kill prostate. lung and breast cancer cells (*Exploitation of Cytotoxicity of Some Essential Oils for Translation in Cancer Therapy. Rossella Russo. et al. Feb 2015).*

GABA and Jasmine

A research study titled: Fragrances in oolong tea that enhance the response of GABAA receptors, and conducted by S.J. Hossain and colleagues that was published in September 2004, looked at the effect of the compounds cis-jasmone, jasmine lactone, linalool oxide and methyl jasmonate in oolong tea and found that these substances significantly enhanced GABA activity, especially cis-jasmone and methyl jasmonate. The study found that inhalation of methyl jasmonate or 0.1% cis-jasmone significantly increased the ability of mice to sleep. It is interesting to note that an early anaesthesia remedy called Hua Tuo's mafeisan powder includes jasmine root in its formula *("Legendary Hwa Tuo's surgery under general anesthesia in the second century China), Chen J (August 2008). "A Brief Biography of Hua Tuo". Acupuncture Today), (Wang Z; Ping C (1999). "Well-known medical scientists: Hua Tuo". In Ping C. History and Development of Traditional Chinese Medicine. ISBN 7-03-006567-0), Huang Ti Nei Ching Su Wen: The Yellow Emperor's Classic of Internal Medicine. Translated by Ilza Veith). ISBN 0-520-02158-4).* Hence, Jasmine may have an effect on Microtubules.

Jasmine

Jasmine has been shown to create enhanced

alertness, perceived vigor and increased blood oxygen saturation, increased breathing rate and changes in systolic and diastolic blood pressure. Emotionally, participants inhaling Jasmine reported feeling more vigorous, more alert and less relaxed compared to a placebo group (*Sayowan et al 2013*).

Reference
The effects of jasmine Oil inhalation on brain wave activities and emotions. Journal of Health Research Vorasith Siripornpanich et al. April 2013.

Foods that contain Nicotine
Potatoes, eggplant, tomatoes and peppers contain nicotine. All these plants belong to the nightshade family.

Algae and Quantum Effects
Green sulfur bacteria and certain algae were found to be able to transfer energy internally via a coherent fashion. This caused an increase in photosynthesis. In algae, a specialized protein exists that is designed to capture sunlight. The protein's job is to move the sunlight to the reaction centre in the cell as fast as possible. This then allows the energy to be converted into chemical energy. Due to quantum coherence, the energy traverses every possible pathway at the same time before it chooses the fastest route to

the reaction center. The algae that does this is known as cryptophytes. They exist in regions of very little light such as at the bottom of deep ponds or under ice. Also two species of Hemiselmis cryptophytes contain an extra amino acid in their proteins.

Cryptophytes contain genes that are able to control coherence and alter mechanisms used in light harvesting. Because coherence occurs much easier during lower temperatures, species of algae in warmer conditions may not show the same effects. This shows a promising field in making organic solar cells that operate during low light conditions or even from the light of stars. Quote from an interview "*This is very exciting in that we are able to uncover the role of quantum coherence in photosynthesis*". (Article Ref - *Algae can switch quantum coherence on and off*. June 18, 2014), (Technical Ref - Stephen J. Harrop et al., Single-residue insertion switches the quaternary structure and exciton states of cryptophyte light-harvesting proteins, Proceedings of the National Academy of Sciences, 2014, DOI: 10.1073/pnas.1402538111).

Circularly polarized light has been shown to cause a reaction in algae (*The Effect of Circularly Polarized Light on the Photosynthesis and Chlorophyll a Synthesis of Certain Marine Algae. G. C. McLEOD. October 1957*).

Blue-Green Algae used to make High-Performance Battery Electrodes

Researchers have heated blue-green algae to temperatures of 700-1000 °C in argon gas which converts them into "hard carbon" which is used for high-capacity sodium-ion (Na-ion) batteries.

Electrodes in Li-ion batteries currently use graphite, which does not fit well. However the electrodes made from blue green algae fit better into the overall battery structure. Most hard carbon is derived from petroleum. After the algae was heated researchers made electrodes out of 80% hard carbon from algae, 10% binder and 10% carbon black (to enhance conductivity). After drying they made it into coin cells with sodium foil used as the counter electrode.

Article   Reference
Scientists convert harmful algal blooms into high-performance battery electrodes. October 9, 2015 by Lisa Zyga. tech explore.com.

Technical   Reference
Da Deng, et al. "Trash to Treasure: From Harmful Algal Blooms to High-Performance Electrodes for Sodium-Ion Batteries." Environmental Science & Technology. DOI: 10.1021/acs.est.5b03882

Algae Makes Better Lithium Ion Batteries

Researchers tried using beach sand and

portabella mushrooms to make better anodes for batteries. After trying this they tried using fossilized remains of single cell algae known as "diatoms". Silicon can store 10 times more energy then graphite. The problem is manufacturing it is expensive and requires a lot of energy.

Diatomaceous earth (DE) is a silicon-rich sedimentary rock composed of fossilized remains of diatoms deposited over millions of years. Using a process known as magnesiothermic reduction, the researchers converted the Diatomaceous earth to pure silicon nano-particles. They discovered that this created a highly porous anode which allowed easy access for the electrolyte.

Reference
Carbon-Coated, Diatomite-Derived Nanosilicon as a High Rate Capable Li-ion Battery Anode. Mihri Ozkan

Further Reading
Quantum mechanical study of the conformational and electronic properties of acetylcholine and its agonists. A quantum theoretical study of themolecular electronic structure of muscarine, nicotine, acetyl- -methylcholine, acetyl- -methylcholine, acetyl- ,-dimethylcholine, and further studies on acetylcholine. Radna RJ,

Beveridge DL, Bender AL. J Am Chem Soc. 1973 Jun 13;95(12):3831-46.

Excess Nicotine and Parkinson's

As we showed earlier, Eggplants and Green Tomatoes contain nicotine *(The Encyclopedia of Healing Foods. Michael T. Murray), (S. Dept of Agriculture).* So if these foods contain nicotine could they increase the risk of Parkinson's? A study looked at the consumption of nicotine-containing foods such as potatoes, peppers and tomato juice in 490 people who had recently been diagnosed with Parkinson's and found that peppers showed an inverse effect and that a protective effect occurred in people who had never smoked cigarettes the last 10 years.

Reference

Nicotine from edible Solanaceae and risk of Parkinson disease (ANA-12-1625). Ann Neurol. Author manuscript; available in PMC 2016 May 12. PMCID: PMC4864980. NIHMSID: NIHMS454255Susan Searles Nielsen et al.

A flower that naturally contains Geraniol and Linalool

We discussed earlier that Geraniol restores dopamine levels. The plant Erysimum cheiri syn. Cheiranthus cheiri (*common name is wallflower*) is a plant with a large bright yellow flower that is part of the Brassicaceae (Cruciferae) family and is native to Europe. It is widely cultivated as a garden plant.

Wallflower contains the following substances: **Geraniol**, Anisicaldehyde, **Linalool**, Nerol, Salicylic Acid, Anthranilic acid and Acetic Acid. It works most effectively when used in very, very small doses and is extremely toxic if taken in large amounts. Because this plant has Geraniol and Linalool within it, further research may show it has extremely powerful natural sedative and relaxing properties when used as a diluted essential oil.

Medical   **Uses**

While I could find no supporting studies, I did find this information from some botanical sites mentioning Wallflower. It is used as a diuretic and emmenagogue and recent research is starting to show it is extremely effective for heart disorders such as strengthening a failing heart, acting in a similar fashion to foxglove (Digitalis purpurea). Usually it is used in essential oil form. It contains substances that cause cardiotonic actions. The seeds contain expectorant, stomachic, aphrodisiac, diuretic and tonic properties and are used to heal injuries to the eyes, dry bronchitis and fevers.

4-Anisaldehyde
4-Anisaldehyde is also called p-anisaldehyde, anise aldehyde or anisic aldehyde and is a compound found in natural and synthetic fragrances. Its molecular structure consists of a benzene ring with a aldehyde and methoxy group. The liquid is clear with a strong aroma. Its scent is sweet, licorice like with a floral note and has a strong aniseed scent.

The molecular structure of anisaldehyde is very similar to vanillin. Anisaldehyde is also used to synthesize compounds used in pharmaceuticals or perfumery. Anisaldehyde is prepared via the oxidation of methoxytoluene (p-cresyl methyl ether) involving **manganese**dioxide. It is also produced by the oxidation of anethole, which is a related fragrance that is found in some alcoholic beverages. 4-Anisaldehyde is commonly available online. Anethole is found in the spices anise and fennel (*Fennel and anise as estrogenic agents. Albert-Puleo M. J Ethnopharmacol. 1980*).

Piperonal
Piperonal is also called heliotropin. It is commonly used as a substitute for anisic aldehyde and is commonly found in fragrances and flavors and mixed with essential oil blends. Its molecule is similar to benzaldehyde and vanillin. Piperonal can be found in various plants such as vanilla, violet flowers, dill and black pepper. Now, dopamine release is not enough. Having healthy functioning neurotransmitters is key to dopamine release. Let's explore this next.

Further Reading
The Notion Of Time In Special Relativity Yefim Bakman* and Boris Pogorelsky *Tel-Aviv University.

Bumblebees are attracted to Nicotine at Low Concentrations
A research study discovered that bumblebees are attracted to flowers with low nicotine levels and deterred by flowers with unnaturally high nicotine concentrations. The study found that nicotine has profound effects on learning when exposed to

Bumblebees in a dose-dependent manner due to the fact that the Bumblebees exposed to nicotine at high concentrations were able to learn the colors of flowers faster. The study also found that flowers containing nicotine in any concentration showed that the bumblebees stayed more faithful to the flowers. The study concluded that nicotine acts as an enhanced pollinator.

Reference
Nicotine in floral nectar pharmacologically influences bumblebee learning of floral features. D. Baracchi, A. Marples, A. J. Jenkins, A. R. Leitch & L. Chittka. Scientific Reports 7, Article number: 1951

This study has similar connotations to the dopamine reward effect, where dopamine receptors are released after certain tasks or goals have been accomplished. Hence nicotine may be acting similar to substances that enhance or create the dopamine reward factor.

Lavender and Nicotine
It may be that lavender allows the body to absorb or hold larger amounts of nicotine. A study found that inhalation of lavender inhibited convulsions that were induced by nicotine, pentylenetetrazol or electroshock in a study involving mice.

Reference
Anticonvulsive effects of inhaling lavender oil vapour. K. Yamada, Y. Mimaki, and Y. Sashida, Biological and Pharmaceutical Bulletin, vol. 17, no. 2, pp. 359–360, 1994. Scholar ·

References. Chapter 13

(1) Seasonal variation in human central dopamine activity. Karson CN, Berman KF, Kleinman J, Karoum F Psychiatry Res. 1984 Feb; 11(2):111-7.
(2) Seasonal Effects on Human Striatal Presynaptic Dopamine Synthesis Daniel P Eisenberg, et al. Nov 2010
(3) Seasonality of Suicidal Behavior Jong-Min Woo, et al. Feb 2012
(4) Dopamine release during stress in the prefrontal cortex of the rat decreases with age. Del Arco et al. Dec 2001
(5) Effect of dietary proteins and carbohydrates on urinary and sympathoadrenal catecholamines. J.C. Agharanya and R.J. Wurtman. 1985
(6) Neuroprotective Effects of Paeoniflorin on 6-OHDA-Lesioned Rat Model of Parkinson's Disease. XS Gu et al. Nov 2016.
(7) The Ginkgo biloba extract EGb 761® and its main constituent flavonoids and ginkgolides increase extracellular dopamine levels in the rat prefrontal cortex T Yoshitake, et al. Feb 2010
(8) Protective effects of geraniol (a monoterpene) in a diabetic neuropathy rat model: attenuation of behavioral impairments and biochemical perturbations. S.N. Prasad SN and Muralidhara. Sept 2014

(9) Protective effect of Geraniol on the transgenic Drosophila model of Parkinson's disease. Y.H. Siddique et al. April 2016
(10) Curcumin ameliorates dopaminergic neuronal oxidative damage via activation of the Akt/Nrf2 pathway.
(11) The neuroprotective effects of cocoa flavanol and its influence on cognitive performance. Astrid Nehlig. March 2013
(12) Pistachio supplementation attenuates motor and cognition impairments induced by cisplatin or vincristine in rats Leila Golchin et al. May 2015
(13) Effects of anthocyanins on psychological stress-induced oxidative stress and neurotransmitter status. MM Rahman et al. Aug 2008
(14) Cytoprotective effects of anthocyanins and other phenolic fractions of Boysenberry and blackcurrant on dopamine and amyloid β-induced oxidative stress in transfected COS-7 cells† Authors Dr Dilip Ghosh et al. June 2007
(15) Naringin protects the nigrostriatal dopaminergic projection through induction of GDNF in a neurotoxin model of Parkinson's disease. E. Leem et al. July 2014
(16) Resveratrol Protects Dopamine Neurons Against Lipopolysaccharide-Induced Neurotoxicity through Its Anti-Inflammatory Actions Feng Zhang, et al. Nov 2010

(17) Dopamine-dependent neurotoxicity of lipopolysaccharide in substantia nigra. R.M. De Pablos et al. March 2005

(18) Effect of probiotics Lactobacillus and Bifidobacterium on gut-derived lipopolysaccharides and inflammatory cytokines: an in vitro study using a human colonic microbiota model. L. Rodes et al. April 2013

(19) Bromelain inhibits lipopolysaccharide-induced cytokine production in human THP-1 monocytes via the removal of CD14. J.R. Huang et al. 2008

(20) Neuroprotective effects of berry fruits on neurodegenerative diseases Selvaraju Subash et al. Aug 2014

(21) Brilliant Blue-G but not Fenofibrate Treatment Reverts Hemiparkinsonian Behavior and Restores Dopamine Levels in an Animal Model of Parkinson's Disease. E.G. Ferrazoli et al. March 2017

(22) Low selenium diet increases the dopamine turnover in prefrontal cortex of the rat. A. Castaño et al. June 1997

(23) Effects of riboflavin and selenium deficiencies on glutathione and its relating enzyme activities with respect to lipid peroxide content of rat livers. M. Taniguchi and T, Hara. June 1983

(24) Evidences for the involvement of monoaminergic and GABAergic systems in antidepressant-like activity of garlic extract

in mice. D. Dhingra and V. Kumar. Aug 2008
(25) Antidepressant-like effect of onion (Allium cepa L.) powder in a rat behavioral model of depression. H. Sakakibara et al. Jan 2008
(26) EFFECT OF CREATINE SUPPLEMENTATION ON BRAIN NEUROTRANSMITTERS AFTER AN EXHAUSTIVE AEROBIC EXERCISE. Mehrzad Moghadasi et al. Jan 2012
(27) Creatine phosphokinase and lactate dehydrogenase levels after ultra long-distance running. An analysis of iso-enzyme profiles with special reference to indicators of myocardial damage. A. J. Kielblock et al. June 1979
(28) Imbalance between thyroid hormones and the dopaminergic system might be central to the pathophysiology of restless legs syndrome: a hypothesis Jose Carlos Pereira, Jr et al. May 2010
(29) Thyroid function and tyrosine metabolism. RIGO J, et al. 1962
(30) Effects of selenium supplementation on iodine and thyroid hormone status in a selected population with goitre in Pakistan. G.A. Kandhro et al. 2011
(31) Reductions in tyrosine levels are associated with thyroid hormone and catecholamine disturbances in sepsis W Khaliq, et al. Oc 2015

(32) Dose-dependent protective effect of selenium in rat model of Parkinson's disease: neurobehavioral and neurochemical evidences. K.S. Zafar et al. Feb 2003

(33) Implicit learning of sequential bias in a guessing task: Failure to demonstrate effects of dopamine administration and paranormal belief. Peter Krummenacher et al. Feb 2007

(34) The effect of an increased ratio of carbidopa to levodopa on the pharmacokinetics of levodopa. S. Kaakkola et al. Oct 1985

(35) Dopamine mediated antidepressant effect of Mucuna

(36) Mucuna pruriens in Parkinson's disease: a double blind clinical and pharmacological study. R. Katzenschlager et al. Dec 2004

(37) Seer Training Module. PITUITARY & PINEAL GLANDS. Natinoal Cancer Institute.

(38) Nicotine enhances the biosynthesis and secretion of transthyretin from the choroid plexus in rats: Implications for beta-amyloid formation MD Li et al .Feb 2000.

(39) Purkey HE, Palaninathan SK, Kent KC, Smith C, Safe SH, Sacchettini JC, Kelly JW (December 2004). "Hydroxylated polychlorinated biphenyls selectively bind transthyretin in blood and inhibit amyloidogenesis: rationalizing rodent PCB toxicity". Chem. Biol. 11 (12): 1719–28.

PMID 15610856.
doi:10.1016/j.chembiol.2004.10.009.
(40) Huang JT, Leweke FM, Oxley D, Wang L, Harris N, Koethe D, Gerth CW, Nolden BM, Gross S, Schreiber D, Reed B, Bahn S (November 2006). "Disease biomarkers in cerebrospinal fluid of patients with first-onset psychosis". PLoS Med. 3 (11): e428. PMC 1630717 Freely accessible. PMID 17090210. doi:10.1371/journal.pmed.0030428.
(41) Chronic lithium administration down regulates transthyretin mRNA expression in rat choroidplexus David J Pulford, et al. Dec 2006
(42) Short-term administration of omega 3 fatty acids from fish oil results in increased transthyretin transcription in old rat hippocampus. L.G. Puskás et al. Feb 2003
(43) Stress and glucocorticoids increase transthyretin expression in rat choroid plexus via mineralocorticoid and glucocorticoid receptors. A. Martinho et al. Sept 2012
(44) Lee YS, Park CO, Noh JY, Jin S, Lee NR, Noh S, Lee JH, Lee KH (Sep 2012). "Knockdown of paraoxonase 1 expression influences the ageing of human dermal microvascular endothelial cells". Experimental Dermatology. 21 (9): 682–7. PMID 22897574. doi:10.1111/j.1600-0625.2012.01555.x

(45) Resveratrol and fish oil reduce catecholamine-induced mortality in obese rats: role of oxidative stress in the myocardium and aorta. Avila PR1, et al. Nov 2013

(46) Dietary fish oil affects monoaminergic neurotransmission and behavior in rats. Chalon S, Delion-Vancassel S, Belzung C, Guilloteau D, Leguisquet AM, Besnard JC, Durand G J Nutr. 1998 Dec; 128(12):2512-9.

(47) hippocampus Beneficial effects of omega-3 fatty acids and vitamin B12 supplementation on brain docosahexaenoic acid, brain derived neurotrophic factor, and cognitive performance in the second-generation Wistar rats. R.S. Rathod et al. Aug 2015

(48) Aspirin and salicylate protect against MPTP-induced dopamine depletion in mice. N. Aubin et al. Oc 1998

(49) Synaptic Mechanisms Underlie Nicotine-Induced Excitability of Brain Reward Areas Author links open overlay panel. Huibert D Mansvelde et al. March 2002

(50) Eagle-eyed visual acuity: an experimental investigation of enhanced perception in autism. E. Ashwin et al. Jan 2009

(51) Dopamine, Paranormal Belief, and the Detection of Meaningful Stimuli Peter Krummenacher, et al. April 2010

(52) Newborns discriminate schematic faces from scrambled faces. M.A. Easterbrook et al. Sept 1999

(53) Reed, P., Wakefield, D., Harris, J., Parry, J., Cella, M., & Tsakanikos, E. (2008). Seeing non-existent events: Effects of environmental conditions, schizotypal symptoms, and sub-clinical characteristics. Journal of Behavior Therapy and Experimental Psychiatry, 39, 276–291. Crossref

(54) Pathophysiology and treatment of psychosis in Parkinson's disease: a review. Zahodne LB1, Fernandez HH. 2008

(55) Van Tol HH, Bunzow JR, Guan HC, Sunahara RK, Seeman P, Niznik HB, Civelli O (1991). "Cloning of the gene for a human dopamine D4 receptor with high affinity for the antipsychotic clozapine". Nature. 350 (6319): 610–4. PMID 1840645. doi:10.1038/350610a0.

Chapter 14. Substances that Enhance the Brain's Neurotransmitters

Dopamine is one of the primary substances that stimulate the brain's neurotransmitters. However, there are also other substances that enhance the speed of the signals of the nerves and these are known as neurotransmitters. Some neurotransmitters affect the body's sensory perceptions which governs how we perceive the world around us.

Bolm and Pribram's theory of Holonmic Brain Functioning[1] is very similar to Orch-OR. This is due to the fact that the effects are very similar to quantum transitions occurring within the brain. Instead of the transitions occurring in the microtubules, they occur at the brain's synapses where neurotransmitters exist as clouds of ions.

What is OR?

Or is short for Objective Reduction and denotes the "collapse of the wave" function that occurs only when a quantum system is large enough to force self-collapse. OR is explained in more detail in the paper published by Penrose titled: *Shadows of the Mind, Penrose* (1994).

OR has also been described by Pearle (1989), (Pearle and Squires, 1994) and Ghirardi et al (1986). Gravitational effects are also attributed to OR and are explained by Károlházy et al (1986),

Diósi (1989), Penrose (1989) and Ghirardi et al (1990).

Linoleic Acid and Neurotransmitter Activity
In one research study the effects of linoleic acid and its effect on neurotransmitter activity was studied in 60 year old male rats (*in human years*) [2]. One group of rats that was the control group were fed peanut plus rapeseed oil. The other group that was the linoleic acid deficient group was fed peanut oil. The rats fed the peanut oil exhibited higher 5-HT2 receptors in the brain's frontal cortex region compared to the control group. The study concluded that dopamine levels governing neurotransmitters increased in the frontal cortex regions in the rats that were deficient in linoleic acid and that they had **lower levels of dopamine** in their brains compared to the control group. They also had low levels of fatty acids. These lower fatty acid levels were restored when the rats were fed more omega 3's [2]. Hence, we can see that healthy neurotransmitters correlate with healthy dopamine levels.

Exercise and Neurotransmission
Physical exercise has been shown to enhance the brain's central dopaminergic systems and that the exercise releases neurotransmitters[3].

Opiates

Opiates are produced by the body after exercise. They act as neurotransmitters and modulate the body's reaction to painful stimulus. Higher levels of endorphins in the body after exercise releases opoids which in turn release dopamine. One such opoid is Kratom.

Myelin

Myelin is a protein that binds itself to zinc in the brain (*Myelin basic protein is a zinc-binding protein in brain: possible role in myelin compaction. Tsang D et al. July 1997*). Myelin can be thought of as the octane fuel responsible for the conduction of nerve signals throughout the body. Myelin consists of a multilayered membrane wrapped around our nerve fibers much like a type of insulated material. **Myelin increases the conduction of the transmissions between nerves up to 50-fold**[4]. To put it simply it accelerates the speed at which our nerves communicate with one another. Myelin has been shown to stablize Microubules (*Myelin basic protein functions as a microtubule stabilizing protein in differentiated oligodendrocytes. Galiano MR et al. 2006 Aug 15;84(3):534-41*).

Mantis   Shrimp

The species of shrimp known as '*smashers*' hit their targets with the force of a rifle bullet. They are known to deliver the fastest punches in the animal kingdom, all with their small brain that is the fraction of a humans. This is due to their fast eyes. Their eyes contain twelve different photoreceptors compared to most animals which only have three. The technology in their eyes is similar to that found in DVD and CD players. They can also **see circularly polarized light** that no other animal is able to[5].

The Sunstone and Polarized Light
Sunstones are used to detect polarized light *(The sunstone and polarised skylight: ancient Viking navigational tools? 1stGuy Ropars et al. Oct 2014).*

When polarized sunlight strikes calcite it splits it in two creating a double image. Each image exists as a different intensity depending upon the orientation and position of the mineral, compared to the light source. When one rotates the calcite mineral, the differences are easily measured, observed. It is possible that the Vikings exploited this method to estimate the position of the sun and determine north when the sky was covered in clouds.

Further Reading
Circularly Polarized Light as a Communication Signal in Mantis Shrimps. Y.L. Gagnon et al. Dec 2015

Substances that Enhance Myelin Growth
Lithium
Lithium has been shown to stimulate expression of myelin genes and restore myelin structure in mice. [6]. Lithium synergizes with Avidin: Biotinylated enzyme Complex (*ABC; Vector Laboratories, Segrate, Italy*) by improving regenerative response after a central nervous system injury[7].

Lithium and Microtubules
A research study found that lithium promotes microtubule assembly and stabilization and that it could be an effective treatment for Alzheimer's[8].

Lithium administered in concentrations of 0.2 – 1.0 mM contributes to tubulin polymerization as long as small amounts of magnesium are present[9]. Also calcium plays a role in regulating the asymmetry of flagellar waveform and that low concentrations of Lithium inhibit microtubule-based movement of reactivated sea urchin sperm flagella[10].

Lithium can be found in eggs, lentils, pepper, cabbage, milk, mushrooms, seaweed, apples, bananas, **tomatoes**, cucumbers, carrots, cauliflower, cinnamon, **lemon**, seafood, sugar cane and seeds. Lithium has also been shown to cause a significant increase of arginine and threonine in the brain[11].

Cholesterol

High cholesterol is also necessary for healthy Myelin growth[12], which is why Fish Oil supplements are an excellent substance to take to enhance the production of myelin.

Lecithin

Lecithin can also help enhance the body's production of Myelin[13].

Foods that promote regeneration of Myelin

Vitamin B12 and Folic acid [14], Pyrroloquinoline Quinone (PQQ), [15] Ashwangdha [16] and Lion's Mane [17] have been shown to promote myelin regeneration.. Also **Zinc** and Vitamin C play a valuable role in maintenance of Myelin.

Pyrroloquinoline helps the body create Myelin which strengthens nerves (*Enhanced rat sciatic nerve regeneration through silicon tubes filled with pyrroloquinoline quinone. Liu S et al. 2005*), Pyrroloquinoline is also used for the early germination of tobacco seeds (*Effect of pyrroloquinoline quinone on pollen germination and pollen tube growth of tobacco. Liu WeiQun; Yang TieZhao; Ding YongLe; Zhao YongFang, 1996.*) and Pyrroloquinoline has also been shown to be a powerful antioxidant for plants (*Pyrroloquinoline quinone is a plant growth promotion factor produced by Pseudomonas fluorescens B16. O. Choi et al. Feb 2008*). Myelin supplements can also be purchased online.

Foods Highest in Pyrroloquinoline from highest to lowest[18]

Parsley 34

Green Pepper 28

Kiwi Fruit 27

Papaya 26

Green Tea 29

Oolong Tea 27

Fermented Soybeans (Natto) 61

Tofu 24

Pyrroloquinoline can also be found in tobacco smoke [19] which could explain the "nerve withdrawal" associated with quitting smoking due to Pyrroloquinoline's strengthening effects of myelin.

Catecholamines
Catecholamines enhance neurotransmitter function by strengthening the signals flowing between the brain's neurons. They also regulate breathing and heartbeat. Catecholamines come

from the amino acid Tyrosine which also elevates dopamine release [20].

Catecholamines enhance the biosynthesis of ethylene in plants [21] and have been found in 44 varieties of plants and are believed to perform a protective role against insects and injuries. A combination of **Resveratrol and fish oil** has been found to enhance Catecholamine functioning in obese rats [22] and Catecholamines have also been shown to create a synergistic effect when interacting with Gibberellins. Gibberellins are not found in adult plants. This could be why young stalks, shoots and buds are not wrinkled, dried and deformed as are adult plants [23]. MAO's prevent catecholamines from breaking down [24].

What are Gibberellins?
Gibberellins are plant hormones that stimulate germination, pollen, stem elongation and flowering.

What is an MAO?
MAO's are found outside of the outer membrane of mitochondria cells types in the human body and are commonly used to treat depression.

Foods that enhance Catecholamines

Bananas

Coffee

Vanilla

Cocoa

Tea

Citrus Fruit

Chocolate

Rhodiola Rosea
We have had very good success taking Rhodiola Rosea extract before bed 48 hours before an ARV session. This is because research studies have found that Rhodiola Rosea enhances neurotransmitter transport in the brain[25].

Theanine
Theanine, which is found in green tea or can be purchased as a supplement, boosts neurotransmitter production. Theanine has been shown to protect neurons in the brain's hippocampus in studies conducted on mice[26]. Theanine has also been shown to ameliorate sleep quality and shown positive symptoms in

patients with schizophrenia by stabilizing glutamatergic concentrations in the brain[27].

PTFE
The substance known as Nafion is a form of Teflon. Nafion film has been shown to enhance electron transfer in dopamine [28].

Aspartate and Glutamate
Research shows that low amounts of Aspartate and Glutamate affect brain neurotransmitters when used in low amounts, however large amounts can cause brain damage in the hypothalamic region[29]. Aspartame synergizes with the food additive Quinoline Yellow[30].

Neurotransmitters and The Spine
The majority of the nerves in the body are contained around the body's spine region. The substance known as Brilliant Blue G (BBG) which is commonly used as a food coloring dye in the chocolate candy M&M's has been shown to heal rats with spinal cord injuries when injected into their spinal region[31].

It is interesting to note that Transthyretin, which we mentioned earlier, is a transport protein that exists in the body's serum and **cerebrospinal fluid**. Its purpose is to transport the thyroid hormone thyroxine (**T4**).

Brilliant Blue Synergizes with Glutamic

Acid[32]. The substance Brilliant blue G has also been shown to enhance neural repair stemming from traumatic optic nerve injury **(33)**.

What is Glutamic Acid?
Glutamic Acid is a precursor of GABA and helps transport potassium into the spinal fluid. It is used to treat Parkinson's disease, epilepsy and muscular dystrophy. Low levels of Glutamic acid are linked to depression and schizophrenia[34]. Theanine enhances GABA. Theanine foods include **bananas**, almonds, brown rice and tree nuts.

What is GABA?
GABA is also called gamma-aminobutyric acid. GABA acts as a neurotransmitter which exists in the neurons of the **cortex**. It governs anxiety, vision, motor control, enhances mental clarity and numerous other cortical functions. GABA exists in high quantities in the brain's **hypothalamus**[34].

Geraniol is used for Spinal Cord Injuries
A study on rats looked at using Geraniol to treat oxidative stress, inflammatory responses and apoptosis in traumatic spinal cord injuries. The study found that oxidative stress and inflammatory responses were significantly reduced when geraniol was administered.

Reference
Protective effect of geraniol inhibits inflammatory response, oxidative stress and apoptosis in traumatic injury of the spinal cord through modulation of NF-?B and p38 MAPK. Jiansheng Wang et al. 2016 Oct 27. doi: 10.3892/etm.2016.3850. Exp Ther Med. 2016 Dec; 12(6): 3607–3613.

Valerian and Dopamine
It has been sown that valerian increases the serotonin concentrations in the brain [35]. There was also a significant increase in 5-hydroxyindol acetic acid and serotonin and GABA levels were significantly decreased (*Effects of Valerian on the level of 5-hydroxytryptamine, cell proliferation and neurons in cerebral hippocampus of rats with depression induced by chronic mild stress. Tang JY et al. 2008 Mar; 6(3):283-8*) [35].

Valerian's Calming Effects on the Nervous System

Anxiety, St. John's Wort and Valerian
When St. John's Wort is combined with Valerian it greatly enhances the effects. A scientific research study found that this combination significantly reduced anxiety and that greater reductions in anxiety were seen with higher doses of valerian. The doses of St. John's Wort remained constant

between the groups being treated, suggesting valerian has more of an effect on anxiety symptoms (*Treating depression comorbid with anxiety--results of an open, practice-oriented study with St John's wort WS 5572 and valerian extract in high doses. Müller D, Pfeil T, von den Driesch V. Phytomedicine. 2003; 10 Suppl 4():25-30*).

Valerian Root is as effective as a Pharmaceutical
A research study discovered valerian root was just as effective as the prescription medication diazepam in reducing anxiety symptoms (*Effect of valepotriates (valerian extract) in generalized anxiety disorder: a randomized placebo-controlled pilot study. Andreatini R, Sartori VA, Seabra ML, Leite JR. Phytother Res. 2002 Nov; 16(7):650-4*).

Further Reading
Anxiolytic effects of a combination of Melissa officinalis and Valeriana officinalis during laboratory induced stress. Kennedy DO, Little W, Haskell CF, Scholey AB. Phytother Res. 2006 Feb; 20(2):96-102.

Nardostachys
Inhallation of valerena-4,7(11)-diene from the plant Nardostachys has been shown to enhance dopamine levels. The research study found the effect was caused by an interaction occurring

with the brain's hypothalamic-pituitary-adrenal axis system (*Inhalation administration of valerena-4,7(11)-diene from Nardostachys chinensis roots ameliorates restraint stress-induced changes in murine behavior and stress-related factors. Takemoto H. et al. 2014*). Nardostachys jatamansi is plant in the Valerian family that grows in the Himalayan Mountains.

Research has also found that valerian extract causes significant increases in delta and theta brainwaves while decreasing beta brainwave activity *(The quantitative EEG as a screening instrument to identify sedative effects of single doses of plant extracts in comparison with diazepam. Schulz H, Jobert M, Hübner WDPhytomedicine. 1998 Dec; 5(6):449-58).*

Some GABA supplements combine GABA with glutamine to help increase the body's production of glutaminc acid.

Further   Reading
Roles of glutamine in neurotransmission. J. Albrecht et al. Nov 2010

Why Older People May be More Intuitive
Research has suggested that older persons experience improvement and have more stability in their emotional processes. Hence, a person that relies more on their intact emotional abilities versus their declining deliberative faculties has

the resources to make better decisions.

An interesting research study published in April 2013 and conducted by J.A. Mikels and colleagues titled: The dark side of intuition: aging and increases in non-optimal intuitive decisions, discovered that older adults were more intuitive on average compared to younger adults[36]. Another deciding factor in enhanced intuition in older individuals is that studies have shown that aged mice show increased levels of GABA (*Taurine improves learning and retention in aged mice.El Idrissi A. May 2008*). And as just stated GABA enhances neurotransmitter activity [37]. This may be why people who are "*middle age*" make good security guards or law enforcement employees.

The Connection between GABA and Enhanced Intuition, Psychic and Precognition
A study showed that GABA increased low-frequency activity (alpha waves) in healthy participants[38]. Another study showed that after participants partook of GABA that they showed significant increases in their alpha brainwave activity[39]. Alpha brainwave activity, besides theta, has been shown to be present during times the mind is in a psychic state of being.

Acetylcholine and GABA both enhance the ability of the mind to generate theta brainwaves. Studies have shown that a reduction in either

Acetylcholine or GABA causes partial but not complete abolishment of theta brainwaves (*Yoder and Pang, 2005; Li et al., 2007*). Above average levels of acetylcholine also enhance memory (*much like niacin*).

Reference
The role of REM sleep theta activity in emotional memory Isabel C. Hutchison and Shailendra Rathore. Oct 2015. Front Psychol. 2015.

When brainwaves are in Alpha, both hemispheres of the brain are in sync. This allows for clearer thinking and better communication. Alpha waves cause left brain activity to slow, allowing right brain activity to express itself through creativity, intuition and new ideas. Professional athletes, singers, artists and writers have been found to have higher alpha brainwave activity compared to the average person.

10 Hz alpha brainwaves are generally known as the peak performance state of mind. As we covered earlier in this book, professional golfers and basketball players have been found to show increased alpha waves in the left sides of their brain hemisphere just before they make a great golf shot or a great free-throw of a basketball. Players that didn't make the good shot showed enhanced beta brainwave activity in their left brain hemisphere.

How to Generate Acetylcholine and GABA in the body

By getting enough choline in the body, acetylcholine is automatically generated. Foods that have choline include brussels sprouts, chocolate, peanut butter and broccoli. Foods that contain GABA (*which forms Glutamine in the body*) include tree nuts, **bananas**, brown rice, oats and almonds.

The Russian Telepathic Experiments

A research project in Russia during 1966 involving long-distance telepathy with two people spaced miles apart from one another consisted of a sender and a receiver. Kamensky was in Moscow and acted as the sender of the telepathic signal and Nikolayev served as the receiver. Nikolayev was stationed at the science research centre in Novosibirsk which is located in western Siberia. Two types of tests took place.

Test 1 was modeled after research by Dr J. B. Rhine from the Parapsychology Laboratory of Duke University. He used cards consisting of five geometric symbols: star, wavy lines, cross, circle and square. The newspaper that carried the Russian telepathic research project did not provide the details on the outcome of the experiment. However, it stated the number of correct symbols was higher than random according to the theory of probability.

Test 2 involved a telepathic transfer of images of objects. The newspaper stated Nikolayev telepathically received clear images of a screw-driver and dumbbells sent from Moscow by Kamensky. The newspaper concluded that the experiment showed the phenomenon is real and pointed to the need for further research.

Extending the 'Split Second' Retrieval of Information from the Future
This next section covers a third experiment conducted in Russia concerning the *'split second'* effect which occurs in these types of experiments. This part is of special significance because during our ARV sessions, the information received occurs in almost a fraction of a second, of which the mind must then interpret correctly. The hard part over the years has been trying to find a way to hold this image or slow down the speed at which this split second information arrives long enough to

make an accurate interpretation.
It just so happens the key to this is = **EMOTION**. Emotion can stretch the time associated with received information during associative remote viewing, even though it lasts only a split second.

In this third experiment we are about to cover, we shall show how the Russian experiments were able to use this technique to enhance the clarity of their telepathic experiments. Book 2 of our associative remote viewing series goes into greater detail about how emotion greatly enhances the clarity of remote viewing, especially emotion that occurs on the future date that is being viewed.

The third experiment took place between Moscow and Leningrad and occurred a year later. The goal was to harness a new type of telepathy known as *'crisis telepathy'* and form it into a single code transmission.

The name of the group was called the Popov group and their goal was to implement the following -

(a) be appropriately suited to the skills of its telepathists

(b) to contain emotional elements

(c) to achieve specific transmission of information

Their problem was that the telepathic transmissions often occurred as 'flashes' and their problem was how to transform split-second impressions into a meaningful message. The solution came from Dr. Genady Sergeyev, a member of the A. A. Uktomskii Physiological Institute in Leningrad and who had been the senior experimenter with Nina Kulagina. Sergeyev decided the solution would be to express a short outburst of emotion during the telepathic transfer of information and that it may have a sufficient enough impact to form the Morse code equivalent of a letter of the alphabet.

Dr. Genady Sergeyev's experiment consisted of a message containing aggressive emotion that would last between 15 and 30 seconds and was to represent a dot in Morse code and an emotional message of 45 seconds was to be the equivalent of a morse code dash. In order to visualize an aggressive emotion, Kamensky was told to imagine that he was administering Mikolayev a severe beating (*length of time being the short or the long period*). Kamensky also sought to enhance his emotions by imagining a flash in own mind whenever he was about to make a telepathic impact on Nikolayev.

The experiment was constructed in such a way that Nikolayev would feel the "morse code beating" on an intellectual level, but would instead register them in his cardiovascular system or brain. To examine these effects of the morse code transmissions, Nikolayev sat in a soundproof chamber in Leningrad University's Physiology Laboratory. The activity of his heart was monitored by an electro-cardiograph and his brain activity was recorded on an electroencephalograph. The test showed that seven changes took place in Nikolayev's heart rhythm which coincided with Kamensky's emotional morse code messages.

During the first experiment, telepathic contact took place at 10 p.m. At this time, Nikolayev tried to turn away from a series of mental blinding flashes. Kolodny stated that when the researchers in the experiment came to the room, that Nikolayev's eyes were red and inflamed.

Phase 2 of this experiment took place two hours later. Kamensky, who was in a test room in Moscow's Poly technical Museum, selected an object from several sealed packages. Kamensky picked a box of empty cigarettes and visualized Nikolayev selecting the cigarette from the empty box. Just at that moment the "receiver" located in Leningrad stated he experienced "the illusion of a cigarette". He also stated he observed "*a lid, yet empty inside,*" and also stated that "*The outside is not cold, but cardboard.*" In conclusion he picked up the complete image of object in his mind, along with the thought image that Kamensky created for himself.

This is interesting because as we show in this book, nicotine greatly affects microtubules and may enhance the clarity of associative remote viewing. Hence, the presence of nicotine during the experiment may have contributed to its success.

When the research team returned to Moscow, Edward Naumov of the Bio-Communication Laboratory was quick to publicize the results of these experiments. They had just succeeded in proving that telepathy could have practical uses in the transmission of thought and that it may have potential for military and academic use. In summary, the Moscow-Leningrad experiment proved that bursts of emotion enhance telepathic communication.

While we don't condone images of violence or summon emotions of beatings to enhance remote viewing sessions, in order to enhance emotion, having feelings of deep profound appreciation works just as well. We also use the technique known as "Smiling into the Heart" during our Heart Math sessions, which sends strong emotional energy into the heart, the body's cells and the immediate surroundings.

Another trick we use is to place the future target date we are remote viewing on a date where emotion is more intensified, such as during a first rainstorm, on a Monday or during a solar eclipses. This works well because these are environments where emotion is naturally intensified. For example if the associative remote viewing session was to take place on a Friday, the target date would be a Monday. The results would be especially strong if there was a solar eclipse on a Monday and it was the first rainstorm in weeks or months.

I had shown in the second ARV book that ants "swarm" before a storm, taking refuge in cupboards etc, just before a major rain. Speaking from personal experience, certain flying insects will also "swarm" mid-air after a major rainstorm and do this if no more rain is due to occur the next few days.

The Underlying Mechanism of Telepathy
An independent research group called the Ai Research Manufacturing Company, which had submitted project reports to the US Central Intelligence Agency in the 1970's, undertook a summarized speculative study involving the fundamental aspects of telepathy and concluded that three methods existed that are compatible with the laws of modern physics.

(1) **Extremely Low Frequency** (ELF) electromagnetic waves and Very Low Frequency (VLF)

(2) **Neutrinos**. Based on the photon theory of neutrinos

(3) **Quantum Mechanical Waves**. Based upon the schizo-physical interpretation of basic quantum mechanics theory.

Improve your Remote Viewing Accuracy Techniques using Quantum Microtubules

It is interesting to note that number (2) states **neutrinos** are a key component. In book 2 of our associative remote viewing series **Remote Viewing. The Complete User's Manual on Experiencing Future Consciousness** we go to great length to show how neutrinos play a major role in the success of remote viewing. What is also even more interesting is that the above 3 points are the same points that came up when we searched for methods to enhance our remote viewing methods. As a matter of fact point (1) ELF was shown in great detail in our first remote viewing book, point (2) in our second book and in point (3) in this edition we cover the quantum effect and the brain's microtubules in great detail. This shows that even though this report was issued in the 1970's that the information in it is very likely valid. The report went on to state that experiments in the USA and Russia in this field point to the ELF/VLF mechanisms, however the other two possibilities can't be ruled out.

This also means that a person's mind is vulnerable to outside monitoring, or that thought transference could occur or be intercepted using extremely low frequency receivers (ELF). For example electronics used in the medical specialties has perfected such apparatuses and some come very close to such uses. It would only be a matter of having the right skill to modify/create more sensitive and more powerful devices responsive to the aforementioned environments.

Kazhinsky's book titled: Thought Transference stated the human nervous system incorporates elements of its own historic evolution. He stated, just as all other parts of the living organism such as the nerve elements and nerve circuits perform protective and adaptive functions (*they adapt the overall body to the effects of environment and its factors*), over the long term they have undergone changes and improvements the last few thousand years. Nature has crafted highly delicate nerve structures which result in greater improvement of vital functions. The electromagnetic transmission of mental information over long distances is a vital component of the human nervous system. Hence, the central nervous system (*especially the brain*) consists of a repository of sophisticated instruments of a form of biological radio communication. Its construction is far superior to the latest instruments of today's primitive technical radio communication.

Hence he may be hinting at the fact that it is mankind's destiny to eventually use telepathy as the body will eventually learn to do so or that it may be taught as a course.

Summary

There may exist 'living' instruments that exhibit technical/biological communication which are still unknown to today's contemporary radio engineering. As the human nervous system continues to evolve new and higher stages of development continue to occur, including biological radio communication, one of the laws of nature follows this path; in that the more a certain capacity is exercised, the keener it will become and hence the greater man's power over nature will be. Kazhinsky stated with a quote from V.I. Lenin: "*Sensation is the resulting effect of matter on our sensory organs.*" (*Materialism and Empirio-criticism, Moscow, 1953*).

A list of former USSR PSI Labs

Many of these labs are now most likely closed, however some may still be operating on a secret level or may have changed their names or moved.

Scott Rauvers

A. S. Popov All-Union Scientific and Technical Society of Radio Technology and Electrical Engineering, Moscow;

Laboratory of Bio-Information, 1965–1975;

Laboratory of Bio-Energetics, established 1978.

Scientific Research Institute of General and Educational Psychology, USSR Academy of Pedagogical Sciences, Moscow.

Baumann Institute of Advanced Technology, Moscow; Laboratory of Dr Wagner.

Institute of Energetics, Moscow; Laboratory of Dr Sokolov. Moscow State University; Laboratory of Prof. Kholodov.

State Instrument of Engineering College, Department of Physics, Moscow.

Moscow Institute of Aviation. I. V. Pavlov Institute, Moscow.

Institute of Reflexology, Moscow.

Moscow University, Department of Theoretical Physics.

Moscow State University, Department of Geology.

Interdepartmental Commission for Coordination of Study on the Biophysical Effect, Moscow (dowsing research).

Improve your Remote Viewing Accuracy Techniques using Quantum Microtubules

Adjunct Laboratory of Medical and Biological Problems, Moscow.

University of Leningrad, Laboratory on the Physiology of Labour; Department of Physiology, Laboratory of Biological Cybernetics.

A. A. Uktomskii Physiological Institute, Leningrad.

Leningrad Polytechnic Institute, Department of Cybernetics.

University of Leningrad, Bekhterev Brain Institute.

Research Institute of Psychology, Ukrainian SSR Academy of Science.

Institute of Problems of Information Transmission of the USSR Academy of Sciences, Moscow.

Pulkovo Observatory, Leningrad.

Filatov Institute, Laboratory of the Physiology of Vision, Odessa.

Scientific-Industrial Unit "Quantum", Krasnodar. State University of Georgia, Tbiblisi (Tiflis). Kazakhstan State University, Alma Ata, Kazakhstan.

Institute of Cybernetics of the Ukrainian SSR, Kiev.• Institute of Clinical Physiology, Kiev.

Scientific Research Institute of Biophysics, Department of Cybernetics, Puschino.

Institute of Psychiatry and Neurology, Kharkov.

Institute of Automation and Electricity, Special Department No. 8 (1965–1969), Siberian Academy of Science, Novosibirsk.

Institute of Clinical and Experimental Medicine, Novosibirsk.

1. **References**
   Moscow daily Komsomolskaya Pravda (July 9, 1966)

2. Psychic Discoveries: The Iron Curtain Lifted. By Sheila Ostrander and Lynn Shroeder
   ISBN 1-56924-750-1

3. Nexus - 0805 - New Times Magazine January 22, 2017 / www.nexusmagazine.com August – September 2001

Nicotine Produces Alpha Brainwaves
A research study examining 16 volunteers who smoked cigarettes, showed increased alpha brainwaves and beta waves. After smoking, their alpha brainwaves decreased ($P < 0.05$) which was followed by an increase in beta waves.

Reference
The Effect of Smoking on Brain Wave Activity in Middle-Aged Men Measured by Electrocorticography So-Hyung KANG. Iran J Public Health. 2015 Sep; 44(9): 1288–1290. PMCID: PMC4645788

Herbs for Healthy Neurotransmission
Proper foods and supplements are just the first part towards healthy neurotransmission. The second key is healthy blood flow and circulation. Herbs such as periwinkle, ginkgo, centella asiatica and huperzia serrata are some of the very best herbs to enhance blood circulation in the brain's neurotransmitters.

Ginkgo and Brain Circulation
Ginkgo stimulates a healthy release of neurotransmitters which keep the mind happy. Centella, as well as Gingko, improve capillary strength and increase blood flow to the brain. Periwinkle is also good for helping restore brain cells deprived of oxygen and is recommended for stroke victims (*do not take periwinkle with some medications*). It is also interesting to note that Vinblastine comes from Periwinkle. Vinblastine acts as a stabilizer/amplifier for microtubule operation.

Cedar essential oil has also been shown to enhance oxygen levels and may do so if added to chlorophyll, although studies are needed to confirm this. We have had success adding a few drops of cedar essential oil to a bowl of algae water during ARV sessions and the beneficial effects may be due to the enhanced oxygen levels.

While I could find no published scientific documented studies on Cedarwood improving oxygen levels, I did find a website that publishes essential oil testimonials. In one testimonial submitted in March 2007 titled: **Oxygen Miracle using Cedarwood and RC**, the person stated after rubbing Cedarwood Essential Oil into a person's feet and lower back that it enhanced their oxygen levels over 90%. The person also used a combination of Cypress, Spruce, and three types of Eucalyptus oils (E. radiata, E. globulus, and E. citriodora in the formula known as "**RC**". Spruce essential oil is reputed to have a very high frequency of 400Hz.

Huperzia Serrata
Huperzia serrata, contains the component huperzine A, which is responsible for the neurotransmitter acetylcholine. Acetylcholine is one of the major neurotransmitters in preganglionic fibers and is a memory and muscle movement neurotransmitter.

What are Preganglionic Fibers?
Postganglionic fibers are a neurotransmitter in the nervous system.

What is Phosphatidylcholine ?
Phosphatidylcholine is a special type of choline that absorbs more thoroughly into the body compared to normal choline.

Bergamot Essential Oil
Essential oil of Bergamot has been shown in mice studies to release neurotransmitters in the brain.

One study found that Bergamot essential oil increased the levels of taurine, glycine and aspartate. Aspartate release was also found to come from the hippocampus region. The study also found that when the monoterpenes and hydrocarbons were removed from the Bergamot essential oil, that the effects ceased, suggesting that monoterpenes and hydrocarbons play a key role in neurotransmitter production and operation. The study concluded that the hydrocarbons were responsible for the release of glutamate *(The essential oil of bergamot enhances the levels of amino acid neurotransmitters in the hippocampus of rat: implication of monoterpene hydrocarbons. Morrone LA et al. Pharmacol Res. 2007 Apr;55(4):255-62. Epub 2006 Dec).*

Bergamot essential oil has also been shown to offer neuroprotection *(Cell signaling pathways in the mechanisms of neuroprotection afforded by bergamot essential oil against NMDA-induced cell death in vitroM T Corasaniti et al. June 2007).*

Monoterpenes

Monoterpenes have the smell of wood. Major monoterpenes include limonene, pinene and pinene. They are commonly emitted from kilns during drying *(McDonald & Wastney, 1995; Wu, 1997).* Monoterpenes are smaller than Sesquiterpenes and have the ability to heal DNA.

Monoterpenes levels in Essential Oils

Grapefruit (94%)
Angelica (79%)
Frankincense (77%)
Cypress (smells like wood) (76%)
Galbanum (75%)
Pine (60%)
Rose of Sharon (55%)
Juniper (54%)
Spruce (50%)
Myrtle (36%)
Hyssop (30%)
Peppermint (11%)

Rosemary and Sandalwood

Rosemary and Sandalwood blends extremely well with bergamot, and bergamot is mixed with cedar and marjoam to induce sleep quickly and peacefully. The ratio is 3 drops of Bergamot, 2 drops of Cedar and 1 drop of marjaom.

Selenium

Selenium has been shown to exhibit protective effects upon the brain's neurons and influence neurotransmitters. It has also been shown to regulate mood function. Research has also shown that brains deprived of selenium show disturbances in the neurotransmitters dopamine,

serotonin and adrenaline *(Selenium and the brain; a review"*
Nutr Neurosci. 2001;4(2):81-97.Selenium and the brain: a review. Whanger PD).* Two of the best neuroprotectors are selenium and gingko.

Vitamin B6
Vitamin B6 also enhances neurotransmitter activity. This could be why Vitamin B6 supplements are used as a treatment for carpal tunnel syndrome.

Root tissues from celery, radish, potato and carrots were found to be the best for promoting the disappearance of pyrene within 40 days. Plant species enhance biodegradation of polycyclic aromatic hydrocarbons. Experiments with A. graveolens demonstrated soil contaminated with pyrene or benzo pyrene was as effective as addition of crushed root tissues. The study found that linoleic acid was the major substance causing PAH degradation *(Biostimulation of PAH degradation with plants containing high concentrations of linoleic acid. Yi H, Crowley DE. Environ Sci Technol. 2007 Jun 15;41(12):4382-8).*

What are PAH's?
PAHs are the starting material for the earliest forms of life. In space, light emitted by the Red Rectangle nebula found spectral imagery suggesting the presence of pyrene and

anthracene and fullerenes (*also called "buckyballs"*), have been found in other nebulae.

Fullerenes also represent a major component in the origin of life. Astronomer Letizia Stanghellini states that buckyballs from outer space may have provided the seeds for life on Earth.

PAHs also contain only carbon and hydrogen and are composed of multiple aromatic rings. PAHs are found in coal and in tar deposits and are also produced by the incomplete combustion of matter such as incinerators, engines and burning biomass in forest fires, etc).

PAHs may be abundant in the universe and seem to have formed as early as a couple of billion years after the Universe's Big Bang. They are also associated with exoplanets and new stars. Over 20% of the carbon in our universe may consist of PAHs.

The Structure of PAH's

Polycyclic aromatic hydrocarbons also called graphenes, include carbon nanotubes, graphite and fullerenes. These all consist of a hexagonal structure that looks like ordinary chicken wire-like. These chicken wire structures also occur in Hydrogen-bonded structures of boric acid, interwoven molecule chains in the inorganic polymer NaAuS and complexes of the protein clathrin. PAHs are lipophilic and nonpolar making them insoluble in water and they are usually colorless.

Alpha Brain Waves

Alpha brainwaves exist as high amplitude brainwaves that are slower then beta. Their frequency is between 9 and 14 cycles per second.

Alpha brainwaves exhibit effects similar to the brain's dopamine reward system. For example a person who has finished a task and then sits down to rest experiences alpha waves or a person meditating, taking time out to reflect is in alpha state. Another example is an individual taking a break from a meeting or conference, then walking out into a garden is in alpha.

Theta Brain Waves

The state after alpha, which are brainwaves of even greater amplitude and slower frequency is theta. Theta ranges between 5 and 8 cycles per second. An example of theta is a person who has just taken time off from performing a task and starts daydreaming is in theta. Another example is someone who has been driving on a freeway for a period of time and suddenly discovers they can't recall the last 7 miles will often be in a theta state, due to the speed and motion of the vehicle.

Alpha Brain Waves and Remote Viewing

It may be that standard remote viewing works well while the brain is in Alpha and Associative Remote Viewing, which involves remote viewing the future works well while the brain is in Theta. This may be

due to the fact that Theta brainwaves are of a stronger amplitude and slower frequency, compared to Alpha. It may also be that Alpha Brainwaves occur in the background while the brain is in Theta.

Properties of Alpha Brain Waves

- Improved Mood and Stability of Emotions
- Better control over emotions

Alpha waves and Heart Math

A research study found that subjects who listened to alpha rhythms using brain entrainment showed enhanced heart rate variability (**HRV**) (*Alpha-rhythm stimulation using brain entrainment enhances heart rate variability in subjects with reduced HRV Francesco Casciaro. Nov 2013*)

Tobacco Enhances Alpha Brainwaves

A research study found that smoking tobacco produced dominant alpha brain waves.

Reference
Tobacco smoking produces widespread dominant brain wave alpha frequency increases. Domino EF. et al. Dec 2009

Theanine Produces Alpha Brainwaves
A research study found that the substance theanine caused significant increases in alpha brainwaves *(L-theanine, a natural constituent in tea, and its effect on mental state. Nobre AC et al. 2008)*. Theanine has been shown to protect neurons in the brain's hippocampus in studies conducted on mice *(covered in another chapter in this book)*.

Ashwagandha is commonly used to reduce anxiety and bring about a state of calmness. Ashwagandha has also been found to enhance the health of the hippocampus *(covered in another chapter in this book)*. As we covered earlier in this book, anxiety was one of the sole reasons professional athletes failed when performing sports related tasks.

Common types of meditation will cause the brain to experience alpha-theta brainwaves. This results in increased awareness of the brainwave state, thus enhancing gamma brainwave activity. Alpha brainwaves also allow the body's sensory inputs to become minimized. Hence the mind is clear of unwanted thoughts and distractions. If the brain shifts gears and focuses on a specific thought, whether it be positive or negative, the alpha oscillations will vanish and higher frequency oscillations will begin.

Mindfulness is one form of meditation that can help enhance alpha and research is showing that mindfulness exercises create alpha waves that allow people suffering from chronic pain to reduce the perceptions of their pain and that people suffering from depression or anxiety can learn to use mindfulness meditation to minimize their symptoms.

Further Reading
Attention Drives Synchronization of Alpha and Beta Rhythms between Right Inferior Frontal and Primary Sensory Neocorx. Journal of Neuroscience.

10 Hz Current Produces Alpha Brainwave Rhythmus
A research study by the University of North Carolina (UNC) School of Medicine showed that low doses of electric current at 10-hertz enhanced alpha brain wave activity. The study also found that it enhanced creativity by 7.4% in adults. The study also suggested that people with depression showed impaired alpha oscillations.

Reference
Functional Role of Frontal Alpha Oscillations in Creativity. Caroline Lustenbergera et al et al. April 2015.

The Schuman Resonance and Alpha Brainwaves

It is very likely that during favorable solar weather conditions that the amplitude/intensity of the Schuman resonance is increased. When the brain experiences alpha brainwaves during this time, such as performing Heart Math (HRV), a synergy may be taking place. For example one part of the Schuman resonance exists at the frequency of 7.8 Hz. 7.8Hz is very close to the brain's alpha rhythm (between 9 and 14 cycles per second).

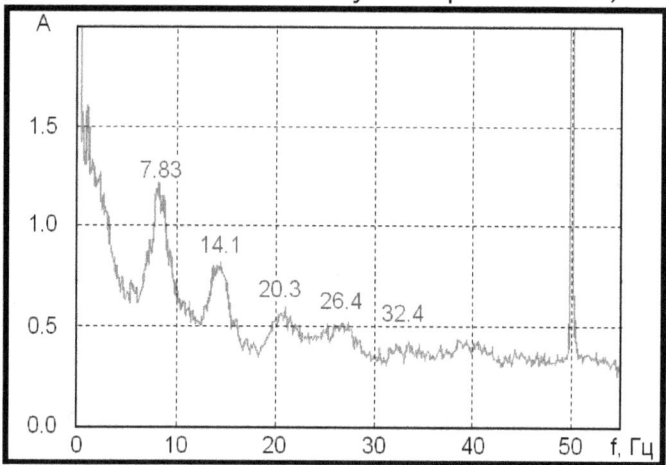

Nicotine increases Alpha Brainwaves
A research study examining 16 volunteers who smoked cigarettes showed increased alpha brainwaves and beta waves. After smoking, their alpha brainwaves decreased ($P < 0.05$) which was then followed by an increase in beta waves.

Reference
The Effect of Smoking on Brain Wave Activity in Middle-Aged Men Measured by Electrocorticography. Iran J Public Health. 2015 Sep; 44(9): 1288–1290. PMCID: PMC4645788 So-Hyung KANG.

Tobacco Enhances Alpha Brainwaves
Another research study found that smoking tobacco produced dominant alpha brain waves.

Reference
Tobacco smoking produces widespread dominant brain wave alpha frequency increases. Domino EF et al. Dec 2009

Stochastic Resonance and Alpha Waves
The effect termed 'stochastic resonance' exists in a wide range of living organisms as well as systems.

Shoichi Kai and Toshio Mori of the University of Kyushu in Japan found that stochastic resonance in information processing exists in the human central nervous system (*T Mori and S Kai 2002 Phys. Rev. Lett. 88 218101*). Most complex systems that have weak periodic signals are able to be strengthened by noise. The process occurs when random peaks in the noisy signal coincide with the regular peaks in the periodic signal. It is most effective when the noisy signal exists at a specific amplitude that is relative to the periodic signal. This effect already exists in our body in the sense of touch and the control of blood pressure in our brain.

During Kai and Mori's research study, they shone light signals into eyes of 5 students while at the same time measuring their brainwaves. Noisy signals were shone into the student's left eyelids and periodic signals shone into their right eyelids while the students were resting. During this time, Kai and Mori measured the intensity of the student's alpha brain waves. As they did this, they found a sharp peak occurring at 5 Hz (frequency of the periodic signal). However after increasing the strength of the noise signal that was relative to the periodic signal, they found that a 'harmonic' peak emerged in the student's alpha brainwaves at 10 Hz.

Reference
Noise-induced entrainment and stochastic

resonance in human brain waves. May 2002. T Mori and S Kai 2002 Phys. Rev. Lett. 88 218101.

Weak Noise Enhances Neural Synchronization
Neural synchronization occurs as a result where specific brain regions establish networks responsible for cognition, perception and action. Adding weak noise (fast random fluctuations) to specific neural systems enhances neural synchronization via stochastic resonance (*also called SR*).

Stochastic Resonance plays important roles in cognition, human perception and action. It may be that cognition, perception and action are dependent upon synchronized oscillations occurring within specific brain networks.

A research study measured 40-Hz responses in the auditory cortex of a person's brain while they listened to brief pure tones. As the person listened to the tones, the researchers added a stream of broadband noise and then measured the person's 40-Hz response via EEG. The study found that the added noise increased their 40-Hz response. The study also discovered that the added noise increased brain synchronization in alpha and gamma during and after the test. The study concluded that regional synchronization of neural activity in the brain can be enhanced by adding moderate amounts of random noise and that stochastic resonance may play a role in optimal brain functioning.

Reference
Stochastic Resonance Modulates Neural Synchronization within and between Cortical Sources. Lawrence M. Ward et al. Dec 2010. PLoS One. 2010; 5(12): e14371. Published online doi: 10.1371/journal.pone.001437.1 PMCID: PMC3002936

Further Reading
Neural synchrony in stochastic resonance, attention, and consciousness. Ward LM et al. Can J Exp Psychol. 2006 Dec;60(4):319-26.

Alpha and Gamma Enhance Creativity
Another research study showed that listening to binaural beats at the alpha (10Hz) and gamma (40?Hz) frequencies enhanced the listener's creativity. The participant's listened to binaural beats for 3 minutes before performing a series of tasks. The study found that listening to binaural beats at both frequencies enhanced the participant's divergent thinking (*the generation of multiple answers to a given problem*). The study

also found that participants who exhibited low eye blink rates benefited most from the alpha binaural-beat tones. However participants with high eye blink rates were unaffected or showed impairments by the alpha and gamma binaural-beat tones. The study also added background white noise to amplify the binaural-beat percept.

Reference
The impact of binaural beats on creativity.
Reedijk SA, Bolders A, Hommel B. Front Hum Neurosci. 2013; 7():786.

Sunlight, Opiates and Exercise
Our remote viewing sessions always worked well doing moderate exercise a few hours before a remote viewing session. This may be due to several factors. A research study found that some people who use tanning salons or sunbathe a lot may become addicted to the heat and warmth. The research study found that sunlight caused the synthesis of endorphins to take place in the skin. These endorphins were of the same opiate type the body produces when it exercises.

Reference
Skin Endorphin Mediates Addiction to UV Light. Gillian L. et al. June 2014. Volume 157, Issue 7, p1527–1534, 19

Essential Oils for a Healthy Autonomic Nervous System

Anxiety

Research by Lehrner et al. looked at the effect of lavender and orange essential oils had on mood, anxiety, alertness and calmness in dental patients. He found that compared to the control group, both scents improved mood and reduced anxiety in patients waiting to see the dentist (*Ambient odors of orange and lavender reduce anxiety and improve mood in a dental office. Lehrner J, Marwinski G, Lehr S, Johren P, Deecke L Physiol Behav. 2005 Sep 15; 86(1-2):92-5*).

Work   Productivity

Studies conducted by Sakamoto et al. looked at if exposure to lavender and jasmine during recess periods would affect work performance. The study found lavender significantly increased concentration levels, however jasmine had little to no effect (*Effectiveness of aroma on work efficiency: lavender aroma during recesses prevents deterioration of work performance. Sakamoto R, Minoura K, Usui A, Ishizuka Y, Kanba S. Chem Senses. 2005 Oct; 30(8):683-91*).

Self   Esteem

Studies by Rho et al. researching lemon, lavender, rosemary and chamomile found it reduced anxiety and enhanced self-esteem in Korean elderly women (*Effects of aromatherapy massage on anxiety and self-esteem in Korean elderly women: a pilot study. Rho KH, Han SH, Kim KS, Lee MS. Int J Neurosci. 2006 Dec; 116(12):1447-55*).

Peppermint Oil and Athletic Ability
A research study looking at volunteer's who inhaled peppermint essential oil fond it significantly increased their running speed, hand grip strength, and number of push-ups. However, it had no effect on their skill-related tasks (*Raudenbush B., Corley N., Eppich W. Enhancing athletic performance through administration of peppermint odor. J. Sport Exerc. Psychol. 2001;23:156– 160. doi: 10.1123/jsep.23.2.156*). This effect has also been found to occur in other studies such as Ho and Spence (*Preliminary investigation of the effect of peppermint oil on an objective measure of daytime sleepiness. Norrish MI, Dwyer KL. Int J Psychophysiol. 2005 Mar; 55(3):291-8*).

Exercise and Jasmine Rose and Lavender Essential Oils
Studies by Nagai et al. looked at the results of inhaling jasmine rose and lavender essential oils and physical exercise in college students. The study found that inhalation suppressed their muscle sympathetic vasoconstrictor activity. Muscle sympathetic vasoconstrictor activity

relates to the arteries. (*Pleasant odors attenuate the blood pressure increase during rhythmic handgrip in humans. Nagai M, Wada M, Usui N, Tanaka A, Hasebe Y. Neurosci Lett. 2000 Aug 11; 289(3):227-9*).

Ylang-Ylang Lengthens Processing Speed
A study by Moss et al. examined volunteer's inhaling the essential oils of peppermint and ylang-ylang and looked at their mood and cognition. The study found that peppermint significantly enhanced their alertness and memory and that ylang-ylang increased calmness and lengthened their processing speed (*Modulation of cognitive performance and mood by aromas of peppermint and ylang-ylang. Moss M, Hewitt S, Moss L, Wesnes K. Int J Neurosci. 2008 Jan; 118(1):59-77*).

Essential Oils for Enhancing Attention
Inhaling essential oils of jasmine, ylang-ylang and peppermint have been shown to significantly enhance attention. This may be due to the presence of the cineole and menthol in these essential oils (*The influence of essential oils on human attention. I: alertness. Ilmberger J, Heuberger E, Mahrhofer C, Dessovic H, Kowarik D, Buchbauer G. Chem Senses. 2001 Mar; 26(3):239-45*).

Topical Application of Ylang Ylang Increases Skin Temperature

Hongratanaworakit and Buchbauer researched transdermal absorption of ylang-ylang essential oil and its physiological effects on humans. The study discovered that Ylang-ylang significantly decreased blood pressure while increasing skin temperature. Also the participants rated themselves as more relaxed and calmer compared to the control group (*Relaxing effect of ylang ylang oil on humans after transdermal absorption. Hongratanaworakit T, Buchbauer G. Phytother Res. 2006 Sep; 20(9):758-63*).

Carvone and Limonene are Chiral Fragrances

Heuberger et al. researched the enantiomers carvone and limonene, both of which are chiral fragrances on the human autonomic nervous system. He discovered prolonged inhalation of these fragrances affected the autonomic nervous system and that the chirality of these odor components may be a major factor in relation to the biological activity caused by these fragrances (*Effects of chiral fragrances on human autonomic nervous system parameters and self-evaluation. Heuberger E, Hongratanaworakit T, Böhm C, Weber R, Buchbauer G. Chem Senses. 2001 Mar; 26(3):281-92*).

Scott Rauvers

Further **Reading**

Quantum effect could explain how chiral molecules interact. Electron spin polarization promotes recognition between molecules of similar chiralityBy Jyllian Kemsley. ww.cen.acs.orgw.

Chirality of Superfluid 3He-A . M. Walmsley and A. I. Golov. November 2012.

References. Chapter 14

(1) Forsdyke D. R. (2009). "Samuel Butler and human long term memory: Is the cupboard bare?". Journal of Theoretical Biology. 258: 156–164. doi:10.1016/j.jtbi.2009.01.028. PMID 19490862.

(2) Chronic dietary alpha-linolenic acid deficiency alters dopaminergic and serotoninergic neurotransmission in rats. Delion S et al. Dec 1994

(3) Exercise and brain neurotransmission. Meeusen R1, De Meirleir K.. Sept 1995

(4) Molecular Cell Biology. 4th edition.. Section 21.2 The Action Potential and Conduction of Electric Impulses. Lodish H, Berk A, Zipursky SL, et al. New York: W. H. Freeman; 2000.

(5) Marshall, N. J. et al. The compound eyes of mantis shrimps (Crustacea, Hoplocarida, Stomatopoda). 1. Compound eye structure—the detection of polarized light. Philos. Trans. Roy. Soc. B 334, 33–56 (1991), (A biological quarter-wave retarder with excellent achromaticity in the visible wavelength region N. W. Roberts. July 2008)

(6) Lithium enhances remyelination of peripheral nerves Joelle Makoukji et al. Feb 2012

(7) Lithium Chloride Reinforces the Regeneration-Promoting Effect of Chondroitinase ABC on

Rubrospinal Neurons after Spinal Cord Injury. Leung-Wah Yick et al. Aug 2004,(Huang X, Wu DY, Chen G, Manji H, Chen DF. Support of retinal ganglion cell survival and axon regeneration by lithium through a Bcl-2-dependent mechanism. Invest Ophthalmol Vis Sci. 2003;44:347-354).

(8) Tau-targeted treatment strategies in Alzheimer's disease. Jürgen Götz et al. March 2012

(9) Stabilization of microtubules by lithium ion. B Bhattacharyya and J Wolff. Dec 1976

(10) Lithium reversibly inhibits microtubule-based motility in sperm flagella. Gibbons BH and Gibbons IR. June 1984

(11) Lithium-induced changes in the brain levels of free amino acids in stress-exposed rats. Eroglu L et al. 1980

(12) Cholesterol metabolism and homeostasis in the brain Juan Zhang and Qiang Liu. Feb 2015

(13) Myelin Recovery in Multiple Sclerosis: The Challenge of Remyelination Maria Podbielska. Aug 2013

(14) Effects of vitamin B12 and folate deficiency on brain development in children Maureen M. Black. June 2008

(15) Enhanced rat sciatic nerve regeneration through silicon tubes filled with pyrroloquinoline quinone. Liu S et al. 2005

(16) Combinations of Ashwagandha Leaf Extracts Protect Brain-Derived Cells against Oxidative Stress and Induce Differentiation Navjot Shah et al. March 2015

(17) Peripheral Nerve Regeneration Following Crush Injury to Rat Peroneal Nerve by Aqueous Extract of Medicinal Mushroom Hericium erinaceus (Bull.: Fr) Pers. (Aphyllophoromycetideae) Kah-Hui Wong et al. Aug 2011

(18) RESEARCH COMMUNICATION Levels of pyrroloquinoline quinone in various foods Takeshi KUMAZAWA,. Biochem. J. (1995) 307, 331-333 (

(19) Cigarette smoke-induced oxidative stress in COPD Hoffmann, Roland Frederik. University of Groningen. 2016

(20) Tyrosine, phenylalanine, and catecholamine synthesis and function in the brain. Fernstrom JD1, Fernstrom MH. June 2007

(21) Catecholamines in plants. A. I. Kuklin B and V. Conger. April 1995

(22) Resveratrol and fish oil reduce catecholamine-induced mortality in obese rats: role of oxidative

stress in the myocardium and aorta. Pricila R M Avila et al. April 2013.

(23) Gibberellic acid in plant Still a mystery unresolved Ramwant Gupta and S K Chakrabarty. June 2013

(24) Catecholamine. Wikipedia

(25) Rhodiola rosea: A Versatile Adaptogen. Farhath Khanum, Amarinder Singh Bawa, and Brahm Singh. Comprehensive Reviews in Food Science and Food Safety Volume 4, Issue 3, Version of Record online: 20 NOV 2006

(26) Neuroprotective effects of the green tea components theanine and catechins. Kakuda T. Dec 2002, (Characterization of l-Theanine Excitatory Actions on Hippocampal Neurons: Toward the Generation of Novel N-Methyl-d-aspartate Receptor Modulators Based on Its Backbone. Sebih F et al. Aug 2017)

(27) Effect of L-theanine on glutamatergic function in patients with schizophrenia. Ota M et al. Oct 2015

(28) Electrooxidation and Determination of Dopamine Using a Nafion®-Cobalt Hexacyanoferrate Film Modified Electrode Suely S. L. Castro. et al. March 2008

(29) Glutamate and GABA imbalance following traumatic brain injury Réjean M. Guerriero et al. May 2016

(30) Synergistic interactions between commonly used food additives in a developmental neurotoxicity test. Lau K et al. March 2006

(31) Spinal cord pathology is ameliorated by P2X7 antagonism in a SOD1-mutant mouse model of amyotrophic lateral sclerosis. Apolloni S et al. Sept 2014

(32) Synergistic interactions between commonly used food additives in a developmental neurotoxicity test. Lau K et al. March 2006

(33) Brilliant blue G treatment facilitates regeneration after optic nerve injury in the adult rat. Ridderström M1, Ohlsson M. Dec 2014

(34) GABA: a dominant neurotransmitter in the hypothalamus. Decavel C, Van den Pol AN. Dec 1990

(35) Towards a glutamate hypothesis of depression An emerging frontier of neuropsychopharmacology for mood disorders Gerard Sanacora et al. Aug 2011

(36) The dark side of intuition: aging and increases in nonoptimal intuitive decisions. Mikels JA et al. April 2013

(37) L-theanine, a natural constituent in tea, and its effect on mental state. Nobre AC1, Rao A, Owen GN. 2008

(38) Alpha-band oscillations, attention, and controlled access to stored information Wolfgang Klimesch. Dec 2012

(39) Relaxation and immunity enhancement effects of gamma-aminobutyric acid (GABA) administration in humans. Abdou AM et al. 2006

Chapter 15. Techniques for Controlling the Signal to Noise Ratio during Associative Remote Viewing

In evolutionary terms, our minds are still evolving. Hence we only use a small fraction of our entire brain. Some people are born with extraordinary faculties, however upon closer inspection, it is an extremely noisy species in more ways than one.

A study titled: Dopamine, Paranormal Belief, and the Detection of Meaningful Stimuli published in August 2010 and conducted by P. Krummenacher and colleagues found that Dopamine improves perceptual and cognitive decisions by increasing the signal-to-noise ratio.

Research studies ascertaining the average ratio of signal/noise occurring among gifted and non-gifted persons with naturally occurring psi perceptions were examined to determine what the signal to noise ratio was. By 1975 research had shown that the average was approximately 15% signal to 85% noise *(1973-1975 H. E. Puthoff)* [1].

The entire process of remote viewing was primarily developed as a means to bring the psychic signal-to-noise ratio under control. This allows accurate target-related information to be extracted via the background static noise. Hence, noise is a part of the process that can never be fully erased. Anyone not having a complete understanding of this concept would function any more reliably than 20% effectiveness. This is what so-called "natural" psychics are able to muster under normal conditions.

The signal to noise ratio is relevant to information as is relevant to anything. Simply put, any form of concentrated mental effort creates a processing of information that is susceptible to the signal-to-noise ratio. Perhaps the noise exists as a safety mechanism which "tunes out" high frequencies that would otherwise overwhelm the mind while it operates on a day to day basis.

What does the Signal to Noise Ratio Mean?
The Signal to Noise Ratio is a concept that describes a form of interference occurring during remote viewing sessions. It can consist of "still small voices", the whinings of our own egos, the overlay of our own imaginations or the urges of our own wants and desires. After a series of remote viewing sessions, one learns to develop spiritual discipline which separates the revelatory signals from our own subjective inner noise. Ingo Swann, one of the first pioneers of remote viewing, stated that one should first learn to notice direct sensory data received during the remote viewing sessions. This can include

impressions and imaginable signals such as feelings.

It takes experience from numerous remote viewing sessions to allow one to gradually separate the "wheat" signal from the noise "chaff". Future research may end up revealing that dopamine reduces the level of "noise" that occurs during remote viewing sessions.

A Scientific look at the Signal to Noise Ratio
A more in-depth understanding of the Signal to Noise Concept comes from Van Nostrand's Scientific Encyclopedia (1968).

SIGNAL: (1) Independent variable; (2) Audible, visual, or other indicator used to convey information; (3) The message, intelligence or effect that is to be conveyed through (or over) a communication system; (4) A signal wave.

NOISE: Any type of undesirable sound(s). By extension, noise exists as any unwanted disturbance consisting of a useful frequency band, such as undesired electric waves in a device or transmission channel. Such disturbances, if produced by other systems or sources (including services), are called interference. Noise also exists as random or accidental fluctuation in electric circuits due to motion of the current flow.

Summary

Information = signal

Mind + Signal Processing = Information

Noise is an adjective to denote unwanted fluctuations in quantities which are desired to remain constant (*or clear and without interference*).

Radio propagation and Signal to Noise Ratio

Radio propagation is a term remote viewing theory borrows from radio propagation, which has precise correlation to the principle of revelation.

Quantum computers operate on a signal-to-noise ratio[2]. In electrical engineering no signal can be protected or enhanced unless the eroding noise sources have been clearly identified. In electrical engineering the signal-to-noise ratio is common when dealing with instruments that measure television, radar, radio and sonar signals. A television set tuned to a non-frequency = noise (*a blank television channel*), however when tuned to a specific frequency (television transmitter) = signal. Signals = equate to accuracy. Noise = inaccuracy. The companion metaphor being the ratio of accuracy to inaccuracy.

In view of the electromagnetic universe,

information is transmitted via precise EM waves and frequencies that are referred to as band(s). Visually our eyes receive signals of very small bands of the EM spectrum. This is known as the light spectrum. Our ears receive other bands in the EM spectrum. Analytic overlay is also known as AOL and is just one of a number of sources known as "mental noise" that is overwhelming to the beginning remote viewer. These are known as physical inclemencies and consist of imaginative, emotional and even telepathic overlays. Because we are obtaining our information by proper modulation and understanding how the signal to noise ratio works, if the parameters of the signal-to-noise issue, including its attendant problems, are not completely understood, remote viewing sessions will not be able to be conducted in any real, functional clarity.

Because electromagnetic wave energy is related to frequency, by lowering our brain wave frequency we can think with less energy. Hence more energy is available for remote viewing tasks making it much more efficient.

The Signal to Noise Ratio and the Pareto Principle

The Signal to Noise Ratio can be more fully understood by the Pareto principle, which is also called the 80/20 rule, the principle of factor sparsity, or the law of the vital few.

The Pareto principle states that for multiple events that occur, roughly 80% of their effects will emerge from 20% of the causes. To put it simply, if you find that you can't get things done, or are failing in a certain activity, spend more time on the 20% of what is getting the results and eliminate or reduce the amount of time on the 80% that is not getting results.

To put this into context of the Signal to Noise Ratio concept, mental noise during a remote viewing session will consist of approximately 80% background noise and approximately 20% information / signal.

Summary
A major portion of our universe exists in a signal-to-noise environment, where the ratios between the noise and the signals are of crucial and critical importance as they are interdependent with one another.

Working with the Signal to Noise during Remote Viewing sessions
The key to working with the signal to noise during remote viewing is not to focus entirely on eliminating the noise, but instead to find the "*sweet spot*" in the signal that exists in the noise.

Experience is the key. The more one remote views, the more one learns to locate this "*sweet *spot*". Repeated accurate feedback from remote viewing sessions can also be used to understand and identify these signals.

The Three Crucial Concepts to Remote Viewing Sessions

1 - model of the mind

2 - the transfer of information

3 - the signal-to-noise ratio

1. Model of the mind is an axinom referring to the mindsets of certain people and how they cultivate and perceive reality. Sensations, mental images, emotions and ideas are all included.

2. Transfer of information describes how telepathic signals break down and become reassembled between the sender and receiver. This is analogous to a fax, where the transmitted information becomes broken down into bits and pieces and then sent through the network continuum.

3. Signal-to-noise ratio is defined as two separate segments.

Signal = information transmitted.

Noise= dilutes, distorts or disturbs clean reception.

Further Reading
Remote viewing with the artist Ingo Swann: Neuropsychological profile, electroencephalographic correlates, magnetic resonance imaging (MRI), and possible mechanisms. Persinger MA, Roll WG, Tiller SG, Koren SA, Cook CM. Percept Mot Skills. 2002;94:927–49.

Methods to Enhance the Signal and Reduce the Noise
Besides experience, we can associate our remote viewing sessions in environments that enhance the accurate signals received during remote viewing.

One trick to greatly reduce mental noise is to perform remote viewing at the LST peak times which are shown in this book. From my own personal experience, I have got the best results performing the sessions during midnight at the LST peak times of 13:30LST and 8:45LST. The LST times are times that less interference occurs during remote viewing which has been obtained from thousands of remote viewing trials (*Apparent Association Between Effect size In Free Response Anomalous cognition Experiments And Local sidereal Times. James P. Spottiswoode Cognitive Sciences*

Laboratory). Also having healthy dopamine levels in the body before a remote viewing session greatly helps to bring the signal to noise problem under control.

Another simple method to reduce the "noise" is to understand that trying to remote view numbers and letters is almost impossible. Remote viewing is like art. One is interpreting emotions and sensations received by the nervous system. In the case of remote viewing the Dow, instead of looking for an exact number, instead use a blank chart of the Dow (*shown at the end of this book*) and fill it in during your remote viewing session. The line/pattern you draw during your associative remote viewing session will give an idea of trading activity for the day.

References. Chapter 15

(1) A Perceptual Channel for Information Transfer over Kilometer Distances: Historical Perspective and Recent Research HAROLD E. PUTHOFF. MEMBER, IEEE, AND RUSSELL TARG. March 1976
(2) Algorithmic cooling and scalable NMR quantum computers P. Oscar Boykin et al. March 2002

Chapter 16. Using Disruptors to Enhance Quantum Coherence in Microtubules

We covered Microtubules and their role in consciousness earlier and in the last few chapters we explored methods, substances and techniques to enhance the clarity of remote viewing. Now let's explore methods that enhance the communication between our consciousness and microtubules. These are known as "microtubule disruptors"[1] which put stress on the microtubules just enough to enhance their performance. Disruptors are akin to what's called "elictors" which are substances that exhibit defense mechanisms in plants[2]. A disruptor is able to penetrate just far enough into a microtubule to "*stress it*", enhancing its performance. Excess use of disruptors can degrade performance, so it is key to know the proper amount ("*sweet spot*") to use.

In a plant research study, tyrosine was administered to mimic stress. This caused reorganization to take place in the plant which encouraged development[3]. It is my hypothesis that just as our nervous system has been proven to exhibit "*pre-sentient*" effects in numerous studies, some plant elictors are the fuel that sharpen the receptivity of the nervous system, allowing it to more rapidly respond to future events, information and/or stimuli. Future research may show that certain specific essential oils exhibit quantum entanglement effects. It may also be that our intention, especially the intention exhibited during remote viewing exhibits quantum effects [4].

Microtubule Disruption
What better way to elicit an elector response then to use a concentrated plant extract, also called an "essential oil". Numerous studies have shown that essential oils affect brain waves, usually by causing increased alertness.

One study, among many, showed that after participants inhaled essential oils of chamomile, lavender, eugenol and sandalwood that it caused enhancement of Alpha brainwave activity [5]. According to Friedman et al. essential oils contain two time dependant actions[6].

1 - compounds which act slowly

2 - compounds which act rapidly

Other studies have found that the essential oils of limonene, citrol (from lemon essential oil) cause a release of monoamines from the brain's of rats[8] (*Monoamines are neurotransmitters*).

Monoterpenes and Citrus
Another study found that Monoterpenes in citrus essential oil also enhanced the release of monoamines in rat brains (*Flavor components of monoterpenes in citrus essential oils enhance the release of monoamines from rat brain slices. S. Fukumoto et al. Nutr Neurosci. 2006 Feb-Apr;9(1-2):73-80*).

Rapid acting essential oils include **geraniol**, carvacrol and cinnamaldehyde which we shall cover in greater detail later on. It may be that the rapid acting essential oils exhibit a catalyst type effect which enhance ARV sessions.

The Action of Essential Oils at the Microscopic Scale
Essential oils exhibit hydrophobicity. Hydrophobicity is the ability for a substance to partition with lipids present in cell membranes of mitochondria and bacteria. As it does so, it renders the membrane more permeable. ie: making it more flexible.

Essential oils also contain hydrocarbons which have the ability to interact with the cell's membrane, allowing for easy penetration of substances into the cell[8]. Pei et al. [8] stated that the synergistic effects of carvacrol/eugenol and thymol/eugenol may be due to the fact that thymol and carvacrol cause disintegration in the outer membrane of E. coli. This allows for eugenol to enter the cytoplasm and then combine with the proteins. It appears that this penetrating effect causes a minor disruption in the functioning of the Microtubules via the monoterpenes present in the essential oil as has been scientifically shown [9]. In certain cases, while a person is in HRV coherence, this disruption causes minor short term stress on the brain's microtubules, enhancing remote viewing clarity.

Essential Oils and their Effects on Microtubules

A detailed research study on the effects of essential oils on microtubules found that Limonene and Citronellal showed strong microtubule efficacy[10]. The essential oils of Citral, Geraniol, Menthone (a monoterpene) and Cavone showed moderate microtubule efficacy[10].

Citral is an essential oil found in specific aromatic plants which has a typical lemon smell. Citral has been shown to disrupt microtubules, yet not cause significant damage. In the study, Citral was biotransfomed into geraniol and nerol. Carvacrol has been shown to cause membrane damage, yet not cause a detectable effect.

Limonene has strong effects and Limonene is transformed into carvacrol[10].

What is Carvone?
Carvone is a terpenoid found in essential oil. It can be obtained from caraway seeds (60% to 70%), spearmint and dill seed oil (40% to 60%)[11]. Carvone is also found in abundant quantities in mint [12] and especially spearmint oil (50% to 80%)[11]. The majority of Carvone is made from synthesizing limonene.

In rainforests in Brazil, the substances Pinene, Myrene and Limonene were found to be emitted by the trees in forests (Kuhn 2002)[13]. Limonene also promotes monoamine release from the brain[14]. A research study found that the theanine in Japanese green tea caused dopamine release via the monoamines in the tea[15].

What is A Monoamine Releasing Agent?
A monoamine releasing agent is a substance which induces the release of a monoamine neurotransmitter in the brain's presynaptic neurons. This causes increased extracellular concentrations of the neurotransmitter.

Geraniol
Geraniol has been found to be one of the best microtubule disruptors. Also allicin from garlic has also been found to disrupt microtubules

(*Prager khoutorsky 2008*) [16] and allicin derivatives were found to disrupt microtubules, but only when administered at higher concentrations[16].

Geraniol is used for Spinal Cord Injuries
A study conducted on rats looked at using Geraniol to treat oxidative stress, inflammatory responses and apoptosis in traumatic spinal cord injuries. The study found that oxidative stress and inflammatory responses were significantly suppressed when the rates were exposed to geraniol.

Reference -
Protective effect of geraniol inhibits inflammatory response, oxidative stress and apoptosis in traumatic injury of the spinal cord through modulation of NF-?B and p38 MAPK. Jiansheng Wang et al. 2016 Oct 27. doi: 10.3892/etm.2016.3850. Exp Ther Med. 2016 Dec; 12(6): 3607–3613.

Citral
Citral has been shown to promote apoptosis in certain cancer cell lines as well as exhibit numerous cancer fighting properties. Also citral has been suggested as one of the better substances that can show healing abilities when inhaled, especially by enhancing sleep [17].
Citral is found in large amounts in

lemongrass essential oil and a complete study of the oils in lemongrass found that besides citral, lemongrass also contained geranial, geraniol, nerol and myrcene along with lower levels of nerol (*Lewinsohn et al., 1998*) [18]. Citral has also been shown to be effective in weed management[19]. Pictured below are the main substances present in lemongrass essential oil

Myrcene *Geraniol* *Nerol*

Geranial *Neral*

The Effect of Citral on Microtubules

Citral has been shown to interact directly with tubulin causing a disruption of microtubules[20].

Citral has also been shown to disrupt microtubules in animals, although to a lesser extent compared to plants[21]. Also studies have shown that the effects on Microtubules exhibited by Citral are dose and time dependant and that citral administered at low concentrations disrupted mitotic microtubules and cell plates. When applied at high doses it caused cell elongation through disruption of the cortical microtubules in seedlings [22]. This effect has also been noted in wheat [23].

Citral Protects the Liver

In tests done on the livers of mice, citral showed protective effects against toxins[24].

Synergy between Citral and Geraniol

Geraniol has been shown to be effective at disrupting microtubules and geraniol has the ability to retain citral's ability to disrupt microtubules. This is interesting because one of the major components in our TXP formula uses Geraniol[25].

Synthetic Microtubule Disruptors

Anesthetics and Microtubules

Anesthetics disrupt the functioning of microtubules, leading to a loss of consciousness (*Hameroff, 1994*)[26]. This temporary impairment may show that microtubules represent the physical structures for emerging consciousness. Thus consciousness exists as a result of collective quantum effects occurring in microtubule networks that exist within the central nervous system (*Koruga, 1995*) [26] [27].

Anesthetics and Microtubules
An anesthetic has similar effects to an opoid, in that it is used to prevent pain during surgery. There are two types of anesthetics. 1 - Causes a reversible loss of consciousness. 2 - Local anesthetics which causes a temporary loss of sensation in a generalized region of the body. Anesthetics have also been shown to eliminate the propagation of action potentials. This could be because general anesthetics readily bind to membranes and hydrophobic regions to form water clathrates and water is an important part of a microtubule's internal structure[28].

Further Reading
Anesthetics act in quantum channels in brain microtubules to prevent consciousness.

What is an Action Potential?
An action potential is a change of electrical activity occurring in the passage of impulses along membranes in muscles or nerve cells.

Researchers at the University of Pennsylvania found that anesthetic compounds work by causing a disruption in the functioning of microtubules[29]. This effect is caused by a dispersal of the electric dipoles that are necessary for consciousness. Also research by Hammeroff found that the dynamics of microtubules would change when they were exposed to anesthetic compounds [26].

As we showed earlier where small amounts of certain essential oils would disrupt microtubule operation, the same may be with certain anesthetic substances, in that there may exist a "sweet spot" that alters a microtubule's function just enough to enhance or create a stronger functioning of the brain regions responsible for remote viewing. Further research is necessary to confirm this hypothesis.

Katanin and its role in the Nervous System
Another substance that is required for optimal nervous system functioning is the AAA protein known as Katanin. Katanin is found in abundant levels in the nervous system and even low levels have been shown to significantly contribute to a depletion in the performance of the body's microtubules[30]. Research has found that seedlings that do not have the protein Katanin are unable to make the microtubules change their direction, thus they are unable to grow towards light sources [31].

Kinesin
Kinesin has been shown to be affected by local

anesthetics by stopping microtubule movement[32].

What is Kinesin?
Kinesin is a mini-ATP motor that "*walks*" along the exterior of the microtubule by working two heads in a sort of "*hand over hand type*" moving motion. As the image below shows, the stick figure with the large head and shoes is walking back and forth along the microtubule exterior.

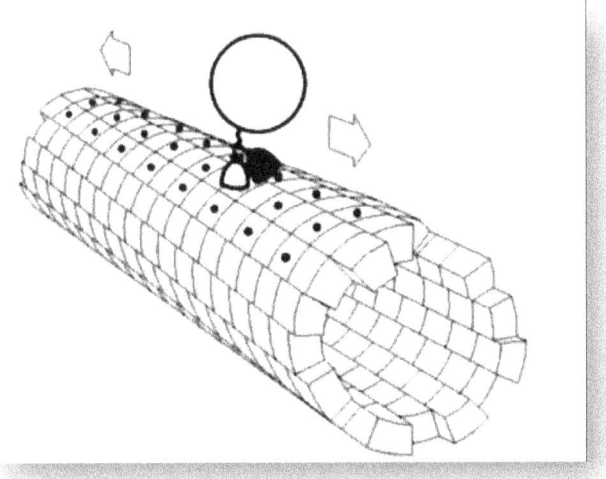

One research study showed that hydrostatic pressure inhibited this activity. The study also showed that high pressure shortened the taxol stabilized microtubules [33]. Perhaps the high pressure causes coherence, especially due to the fact that microtubules contain an inner cavity that exhibits coherent effects.

Lidocaine

Lidocaine is used as an anesthetic and it has been shown to inhibit synaptic transmissions through the membranes of the brain's presynaptic neurons[34].

Substances with similar effects to Lidocaine

The substances known as Linalool and Fenchone were found in a study to have similar effects to Linalool, exhibiting the effects of a local anesthetic. Linalool was shown to be effective even at low concentrations and that lidocaine and linalool were concentration dependant, but not for fenchone, which was inactive at low concentrations[35].

QX-314

QX-314 (*a derivative of lidocaine*) has been shown to produce long-lasting local anesthesia in animal studies [36]

Limonene

The monoterpene limonene (*at 1%*) has been shown to dramatically enhance the absorption of trimecaine (*anesthetic*). When mixed together, Trimecaine synergizes with Lidocane [37].

Anesthesia for Stress Relief

A research study on rainbow trout showed that handling fish caused stress which caused increased stimulation of the hypothalamic-pituitary inter-renal axis and increased levels of blood catecholamines. After administering the rainbow trout the anesthetic (phenoxyethanol) into the fish tanks, it lowered their adrenaline levels and was effective in reducing the stress of fish when they were handled[38].

What is Phenoxyethanol?

Phenoxyethanol is used as an insect repellent; an antiseptic; a perfume fixative; a solvent for cellulose acetate, a preservative for pharmaceuticals, in organic synthesis, cosmetics dyes, inks, and resins; lubricants and as an anesthetic in fish aquaculture.

Phenoxyethanol can be obtained via the hydroxyethylation of phenol. Phenoxyethanol has also been shown to be effective against both gram-positive and gram-negative bacteria and the yeast Candida albicans. Phenoxyethanol is readily available online.

Synthetic Disruptors

Synthetic disruptors are also known as antimitotics. Common antimitotics include vincristine, taxol (**paclitaxel**), colchicine and vinblastine. There is also a fifth and not commonly known synthetic microtubule disruptor called nocodazole[44].

Vincristine as a Microtubule Disruptor
Vincristine is also called leurocristine and is available under the brand name Oncovin. It is commonly used as a chemotherapy medication to treat a number of types of cancer, such as acute myeloid leukemia, Hodgkin's disease, lymphocytic leukemia, neuroblastoma and small cell lung cancer among others[41]. Vincristine is created by a semi-synthesis process of indole alkaloids vindoline and catharanthine found in the Vinca Plant[42]. Vinblastine binds to tubulin which inhibits the assembly of microtubules. Vinblastine is used for patients undergoing chemotherapy medication [43] and for people who are being treated for bladder cancer, brain cancer, melanoma, Hodgkin's lymphoma, non-small cell lung cancer and testicular cancer.[43] Vinblastine is obtained from the Madagascar periwinkle[44].

The Vinca Plant
The Vinca Plant, also called Vinca Rosea, **Madagascar Periwinkle**, Rosy Periwinkle or Teresita, is commonly found in the Caribbean Basin. It is also found in the southern part of Mexico, specifically in Campeche, Champotón and Mérida. The vinca plant exists as an

herbaceous plant or evergreen shrub growing 1 meter in height which is similar in height to the Pacific Yew (*also an evergreen shrub*). The vinca plant has more than 400 alkaloids, and some are used as antineoplastic agents for treating malignant lymphomas, leukemia, Hodgkin's disease, neuroblastoma, Wilms' tumor, rhabdomyosarcoma and other cancers. The extracts of its shoots and roots are poisonous. It has also been used as a memory enhancer and used to alleviate dementia and Alzheimer's disease[45]. European herbalists utilized it for conditions from headache to diabetes. Traditional Chinese medicine utilizes its extracts to heal malaria, diabetes and Hodgkin's lymphoma. Extracts from the leaves and flowers (wet or dry) are used as a paste on wounds and the fresh juice from the flowers of C. roseus are made into a tea used by Ayurvedic physicians in India to treat eczema, acne, dermatitis and other skin issues. The two main components in Vinca are tannins and alkaloids (the main ones being vincamine and vinpocetine).

Extracts of Vinca demonstrate significant anticancer activity. Vinca should be used with caution as if not properly prepared or used can be highly toxic. Hence, small amounts yield powerful effects.

Vinpocetine can be extracted from the seeds

of the periwinkle plant (Vinca minor). It is also used for improving circulation in the cerebral region of the brain as well as memory. Alkaloids of Vinca are the second most-used class of cancer drugs and are estimated to remain so well into the future[46].

One research study found that people treated with 40mg of vinpocetine had improvements in short-term memory[47]. Another study found that microtubules are part of an essential process of learning and memory[48].

Further Reading
Anesthetics act in quantum channels in brain microtubules to prevent consciousness. Craddock TJ et al. 2015

Oryzalin as a Microtubule Disruptor
Oryzalin is a herbicide and acts through the disruption (depolymerization) of microtubules. This allows it to block the anisotropic growth of plant cells[49]. Oryzalin is also used to induce polyploidy in plants. It is a common alternative to colchicine.

Griseofulvin as a Microtubule Disruptor
Griseofulvin is antifungal medication used for fungal infections of the skin and nails when antifungal creams have not proved effective. It is commonly taken by mouth and is ineffective if

used on the skin. It is commonly reserved for special cases involving hair, nail or large body surfaces[50].

Nocodazole as a Microtubule Disruptor
Nocodazole is an antineoplastic agent. It exerts its effect by interfering with the polymerization of microtubules[51]. This is an important observation because antineoplastic agents enhance the flexibility of membranes and a term known as neuroplasticity, which I state in many of my anti-aging books, is an effect that causes aging, hence substances that protect the flexibility of the brain's neurons enhance lifespan due to the ability of them to reduce the "brittleness". Also Selenium has been shown to inhibit the polymerization of tubulin[52] and studies have shown that selenium levels decrease with age and that a decline in the levels of selenium in a person's body corresponded with a decline in cognitive functioning[53].

Further Reading
The aging mind: neuroplasticity in response to cognitive training Denise C. Park, PhD. March 2013.

RH4032
RH4032 acts as a microtube stabilizer [55]and has

shown significant antimicro effects in tobacco[55]. It is commonly found in tobacco. RH4032 also binds strongly to tubulin and strongly affects the roots of tobacco plants[55].

Acetylsalicylic Acid (Aspirin)
Acetylsalicylic Acid has been found to increase a tobacco plant's resistance against TMV and cause an accumulation of PR proteins (*White 1979*)[56].

Another research study found that a 5 mg/kg dose of aspirin significantly enhanced memory [57]. Nicotine has also been used to offset the side effects of excess aspirin intake.

Rotenone
Rotenone is a microtubule disruptor and is a microtubule-depolymerizing agent that affects dopaminergic neurons in the midbrain. This region of the brain has been linked to Parkinson's disease.

The toxicity of rotenone has been shown to be significantly reduced by the microtubule-stabilizing substance taxol[59]. Curcumin, found in turmeric, has been shown to protect against excess Rotenone intake [60].

There is a remarkable connection between Parkinson's disease, depression and microtubules. Research studies found that people with Parkinson's disease learn better from negative feedback, compared to people without

Parkinson's disease who learn better from positive feedback (*Frank 2004, Hearst 1991, Kim, Skimojo and O, Doherty 2006, Wachter 2009*)[61].

Other research has shown that people with Parkinson's have less dopamine [62]. Dopaminergic neurons in the midbrain are vulnerable to rotenone, a microtubule-depolymerizing agent, which has been linked to Parkinson's disease.

The toxicity of rotenone has been shown to be significantly reduced by the microtubule-stabilizing substance **taxol**[63], which mimics the behavior of the microtubule-depolymerizing agents such as nocodazole and colchicine. The application of antioxidants will also significantly reduce the effects of rotenone or colchicine[64]. Taxol has been shown to reverse the effects of Rotenone[64a].

Nicotine as a Central Nervous System Antioxidant

A research study found that nicotine may exhibit antioxidant properties in the body's central nervous system and be used to treat neurodegenerative diseases[65]. This is significant because the body's nervous system plays a key role in the success of remote viewing. Also vitamin C will block the effects of Nicotine. This is why we try to avoid excess vitamin C before an ARV session [66].

Psychotropics and Microtubules
Research studies have found that Microtubules can be altered by lysergic acid diethylamide (LSD)[67]. These may also cause some forms of psychosis.

Colchicine
Colchicine is commonly used to treat gout [68] and comes from the plant Colchicum (autumn Crosus), which is also commonly called Meadow Saffron). Colchicine is available from online pet pharmacies.

Meadow saffron is a very toxic poison and is commonly mistaken for wild garlic. The density of microtubules has been shown to be reduced by colchicine[69]. Colchicine has been shown to exert anti-inflammatory properties in various cardiovascular settings and studies show that it has been used for conditions of the heart. Two studies [70] [71] have already showed that Colchicine is recommended for heart conditions and studies have shown that Microtubules are affected by heart beats. Colchicine has been found to be a stronger disruptor of microtubules compared to citral and that 0.1 um colchicine was enough to disrupt microtubules just enough without affecting acting fibers[73]. Other

Microtubule altering substances include Oryzalin[74] and Propyzamde[75].

High Pressure and its Effects on Microtubule Functioning

I show in my second remote viewing book **Remote Viewing. The Complete User's Manual on Experiencing Future Consciousness**, that high barometric pressure is one of the main components for a successful remote viewing session. A barometric high air pressure system is usually a period of drier weather a few days before the weather turns cooler and rain begins.

An interesting fact is one of the world's greatest Inventors Dr. Yoshiro Nakamats, states the way he gets his brilliant insights is by swimming underwater at the bottom of a pool where the high pressure is strongest and while swimming, he holds his breath for as long as possible while writing down his idea on a waterproof notepad. Hence, if high pressure is giving this inventor brilliant insights, it may prove the link between high pressure and its ability to influence the brain's microtubules / consciousness.

Also we found that when barometric pressure was highest or peaking that our ARV sessions were always more accurate.

Summary
Higher pressure has the ability to preserve and maintain coherence for longer periods of time compared to low pressure.

High Pressure and Microtubule Depolymerization
Research studies have shown that high pressure induced microtubule depolymerization by perturbing the binding of nucleotides to myosin and that it caused changes in the stepping reactions of the kinesin motors within the microtubules[76]. The research study showed that hydrostatic pressure inhibited this activity. The study also showed that the high pressure shortened the taxol stabilized microtubules[76]. Perhaps the high pressure caused microtubule disruption.

High Pressure and X-rays
Coherent x-rays in a high pressure environment have been shown to preserve the coherence when the X-ray beam passes through a diamond cell[77].

Coherence and High Pressure
An effect occurs when several maser microwaves at varying frequencies are formed in a resonant cavity inside high pressure gases. This causes a

condition known as the so called coherence effect[78].

The Effects of High Pressure on Muscles
One study found that hydrostatic high pressure caused changes in the behavior of muscle fibers [79].

Sensory Deprivation and Remote Viewing
These sensory deprivation techniques also help induce lucid dreaming, which can come in handy a few days before an ARV session. One popular method is to use a flotation tank and it may require several sessions before lucid dreams start occurring.

A flotation tank consists of an 8 ft x 4 ft tank that holds 10 inches of a saturated solution of magnesium sulphate (BP or USP grade), at a temperature of 95 degrees F (35 degrees C) which is the skin temperature of the human body. This creates an ambience sensation throughout the body where sensations of touch vanish. The room should be soundproof and dark so all five senses receive no normal input. This allows the brain to "turn up its automatic gain control". (*Hutchison, 1984*) Electroencephalograph's (EEG) of people in flotation tanks exhibit bi-laterally symmetrical beta, theta and alpha brainwaves.

A research study published in 1984 titled:

lower systolic and diastolic pressure and heighten their subjective perception of relaxation. and conducted by G.D. Jacobs and colleagues found that participants in a flotation tank showed lower diastolic and systolic blood pressure and heightened subjective perception of relaxation.

Experiences in a flotation tank include spontaneous bright images as if in a vivid dream, although one is fully awake and self-conscious. Also scenes from long-past memories may reappear. Another study found that flotation tanks do not increase endorphin levels, but do **lower anxiety** and depression[81].

Another trick to induce lucid dreaming is to look at your hands very frequently throughout the day and visualize yourself trying to make them disappear. This is done as a test of whether you are awake or dreaming. If made into a habit, it will automatically continue as you dream, with one's hands disappearing, triggering a realization one is dreaming. Hence at the moment of self consciousness, the dream becomes lucid.

This process encourages the brain's left hemisphere to become active during dreaming, allowing for conscious detection of the dream. Also Apps are now available for the I-phone and I-pod Touch that wake one up during rapid eye movement (REM) sleep, which occurs during lucid dreaming. One App is "Sleep Cycle Alarm Clock" and will wake you up during dreams. The App works by detecting dreaming movement then sounding an alarm. An excellent research study discussing sleep technologies can be found in the research paper titled: Consumer Sleep Technologies: A Review of the Landscape and published in December 2015 by Ping-Ru T. Ko, MD and colleagues.

One method used by the author of the book **Exploring the Sub-conscious using New Technology** (by Michael Gwyn Hocking) states that a vegetarian dish ate in the evening known as Lo Han ("Buddha's delight") with soy sauce, allows vivid dreams to occur. The dish, listed lily buds and soy sauce. Lily contains the substance **glutamine**, known to cause lucid dreams. Extracts from Red Spider Lily (Lycoris radiata) has a long history of people using it for dream enhancement and for memory recall. Soy Sauce just happens to contain the strain Aspergillus oryzae, which **produces GABA** [82].

Also extracts from the Snowdrop plant have been used for memory loss (*Duvoisin, 1983*). And the substance galantamine, which comes from Snowdrop [84] has been shown to improve the memory of those suffering from Alzheimer's disease, as well as to prevent or treat neurocognitive decline [83]. A research study found that people taking Vitamin B6 had significantly enhanced dream recall[84]. Pictured is the Snowdrop plant.

Further Reading
Pre-sleep treatment with glutamine increases the likelihood of lucid dreaming. Poster session, presented June 25, 2012 at the Annual conference for the International Association for the Study of Dreams, Berkeley, CA. La Marca, K. and Laberge, S. (2012).

The Hippocampus And Extremely Low-Frequency Electromagnetic Fields
While our research has shown that low to mild geomagnetic activity (*similar to low frequency electromagnetic fields*) levels create the best opportunity for remote viewing sessions, it may

be due to the fact that the right levels of geomagnetic activity is affecting the Hippocampus region of the brain. This has been looked at in one study titled An interaction of a NR3C1 polymorphism and antenatal solar activity impacts both hippocampus volume and neuroticism in adulthood[88]. Another study found that extremely low-frequency electromagnetic fields were associated with enhanced spatial learning and memory and that the extremely low-frequency electromagnetic fields did so by enhancing the survival of newborn neurons in the hippocampus of mice[89]. It is interesting to note that work done by Puthoff and Stanford Researcher Marcia Adams found that tress act as powerful receivers of low frequency waves[90].

Further Reading
Modification of semantic memory in normal subjects by application across the temporal lobes of a weak (1 microT) magnetic field structure that promotes long-term potentiation in hippocampus slices. Richards P. M., Persinger M. A., Koren S. A. (1996) Electro- and Magnetobiology, 15, 141–148

Endogenous potentials generated in the human hippocampus formation and amygdala by infrequent events. Halgren E., Squires N. K.,

Wilson C. L., Rohrbaugh J. W., Babb T. L., Crandall P. H. (1980) Science, 210, 803–805

Temporally structured replay of awake hippocampus ensemble activity during rapid eye movement sleep. Louie K., Wilson M. A. (2001) Neuron, 29, 145–156.

Moon Phase and Geomagnetic Activity
A research study found that during descending solar cycles, stronger geomagnetic activity occurred around the time of the new moon. As the solar cycle began increasing, reaching solar maximum, geomagnetic activity was weaker/lower several days before full moon, with maximum geomagnetic activity occurring several days after the full moon.

This means that ARV sessions conducted when the solar cycle is increasing or approaching solar maximum, favorable geomagnetic activity that is conductive to successful remote viewing sessions would occur more often several days before full moons.

Another study looking at 31 years of geomagnetic activity found that geomagnetic activity increases about 4% after a full moon which lasts on average seven days. The study also found that geomagnetic activity decreased about 4% seven days before full moons. The probability that these variations occur by random chance is less than 5%. The effect is associated with geomagnetic activity data derived from periods of quiet geomagnetic conditions and is not evident in the data during disturbed periods.

Reference
Variations of geomagnetic activity with lunar phase. Harold L. Stolov and A. G. W. Cameron. Dec 1964

References. Chapter 16

(3) Microtubule disruptors and their interaction with biotransformation enzymes. Modrianský M1, Dvorák Z. Dec 2006

(4) Elicitors and priming agents initiate plant defense responses. Paré PW. Aug 2005

(5) Nitrosative stress triggers microtubule reorganization in Arabidopsis thaliana Elisabeth Lipka Sabine Müller. Aug 2014

(6) Attention, Intention, and Will in Quantum Physics Henry P. Stapp Lawrence Berkeley National Laboratory University of California Berkeley, California 94720

(7) Effects of inhalation of essential oils on EEG activity and sensory evaluation. Masago R. et al. Jan 2000

(8) Antibacterial activities of plant essential oils and their components against Escherichia coli O157:H7 and Salmonella enterica in apple juice. Friedman M et al. Sept 2004

(9) Flavor components of monoterpenes in citrus essential oils enhance the release of monoamines from rat brain slices. Fukumoto S. et al. Apr 2006

(10) Essential Oils in Combination and Their Antimicrobial Properties Imaël

Improve your Remote Viewing Accuracy Techniques using Quantum Microtubules

Henri Nestor Bassolé 1,* and H. Rodolfo Juliani. April 2012
(11) Herbicidal Activity of Monoterpenes Is Associated with Disruption of Microtubule Functionality and Membrane Integrity No Access David Chaimovitsh Alona Shachter, Mohamad Abu-Abied, Baruch Rubin, Einat Sadot and CLOSE Nativ Dudai Unit of Medicinal and Aromatic Plants, ARO, Newe Ya'ar, P.O. Box 1021, Ramat Yishay 30095, Israel Correspondence: nativdud@gmail.com Nativ Dudai. August 2016
(12) January 2017 , pp. 19-30 Herbicidal Activity of Monoterpenes Is Associated with Disruption of Microtubule Functionality and Membrane Integrity David Chaimovitsh (a1) (a2), Alona Shachter (a1), Mohamad Abu-Abied (a3), Baruch Rubin
(13) De Carvalho, C. C. C. R; Da Fonseca, M. M. R. (2006). "Carvone: Why and how should one bother to produce this terpene". Food Chemistry. 95 (3): 413–422. doi:10.1016/j.foodchem.2005.01.003.
(14) Chemical composition of essential oil from several species of mint (Mentha spp.)
(15) Isoprene and monoterpene emissions of Amazônian tree species during the

wet season: Direct and indirect investigations on controlling environmental functions Authors U. Kuhn, et al. Sept 2002

(16) Flavor components of monoterpenes in citrus essential oils enhance the release of monoamines from rat brain slices. Fukumoto S et al. Apr 2006

(17) Effect of theanine, r-glutamylethylamide, on brain monoamines and striatal dopamine release in conscious rats. Yokogoshi H et al. May 1998

(18) Microtubules are an intracellular target of the plant terpene citral David Chaimovitsh et al. 2010. The Plant Journal.

(19) Central effects of citral, myrcene and limonene, constituents of essential oil chemotypes from Lippia alba (Mill.) n.e. Brown. do Vale TG et al. Dec 2002

(20) Histochemical Localization of Citral Accumulation in Lemongrass Leaves (Cymbopogon citratus (DC.) Stapf., Poaceae) EFRA IM LEW INSOHN et al. Sept 1997. Annals of Botany 81: 35-39, 1998. Katsukawa

(21) Microtubules are an intracellular target of the plant terpene citral. Chaimovitsh D et al. Feb 2010

(22) Microtubules are an intracellular target of the plant terpene citral

Authors David Chaimovitsh, et al. Nov 2009

(23) Microtubules are an intracellular target of the plant terpene citral. Chaimovitsh D et al. Feb 2010

(24) The relative effect of citral on mitotic microtubules in wheat roots and BY2 cells. D. Chaimovitsh et al. March 2012

(25) The relative effect of citral on mitotic microtubules in wheat roots and BY2 cells. Chaimovitsh D et al. Mar 2012

(26) Hepatoprotective Effect of Citral on Acetaminophen-Induced Liver Toxicity in Mice. Uchida NS et al. June 2017

(27) Microtubules are an intracellular target of the plant terpene citral Authors David Chaimovitsh, et al. Nov 2009

(28) Anesthetics Act in Quantum Channels in Brain Microtubules to Prevent Consciousness Travis J. A. Craddock et al. 2015

(29) A Quantum Biomechanical Basis for Near-Death Life Reviews Thomas E. Beck, Ph.D and Janet E. Colli Seattle.

(30) The Role of Structured Water in Mediating General Anesthetic Action on α4β2 nAChR Dan Willenbring, et al. July 2010

(31) Direct Modulation of Microtubule Stability Contributes to Anthracene

General Anesthesia Daniel J. Emerson, et al. Mar 2013

(32) Regulation of Microtubule Severing by Katanin Subunits during Neuronal Development Wenqian Yu et al. Sept 2005

(33) . J. Lindeboom, M. Nakamura, A. Hibbel, K. Shundyak, R. Gutierrez, T. Ketelaar, A. M. C. Emons, B. M. Mulder, V. Kirik, D. W. Ehrhardt. A Mechanism for Reorientation of Cortical Microtubule Arrays Driven by Microtubule Severing. Science, 2013; DOI: 10.1126/science.1245533

(34) Direct inhibition of microtubule-based kinesin motility by local anesthetics. Miyamoto Y et al. Feb 2000

(35) Pressure-induced changes in the structure and function of the kinesin-microtubule complex. Nishiyama M et al. Feb 2008

(36) The effect of lidocaine on cholinergic neurotransmission in an identified reconstructed synapse. Onizuka S et al. Oct 2008

(37) Assessing the local anesthetic effect of five essential oil constituents. Zalachoras I et al. Octo 2010

(38) The quaternary lidocaine derivative, QX-314, produces long-lasting local anesthesia in animal models in vivo. Lim TK et al. Aug 2007

(39) Percutaneous absorption of disopyramide, lidocaine and trimecaine. Príborský J et al. 1998
(40) Modulation of stress hormones in rainbow trout by means of anesthesia, sensory deprivation and receptor blockade. L. Gerwick et al. Nov 1999
(41) Phenoxyethanol. Wikipedia
(42) Microtubule disruptors and their interaction with biotransformation enzymes. Modrianský M1, Dvorák Z.. Dec 2005
(43) NEW AGENTS AND COMBINATION CHEMOTHERAPY OF NON-HODGKIN'S LYMPHOMA A. T. SKARIN, et al. 1975
(44) Vincristine. Wikipedia
(45) "Vinblastine Sulfate". The American Society of Health-System Pharmacists. Archived from the original on 2015-01-02. Retrieved Jan 2, 2015.
(46) Liljefors, Tommy; Krogsgaard-Larsen, Povl; Madsen, Ulf (2002). Textbook of Drug Design and Discovery, Third Edition (3 ed.). CRC Press. p. 550. ISBN 9780415282888. Archived from the original on 2016-12-20
(47) Catharanthus roseus flower extract has wound-healing activity in Sprague Dawley rats BS Nayakcorresponding author1 and Lexley M Pinto Pereira Dec 2006

(48) Cytotoxicity of Vinca minor. Khanavi M1 et al. Jan 2010
(49) Vinpocetine effects on cognitive impairments produced by flunitrazepam. Bhatti JZ1, Hindmarch Otc 1987
(50) Of microtubules and memory: implications for microtubule dynamics in dendrites and spines Erik W. Dent. Jan 2017
(51) Taiz, L., Zeiger, E. Plant Physiology, 5/e. 2010. p. 443-4.
(52) "Griseofulvin". The American Society of Health-System Pharmacists. Archived from the original on 20 December 2016. Retrieved 8 December 2016.
(53) Nanomolar concentrations of nocodazole alter microtubule dynamic instability in vivo and in vitro. Vasquez RJ et al. June 1997
(54) Selenium: inhibition of microtubule formation and interaction with tubulin. Leynadier D et al. 1991
(55) Neuroplasticity and Successful Cognitive Aging: A Brief Overview for Nursing David E. Vance, Ph.D et al. August 2012
(56) Covalent binding of the benzamide RH-4032 to tubulin in suspension-cultured tobacco cells and its application in a cell-based competitive-

binding assay. Young DH1, Lewandowski VT. Sept 2000

(57) Covalent Binding of the Benzamide RH-4032 to Tubulin in Suspension-Cultured Tobacco Cells and Its Application in a Cell-Based Competitive-Binding Assay David H. Young* and Veronica T. Lewandowski. Sept 2000

(58) Acetylsalicylic acid (aspirin) induces resistance to tobacco mosaic virus in tobacco. White RF. Dec 1979

(59) Memory-enhancing effect of aspirin is mediated through opioid system modulation in an $AlCl_3$-induced neurotoxicity mouse model SAIMA RIZWAN, et al. May 2016

(60) Effect of nicotine, alcohol and caffeine pretreatment on the gastric mucosal damage induced by aspirin, phenylbutazone and reserpine in rats. Parmar NS, Tariq M, Ageel AM. Sept 1985

(61) Microtubule: a common target for parkin and Parkinson's disease toxins. Feng J. Dec 2006

(62) Protective Effects of Curcumin Against Rotenone and Salsolinol Induced Toxicity: Implications for Parkinson's Disease Zakiya Qualls, et al. Jan 2014

(63) Dopaminergic Medication Modulates Learning from Feedback and Error-Related Negativity in Parkinson's Disease: A Pilot Study Chiara Volpato, et al. Oct 2016

(64) Serotonergic markers in Parkinson's disease and levodopa-induced dyskinesias. Cheshire P et al. May 2015

(65) Neurotrophic factors stabilize microtubules and protect against rotenone toxicity on dopaminergic neurons. Jiang Q et al. Sept 2006

(66) Rotenone selectively kills serotonergic neurons through a microtubule-dependent mechanism Yong Ren*, and Jian Feng. Journal of Neurochemistry, 2007, 103, 303–311

(64a)Rotenone selectively kills serotonergic neurons through a microtubule-dependent mechanism. Ren Y1, Feng J. Octo 2007

(67) Nicotine's oxidative and antioxidant properties in CNS. Newman MB et al. Nov 2002

(68) Role of vitamin C and selenium in attenuation of nicotine induced oxidative stress, P53 and Bcl2 expression in adult rat spleen Author links open overlay panelMarwa A.Ahmeda et al. Sept 2014

(69) Computational study on the binding affinity between microtubules and

consciousnessaltering substances R. Pizzi, T. Rutigliano, A. Ferrarotti and M. Pregnolato. NTERNATIONAL JOURNAL OF BIOLOGY AND BIOMEDICAL ENGINEERING Volume 9, 2015

(70) Chen LX, Schumacher HR (October 2008). "Gout: an evidence-based review". J Clin Rheumatol. 14 (5 Suppl): S55–62. PMID 18830092. doi:10.1097/RHU.0b013e3181896921.

(71) Colchicine, a microtubule depolymerizing agent, inhibits myocardial apoptosis in rats. Saji K. Oct 2007

(72) Volume 2, Issue 2, April 2014 DOI: 10.1016/j.jchf.2014.02.001. Colchicine and the Failing Heart A "FINER" Anti-Inflammatory Agent?

(73) Colchicine, a microtubule depolymerizing agent, inhibits myocardial apoptosis in rats. Saji K1 et al. Oct 2007

(74) May 3, 2016 Microtubules' role in heart cell contraction revealed. NIH Research article

(75) Microtubules are an intracellular target of the plant terpene citral Authors David Chaimovitsh, et al. Nov 2009

(76) Rapid and Reversible High-Affinity Binding of the Dinitroaniline Herbicide

Oryzalin to Tubulin from Zea mays L. Hugdahl JD1, Morejohn LC. Jul 1993
(77) Low concentrations of propyzamide and oryzalin alter microtubule dynamics in Arabidopsis epidermal cells. Nakamura M . Sept 2004
(78) Pressure-Induced Changes in the Structure and Function of the Kinesin-Microtubule Complex Masayoshi Nishiyama, et al. Feb 2009
(79) Combining high pressure and coherent diffraction: a first feasibility test D. Le Bolloc'h, et al. Oct 2009
(80) Coherence effect in maser (microwave) source/ Mladen M. Kekez High-Energy Frequency Tesla Inc., Ottawa, Ontario, Canada, K1H 7L8
(81) Pressure-Induced Changes in the Structure and Function of the Kinesin-Microtubule Complex Masayoshi Nishiyama, Feb 2009
(82) Effects of Flotation REST and Visual Imagery on Athletic Performance: Tennis. Patrick McAleneyArreed Barabasz.
(83) Effects of flotation-REST on muscle tension pain. Kjellgren A. et al. Winter 2001
(84) Evaluation of commercial soy sauce koji strains of Aspergillus oryzae for γ-aminobutyric acid (GABA) production. Ab Kadir S et al. Oct 2016

(85) Galanthamine from snowdrop--the development of a modern drug against Alzheimer's disease from local Caucasian knowledge. Heinrich M1, Lee Teoh H. June 2004

(86) Traditional used Plants against Cognitive Decline and Alzheimer Disease Gunter Peter Eckert. Dec 2010

(87) Effects of pyridoxine on dreaming: a preliminary study. Ebben M et al. Feb 2002

(88) An interaction of a NR3C1 polymorphism and antenatal solar activity impacts both hippocampus volume and neuroticism in adulthood Christian Montag, et al. June 2013

(89) Extremely low-frequency electromagnetic fields enhance the survival of newborn neurons in the mouse hippocampus. Podda MV et al. March 2014

(90) Remote Viewers: The Secret History of America's Psychic Spies By Jim Schnabel

Chapter 17. Substances that Strengthen and Enhance the Operation of Microtubules

While it is good to create the appropriate "stress" upon Microtubules to enhance the clarity of remote viewing, it can also help to keep Microtubules in good functioning condition. Let's look at a few substances that do this.

Ashwagandha as a Microtubule Stabilizer
Ashwagandha has also shown antineoplastic properties[1]. Extracts of Ashwagandha have been found to be highly effective against oxidative stress and glutamate toxicity when conducted in a study examining Ashwagandha's impact on the brain's microtubules and cytoskeleton components[2]. Ashwagandha is commonly used to reduce anxiety and bring about a state of calmness. Ashwagandha has also been found to enhance the health of the hippocampus[3].

The Pacific Yew, Taxol and its Microtubule Stabilizing Effects
The manufactured substance known as Taxol, which comes from the Pacific Yew Tree or its derivatives, has been shown to be a significant microtubule stabilizer[4]. Taxol is used as a microtubule builder. It works with insulin via hydrogen bonds and the **van der wall's** force effect[5]. Testing is underway to see how it works

in treating multiple sclerosis[6]. Taxol is also called Taxon or **Paclitaxel** and is also found in the needles of the European Yew[7]. The plant Anglojap Yew also contains Taxol and is commonly used as a household **garden shrub**[8]. Taxol is also found in certain species of Fungi [8]. The seeds of the Pacific Yew are commonly eaten by wildlife, but are extremely toxic if ate by humans.

The tree Hazel (Corylus Cornuta) has also been found to contain Taxol (*Angela Hoffman 1998*)[9].

Further Reading
In vitro cell cultures obtained from different explants of Corylus avellana produce Taxol and taxanes. Bestoso F. et al. Dec 2006

Taxanes from Shells and Leaves of Corylus avellana. Ottaggio. L et al. Jan 2008

Taxol as a powerful Anti-Cancer Substance
Taxol has been found to be one of the most successful substances used to treat a variety of cancers by killing cancer cells immediately upon contact [10]. Taxol is also used as a powerful leukemia and overall cancer preventative[10].

When Taxol was first discovered, the bark from six 100 year old pacific yew trees was used to treat a single cancer patient at a cost of about

$17,000 an ounce. However one scientist discovered that if he combined the fungus that grew under the bark of the Pacific Yew and combined it with the landscape species of the yew tree that it created paclitaxel (Taxol). This discovery allowed for much more Taxol to be produced, thus reducing the cost of treating cancer patients dramatically.

Taxol Stabilizes Microtubules
Microtubules have been found to be stabilized with Taxol[11].

Where to Find the Pacific Yew Tree
The Pacific Yew grows in the pacific northwestern United States and is found in abundance in Packwood, Washington. Trees that are commonly found growing with the Pacific Yew in Southwestern Oregon and California include Hazel (*Corylus Cornuta*), Northern Twinflower, Salal (*Gaultheria Shallon*), Vine Maple (*Acer Circinatum*) and Dwarf Oregon Grape (*Mahonia Nervosa*).

The Yew lives at elevations of between 2,000 to 8,000 feet, living at lower elevations in Oregon. It likes moist, deep and well drained soils and handles acidic soil well. Pacific Yews growing in Montana and Idaho contain soils with high levels of nitrogen and it also prefers soils high in granite, serpentine, schists and diorite and gabbro. Its average lifespan is 500 years. Pacific Yew also grows in the northern Rocky Mountains[12].

Pacific Yew oil tincture / extracts are available online and a common internet search for "yew tincture" yields many suppliers. Pacific yew oil, vital yew tincture are just 2 common keyword names to use when searching for pacific yew tincture. Below is a picture of a few leaves of the Pacific Yew

Deuterium and Taxol

Taxol has been shown to react with deuterium causing a marked reduction in deuterium in tubulin when Taxol was present[16] and Linoleic Acid synergizes with Taxol[17].

Reversing Aging using Quantum Coherence
Because quantum effects interact with consciousness, it may be that certain substances that enhance the effects of quantum interactions strengthen the brain's neuron's/consciousness, which in turn lengthens lifespan.

A few months ago I published an article on my website showing that 3 people who lived to over 100 years of age attributed their long lifespan to consuming and rubbing 1 teaspoon of olive oil into the joints of their body, which eliminated their arthritis[18]. Olive oil contains oleic acid and sunflower and flaxseed oil are rich in linolenic acid[19].

One study titled Neuroprotective effect of olive oil in the hippocampus, published in April 2013 by M. Zamani and colleagues found when mice were fed olive oil that it significantly reduced the cell death and decreased memory loss by affecting **activity in the hippocampus** region of the brain.

Oleic Acid is a pale yellow oily liquid that is found widely in nature. It is found in large quantities in olive oil, avocados and canola oil.

Linoleic and Oleic acids exist as saturated fatty acids which help the body make many of its natural oils. Linoleic Acid is a unsaturated omega-6 fatty acid and is colorless at room temperature. It is found in the lipids of cell membranes and is found in high levels in sunflower, safflower and corn oil. Rice bran oil contains 30% linoleic acid and 44% oleic acid. When heated, it exhibits a drastic increase its yellow color due to oxidation [20].

Further Reading
Linoleic acid: Is this the key that unlocks the quantum brain? Insights linking broken symmetries in molecular biology, mood disorders and personalistic emergentism Massimo Cocchi et al. Apr 2017.

Linoleic Acid and Lifespan
Fish oil is a good source of linoleic acid and fish oil has been shown to extend lifespan[21] and linoleic acid has been shown to extend the lifespan in worm studies [22]. Studies have also found that a deficiency of Omega 3 oils may contribute to major depression partly due to its effects on the dopaminergic system.

One study concluded that combining fish oil with anti-depression medication may contribute to enhanced therapeutic benefits[23]. Fish oil has also been shown to exert important antineoplastic effects[24].

Acetylation and its Effects on Microtubules
Acetylation has been shown to protect microtubules against aging and to stabilize microtubules. Acetylation enhances flexibility, which causes protection against mechanical stress[25].

What does Acetylation Mean?
Acetylation is a reaction that causes an acetyl function to group into a special chemical compound. It is a process commonly used to synthesize aspirin.

The rare earth element Lanthanum Acetate enhances flexibility of the Arteries
Lanthanum Acetate has been shown to decrease serum phosphorus levels in people who have chronic renal disease. A research study showed that when Lanthanum Acetate was administered to mice that it decreased the degree of their irregular elastic fibers in their aortas. This showed better results compared to mice that were treated with both nicotine and vitamin D3. The study concluded that Lanthanum acetate may inhibit the pathogenesis of vascular calcification (*hardening of the arteries*).

This study has similar connotations to the "flexibility" or MOE (modus of elasticity) materials which were shown to exhibit space/time effects by Kozyrev. It just so happened that the metal

Tungsten, which has one of the highest Modus of Elasticity values (besides carbon nanotubes and diamond) was discovered by Kozyrev to exhibit significant space time effects (greater detail is shown in book 2 of the associative remote viewing series). Perhaps elasticity enhances resonance.

Properties of Lanthanum
Lanthanum oxide absorbs moisture over time and has p-type semiconductor features. During low temperatures it has an hexagonal crystal structure and its resistivity decreases as temperature increases. Lanthanum oxide is used for developing ferroelectric materials and it is used to make optical
glasses, causing increased density, a higher refractive index and increased hardness. When oxides of tantalum, tungsten and thorium are added, Lanthanum increases the resistance of glass to attack by alkali. It is also used to manufacture thermoelectric and piezoelectric materials. Lanthanum Tungsten electrodes are now being used in gas Tungsten arc welding (TIG) (*Wikipedia Lanthanum*).

Lanthanum Enhances Photosynthesis
Research has shown that low amounts of Lanthanum applied to plants stimulate photosynthesis and increase chlorophyll levels. It

also leads a higher incidences of binucleate cells, which cause a small increase in roots and shoot biomass. When Lanthanum is added in higher amounts, growth becomes reduced (*Bioaccumulation and effects of lanthanum on growth and mitotic index in soybean plants. De Oliveira C et al. Ecotoxicol Environ Saf. 2015 Dec;122:136-44. doi: 10.1016/j.ecoenv.2015.07.020. Epub 2015 Jul 29*).

Lanthanum carbonate (Fosrenal) is a phosphate binder and blocks the absorption of phosphorous in the body and raises blood pressure *(www.ehealthme.com)*.

It is interesting to note that **US Patent #20170136050** A1 (*Anti-glycation methods and compositions) uses a compositions to provide anti-glycation activity in the human body*) shows a mineral extract composition that includes Manganese and no less that 0.0001 ppm of the rare earths (Micro Minerals) consisting of Molybdenum, Neodymium, Nickel, Niobium, Palladium, Platinum, Praseodymium, Rhenium, Rhodium, Rubidium, Ruthenium, Samarium, Scandium, Selenium, Silver, Strontium, Sulfur, Tantalum, Terbium, Tellurium, Thallium, Thorium, Thulium, Tin, Titanium, Tungsten, Vanadium, Ytterbium, Yttrium, Zinc, Zirconium, Aluminum, Antimony, Arsenic, Barium, Beryllium, Bismuth, Boron, Bromine, Cadmium, Cerium, Cesium, Chromium, Cobalt, Copper, Dysprosium, Erbium,

Europium, Fluorine, Gadolinium, Gold, Hafnium, Holmium, Iodine, Indium, Iridium, Iron, Lanthanum, Lead, Lithium, Lutetium and Mercury all blended together. This is all mixed together in one liter of Processed Orange.
Glycation is a process where proteins are damaged due to sugar molecules and it is of importance to skin care experts. It has been a strong focus of study in people who have diabetes and also plays a role in the process of aging skin.

Further **Reading**
Advanced glycation end products. Key players in skin aging? Paraskevi Gkogkolou and Markus Böhm. July 20112

References. Chapter 17.

(1) Ashwagandha Root Powder extract. NCI Drug Dictionary. National Cancer Institute
(2) Combinations of Ashwagandha Leaf Extracts Protect Brain-Derived Cells against Oxidative Stress and Induce Differentiation Navjot Shah, et al. March 2015
(3) Neuritic regeneration and synaptic reconstruction induced by withanolide A Tomoharu Kuboyama, et al. Feb 2005
(4) Selective Vulnerability of Dopaminergic Neurons to Microtubule Depolymerization. Yong Ren et al.
(5) Inhibition study on insulin fibrillation and cytotoxicity by paclitaxel. Kachooei E et al. June 2014
(6) Paclitaxel (Taxol) attenuates clinical disease in a spontaneously demyelinating transgenic mouse and induces remyelination. Moscarello MA et al. April 2002
(7) Taxus baccata (European yew). Cambridge Universtiy Online Botanical Garden.
(8) Diversity of endophytic fungi and screening of fungal paclitaxel producer from Anglojap yew, Taxus x media. Xiong ZQ et al. March 2013

(9) Hazel (Corylus avellana L.) as a New Source of Taxol and Taxanes. Ardeshir Qaderi et al. Feb 2012
(10) How Taxol/paclitaxel kills cancer cells Beth A. Weaver. Sept 2014
(11) Taxol stabilizes synaptosomal microtubules without inhibiting acetylcholine release. Burgoyne RD, Cumming R.. Nov 1983
(12) Tirmenstein, D. A. 1990. Taxus brevifolia. In: Fire Effects Information System, [Online].
(13) U.S. Department of Agriculture, Forest Service,
(14) Rocky Mountain Research Station, Fire Sciences Laboratory (Producer).
(15) Available: http://www.fs.fed.us/database/feis/ [2017, October 9]
(16) Insights into the mechanism of microtubule stabilization by Taxol. Xiao H et al. July 2006
(17) The therapeutic efficacy of conjugated linoleic acid - paclitaxel on glioma in the rat. Ke XY et al. Aug 2010
(18) http://www.ez3dbiz.com/olive_oil_extends_lifespan.html
(19) The effect of dietary oleic, linoleic, and linolenic acids on fat oxidation and energy expenditure in healthy men. Jones PJ et al. Setp 2008

(20) Physico-chemical changes in rice bran oil during heating at frying temperature Rangaswamy Baby Latha and D. R. Nasirullah. Aug 2011
(21) Fish oil changes the lifespan of Caenorhabditis elegans via lipid peroxidation. Sugawara S et al. March 2013
(22) The ω-3 fatty acid α-linolenic acid extends Caenorhabditis elegans lifespan via NHR-49/PPARα and oxidation to oxylipins. Qi W et al. Oct 2017
(23) Polyunsaturated Fatty Acid Associations with Dopaminergic Indices in Major Depressive Disorder M. Elizabeth Sublette, MD and PHD et al. Dec 2013
(24) The effects of fish oil, olive oil, oleic acid and linoleic acid on colorectal neoplastic processes. Llor X et al. Feb 2003
(25) Tubulin acetylation protects long-lived microtubules against mechanical aging Didier Portran, et al. April 2017

Chapter 18. How Plants 'See'

Numerous studies have shown that the photosynthesis process taking place in plants undergoes quantum effects[1]. What is even more interesting is that some species of plants such as algae have "eyes".

The region of the plant that contains the "eye" is called the "eyespot apparatus"[2]. The eyespot on a plant is reputed to be the most photosensitive parts of the plant and is located at the anterior portion of the cell that is close to the bottom of the flagellum. Flagellates have been shown to be reactive to light, even when eyespots were not present (*Engelmann 1882*)[3] [4].

The illustration on the next page shows a more detailed description.

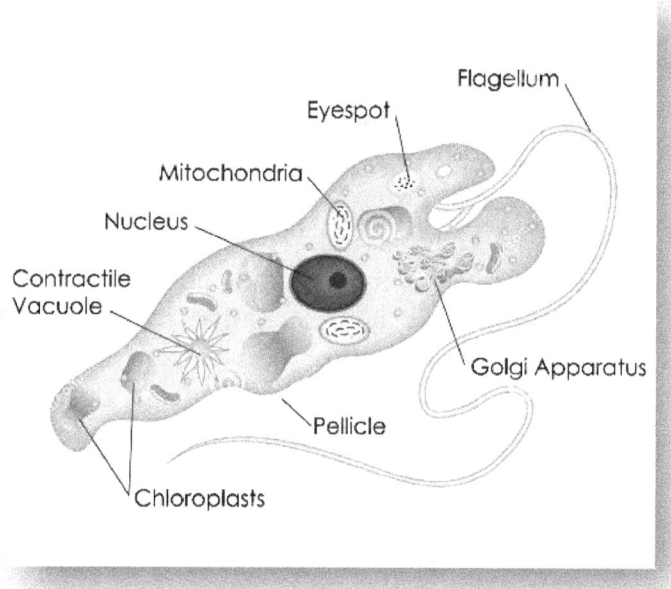

What are Flagellates?

Flagella are hair like creatures that consist primarily of protein. They can be found on the surfaces of cells and are utilized for movement by microorganisms and certain specialized cells in certain plants.

Eyespot Proteins

Eyespot proteins act as photoreceptor proteins which are sensitive to light. There are two forms of Eyespot Proteins.

Improve your Remote Viewing Accuracy Techniques
using Quantum Microtubules

1 - Flavoproteins

2 - Retinylidene

What is a Flavoprotein?
Flavoproteins contain the nucleic acid called riboflavin. They help clear out toxins from the cells and allow nutrients to enter the cells through a respirative type effect, much like our lungs breathe in and filter out bad air[5]. Flavoproteins are yellow in color. The word "Flavus" in latin means "Yellow".
Flavoproteins are found in abundant levels in the body's heart and skeletal tissues[6]. Flavoproteins are a protein containing nucleic acid, which is a derivative of riboflavin. Flavoproteins are responsible for bioluminescence, DNA repair and photosynthesis. Flavins come from the B vitamin ribo**FLAVIN**. When a flavin becomes oxidized such as from heat, it turns a strong yellow. It is blue in its neutral state and red in its anionic state. The discovery of Flavoprotein took place when the closer examination of cow's milk in 1879 was found to contain flavin, which existed as a bright yellow pigment. At this time it was named Lactrochrome[7]. As a side note, Geraniol rapidly oxidizes when exposed to air[7a] [7b].

E106
E106 is a very closely related food dye, also commonly called riboflavin 5 phosphate sodium salt. After it is ingested, it is turned into riboflavin. It is commonly found in baby foods and milk products. Riboflavin has been shown to inhibit the release of glutamate[8].

Algae, Quantum Effects and Photosynthesis
Research has shown that cells in chalmydonionas will swim towards a source of light that is of moderate intensity. If however the cells are exposed to strong light, they will swim away from it[9]. Because light is part of the process of photosynthesis, and photosynthesis has been shown to exhibit quantum effects, there may be a connection between photosynthesis and some kind of communication.

What are Chalmydonionas?
Chlamydomonas are a genus of green algae that contain unicellular flagellates. They are found in abundance in damp soil and stagnant water[10]. They also exist in seawater, freshwater and snow (commonly called "*snow algae*"). Chlamydomonas show the unique trait in that it has ion channels which become activated by light[11]. It is a fact that algae and cyanobacteria both undergo oxygenic photosynthesis[12]. If algae utilize photosynthesis, which has been shown to exhibit

quantum effects[1], then algae may be exhibiting quantum effects.

Hence we see another clue to quantum anti aging, where the red algae **Astaxanthin**, which is one of the most powerful lifespan supplements, may be deriving its anti-aging effects from its powerful UV protective aspects [13] and it has been shown to enhance the retinas of the eyes [13]. Also studies have shown that a higher intake of linoleic acid has been linked to macular degeneration[13]. Hence it is advisable to use linoleic acid sparingly.

Further Reading
Microalgae Biophotonic Optimization Of Photosynthesis By Weakly Absorbed Wavelengths. Erico Rolim De Mattos.

By examining how light and geomagnetic energy interact with one another, we can get a better idea of how geomagnetic energy affects our remote viewing. This is because numerous studies have shown that plants can detect danger before it occurs in the process known as **Plant Perception**. For example Arabidopsis thaliana (pictured), which has almost translucent roots, is extremely sensitive to light. It is able to perceive magnetic fields utilizing cryptochromes[14].

Because plants produce several proteins that have a similar effects as the human nervous system such as glutamate receptors, GABA receptors, endocannabinoid signaling components and acetylcholine esterase, the chemicals they produce as a means of defense may enhance the human nervous system, thus enhancing the pre-sentient effects observed in remote viewing studies.

For example the monoterpenes such as ocimene, limonene, linalool and the sesquiterpenes such as caryophyllene, bergamotene and farnesene are released in some plants when they attacked by insects [16]. One patent states that a monoterpene is to be used for the regression of mammalian nervous systems[17]. Also a study found lemon essential oil, which is high in the monoterpene limonene, relieved physical and psychological stress. The study concluded that the components in the lemon essential oil (limonene and citral) exhibited potent stress-alleviating properties [19]. The plant Angelica archangelica, also called garden angelica or wild celery has been shown to have ant-seizure effects due to the terpenes in the essential oil[20]. This is a key finding because in our second book on remote viewing we state that substances that alleviate seizures in epileptics enhance remote viewing, giving the brain better concentration.

Further Reading
Actions of essential oils on the central nervous system: An updated review. Clara Dobetsberger and Gerhard Buchbauer. November 2010

Where are the Eyes located in Green Algae?
The region of an algae's eye is found in the region of algae known as the flagellate. This region allows the cells to sense where the light is coming from and also how bright the light is. This allows the cells to swim towards it.

How Plants Utilize Quantum Coherence for their Photosynthesis

It is amazing just how much our nervous systems have in common with plants. Just as the microtubules, which exist in human, animal and plants utilize the quantum field, the process of plant photosynthesis also utilizes the quantum realm.

Research on organisms living deep in the ocean or in sulphur rich environments show that they are able to create energy via chemosynthesis instead of photosynthesis [21].

A study showed that molecular vibrations present in quantum fluctuations perpetuate and even regenerate coherence [22]. This discovery shows how the majority of light absorbed by photosynthesis cells can be smoothly transferred to distant locations in cells wirelessly where the electric energy is than converted into chemical energy. This could have huge implications for wireless battery charging stations.

Detection of Quantum Fluctuations

Quantum mechanics states that space is filled with particles and quantum energy blinking in and out of our existence in moments that are fleeting. These are referred to as quantum fluctuations. In 2015 scientists detected the theoretical fluctuations directly. Quantum fluctuations produce randomly fluctuating electric fields which affect electrons. This is how scientists indirectly demonstrated their existence in the 1940s.

In 2015 a research team led by Alfred Leitenstorfer from the University of Konstanz in Germany detected these fluctuations, while observing their influence on a wave of light. In order to do this they fired short laser pulses lasting just a few femtoseconds (a millionth of a billionth of a second) into a vacuum. After doing this they saw subtle changes in the **polarization** of the light which were caused directly by the quantum fluctuations.

Could these quantum fluctuations be behaving like the tides of our earth, exhibiting stronger quantum fluctuations during times the moon's light is polarized (first quarter moon)? or during times the moon is in perigee? As these quantum tides 'rise and fall' they may exhibit highs and lows in the quantum noise. Hence, there may exist a quantum noise "*sweet spot*" that achieves the best associative remote viewing results.

Reference
Article 1 - Physicists Say They've Manipulated 'Pure Nothingness' And Observed The Fallout. Fiona Macdonald. Jan 2017. Science Alert.com

Technical Reference
Subcycle quantum electrodynamics. C. Riek, P. Sulzer, M. Seeger, A. S. Moskalenko, G. Burkard, D. V. Seletskiy & A. Leitenstorfer. Corresponding author. Nature 541, 376–379 (19 January 2017)

Further Reading
Energy transfer: Resonance is the key for coherence. Daniel B. Turner. Feb 2017

Using coherence to enhance function in chemical and biophysical systems. Gregory D. Scholes et al. doi:10.1038/nature 21425

The Quantum Process of Photosynthesis
As sunlight strikes the surface of a plant it transfers its energy along chains of pigments into a reaction center. At this reaction center it is than converted into chemical energy. These pigments are held together by proteins, which create pigment–protein complexes, also known as PPCs. PPCs act as corridors, funneling the energy in the form of molecular excited states, which are called "excitons". These excitons travel along the PPC by "hopping" from one molecule to another.

In 2007 Graham Fleming and his research team showed that excitons exhibit quantum coherence and that the quantum coherence exhibited resonance effects[23][24].

Further Reading
Quantum Coherent Effects in Photosynthesis. K. Birgitta Whaley et al. Berkeley Quantum Information and Computation Center. Darpa.

Quantum Coherence and Energy Landscapes in Photosynthetic Systems Investigated with Two-dimensional Electronic Spectroscopy Author: Calhoun, Tessa Rae. 2010. Berkeley CA.

This presence of quantum effects occurring in photosynthesis stunned both biologists and physicists and left them wondering how such a fragile quantum state is able to cope in such a sophisticated living organism. Simply put the length of coherence is greatly enhanced due to the presence of quantum waves in photosynthesis. This causes the wave states to exhibit a resonance long enough to create a safe passage of nearly 100% of photon energy in which the organisms absorb. All of this takes place without a major loss of energy in the process, much like a superconductor. Much of this process is made possible due to pigment–protein complexes[25].

Further Reading #2
When It Comes to Photosynthesis, Plants Perform Quantum Computation. David Biello. Scientific American. April 13, 2007

Microscopic quantum coherence in a photosynthetic-light-harvesting antenna Jahan M. Dawlaty, Akihito Ishizaki, Arijit K. De, Graham R. Fleming. July 2012.

Summary

The process of photosynthesis in plants acts much like a quantum processor. The process occurs when sunlight strikes electrons causing them to enter into a state of quantum superposition. This places the sun's electrons in two places at the same time while allowing the light-gathering molecules to flow to the reaction center where photosynthesis takes place. This takes place for just a few hundred femtoseconds. During this process, the electrons are taking all paths between the two places simultaneously[26]. For a more complete understanding of this process see: New Scientist. *Quantum life: The weirdness Inside Us*. Written by Michael Brooks. September 2011

Bacteria and Quantum Photosynthesis
A plant soaks up to 1017 joules of energy that comes from the sun every second. This process involves a transformation of sunlight into carbohydrates and it takes less than one million billionth of a second. This rapid transfer of energy is what stops the energy from dissipating as excessive heat and it protects the plant from

sunburn. The energy transfer is akin to the way energy flows in semiconductors. The completion of this circuit of energy ends up with plants retaining as much as 95 percent of the energy absorbed by the sun, and all without overheating or burning up. This also means plants or trees could be used as electrical outlets, provided quantum technology becomes available to harness this immense amount of energy harnessed by the plants. As we showed earlier, trees act as receivers for low frequency energy.

Further   Reading
Quantum entanglement in photosynthetic light-harvesting complexes. Mohan Sarovar et al. April 2010.

Quantum coherence in photosynthesis for efficient solar-energy conversion. Elisabet Romero, et al. July 2014

Plants   as   Quantum   Computers
If the microtubules in our brain behave in similar fashion to a quantum computer and plants also contain microtubules, could this mean plants also have quantum computer properties?
Research by Biophysicists at the University of California, Berkeley, showed that plants use quantum computing to achieve photosynthesis. Biophysicist Gregory Engel and his team cooled

the sulphur bacteria known as chlorobium tepidum, which is one of the oldest bacteria on earth, and then pulsed it with short bursts of laser light in order to track the flow of energy in its photosynthetic system. The study found that instead of the light coming to an intersection and going right or left, that it was going in both directions at the same time, exploring numerous paths along its route[27].

Summary
Bacteria are utilizing quantum behavior in the process of photosynthesis.

Further Reading
Proposal for quantum many-body simulation and torsional matter-wave interferometry with a levitated nano-diamond Yue Ma et al. Aug 2017.

References. Chapter 18

(1) The quantum physics of photosynthesis. Ritz T1, Damjanović A, Schulten K. March 2002
(2) Proteomic Analysis of the Eyespot of Chlamydomonas reinhardtii Provides Novel Insights into Its Components and Tactic Movements[W] Melanie Schmidt, et al. Aug 2006
(3) THE BOTANICAL REVIEW. VOL. 26 APRIL-JUNE, 1960. PHOTOTAXIS'. SELINA W. BENDIX
(4) Contributions of Theodor Wilhelm Engelmann on phototaxis, chemotaxis, and photosynthesis Gerhart Drews. Photosynthesis Research (2005) 83: 25–34
(5) Abbas, Charles A.; Sibirny, Andriy A. (2011-06-01). "Genetic Control of Biosynthesis and Transport of Riboflavin and Flavin Nucleotides and Construction of Robust Biotechnological Producers". Microbiology and Molecular Biology Reviews. 75 (2): 321–360. ISSN 1092-2172. PMC 3122625 Freely accessible. PMID 21646432. doi:10.1128/MMBR.00030-10.
(6) Isolation and properties of a flavoprotein from heart muscle tissue Ferenc Bruno Straub. May 1939

(7) Massey, V (2000). "The chemical and biological versatility of riboflavin". Biochemical Society Transactions. 28 (4): 283–96. PMID 10961912. doi:10.1042/0300-5127:0280283.
(7a) Cross-reactivity between citral and geraniol - can it be attributed to oxidized geraniol? Hagvall L and, Bråred Christensson J.
(7b) Geraniol. Wikipedia
(8) Vitamin B2 inhibits glutamate release from rat cerebrocortical nerve terminals. Wang SJ et al. Aug 2008
(9) How Chiamydomonas Keeps Track of the Light Once It Has Reached the Right Phototactic Orientation Klaus Schaller et al. Biophysical Journal Volume 73 September 1997 1562-1562
(10) Hoham, R.W., Bonome, T.A., Martin, C.W. and Leebens-mack, J.H. 2002. A combined 18S rDNA and rbcL phylogenetic analysis of Chloromonas and Chlamydomonas (Chlorophyceae, Volvocales) emphasizing snow and other cold-termperature habitats. J. Phycol., 38: 1051–1064
(11) A Falciatore, L Merendino, F Barneche, M Ceol, R Meskauskiene, K Apel, JD Rochaix (2005). The FLP proteins act as regulators of chlorophyll synthesis in response to light and plastid signals in Chlamydomonas. The

red eye spot in chlamydomonas is sensitive to light and hence determines movement. Genes & Dev, 19:176-187

(12) Cyanobacterial Oxygenic Photosynthesis is Protected by Flavodiiron Proteins Yagut Allahverdiyeva, et al. March 2015

(13) Astaxanthin affects oxidative stress and hyposalivation in aging mice Manatsu Kuraji, et al. July 2016

(14) Paris, september 7, 2006 The "sixth sense" of plants

(15) Baluška F, Volkmann D, Mancuso S (2006) Communication in Plants: Neuronal Aspects of Plant Life. Springer Verlag. ISBN 978-3-540-28475-8

(16) Their role in plant defense for pest management Abdul Rashid War, et al. Dec 2011

(17) Patent # Clovis Pereira Da Fonseca. US20040087651 A1

(18) S. Fukumoto, A. Morishita, K. Furutachi, T. Terashima, T. Nakayama, H. Yokogoshi, Stress Health 2007, 24, 3

(19) Effect of flavour components in lemon essential oil on physical or psychological stress Authors Syuichi Fukumot et al. Oct 2007

(20) Evaluation of Antiseizure Activity of Essential Oil from Roots of Angelica

archangelica Linn. in Mice Shalini Pathak et al. June 2010

(21) Microbial Communities and Chemosynthesis in Yellowstone Lake Sublacustrine Hydrothermal Vent Waters Tingting Yang, et al. June 2011

(22) Origin of long-lived quantum coherence and excitation dynamics in pigment-protein complexes Zhedong Zhang1,2 and Jin Wang. Nov 2016

(23) Evidence for wavelike energy transfer through quantum coherence in photosynthetic systems Gregory S. Engel et al. April 2007

(24) Dynamics of light harvesting in photosynthesis. Cheng YC1 and Fleming GR. 2009

(25) Microscopic quantum coherence in a photosynthetic-light-harvesting antenna Jahan M. Dawlaty, et al. July 2012

(26) Coherently wired light-harvesting in photosynthetic marine algae at ambient temperature Elisabetta Collini et al. Feb 2010

(27) When It Comes to Photosynthesis, Plants Perform Quantum Computation. David Biello on April 13, 2007. Scientific American.

Chapter 19. Monoterpenes and Photosynthesis

Because plants utilize quantum fields to assist with their photosynthesis, and as we showed earlier, essential oils disrupt microtubules, perhaps there is a connection between monoterpenes (*found in essential oils*) and quantum phenomena. We showed earlier that when the monoterpenes were removed from Bergamot essential oil, that the effects ceased.

Monoterpenes are Produced by Trees During Photosynthesis

Research has found that trees produce monoterpenes during photosynthesis (*Unger 2014*)[1][2] which also exhibits seasonal variations [3]. Monoterpenes have also been shown to cause decreases in the concentrations of beta carotene and chlorophyll (*Effects of menthone and piperitone on growth, chlorophyll a and beta-carotene production in Dunaliella salina. 1stMina Zarei et al. Jan 2016*). Monoterpenes can be found in the substances Mycerene and Geraniol[4]. The algae called Halomon is a monoterpene[5] and is produced by red algae and forests have been found to be major emitters of monoterpenes [6].

The monoterpenes emitted by forests help create bright clouds. One study found that monoterpenes existed at the upper canopy of trees in an Amazon rainforest and that the monoterpenes play an important role in photosynthesis. Low levels of monoterpenes such as myrcene, terpinolene, terpinene and phellandrene have also been detected in the Amazon rainforest (*Helmig 1998, Rinne 2002, Kuhn 2007*)[6]. Also monoterpenes have been found to protect the process of photosynthesis during periods of abiotic stress (*Penuelas and Llusia 2002, Vickers 2009*)[6]. There may also exist a synergistic effect between linoleic acid and monoterpenes. A research study found that mixtures of essential oils containing monoterpene hydrocarbons, exhibited the highest antiradical activity (*which act similar to antioxidants*)[7].

The Monoterpene Linalool

Linalool is a monoterpene alcohol that readily binds to phosphates. Linalool is found in bergamot, basil, clove, lavender, cinnamon and black pepper[8]. Basil essential oil contains approximately 54% linalool[8]. Because linalool is a terpene, it exhibits odors similar to some fungi (*Breher 1997, Malheiro 2013*)[9] and linalool has shown promise in treating opiod addiction[10]. Linalool has also been found to inhibit the binding of glutamate and reduce potassium glutamate release and glutamate update[11].

Linalool has been used traditionally as a sedative and has been shown to reverse behavioural and neuropathological impairments in old mice[12]. Also monoterpenes found in citrus essential oils release monoamine from the brain[13]. Linalool also exhibits fungicidal and bactericidal effects (*Pattnaik 1997*)[14].

What are Monoamines?
Monoamines are also known as monoamine neurotransmitters. They include, noradrenaline, adrenaline, dopamine and serotonin. These are released from neurones in the peripheral nervous system as well as the brain. They are thought to play important roles in emotion, arousal and cognition.

Neurotransmitters for Calm Moods and Emotions
GABA is a neurotransmitter that helps deal with stressful situations. It behaves as a calming substance as the body's stress levels increase. Many people who feel anxious or irritable are deficient in GABA. Serotonin is also a neurotransmitter that regulates anxiety and mood and 90% of the serotonin in the body is found in the stomach and microtubules have been found in the stomach of the human body.

Pirenzepine
Pirenzepine is used to treat peptic ulcers and is a

neurotransmitter of the parasympathetic nervous system.

Foods and Serotonin Levels
Foods with high serotonin concentrations include the following:

plantain 30.3 - **pineapple** 17.0 - **banana** 15.0 - Kiwi fruit 5.8 - plums 4.7 - tomatoes 3.2

Nuts in the hickory or walnut family have high serotonin concentrations via their micrograms/g weight;

butternuts 398 - black **walnuts** 304 - English walnuts 87 - shagbark hickory nuts 143 - mockernut hickory nuts 67 - pecans 29 - sweet pignuts 25

Reference
Serotonin content of foods: effect on urinary excretion of 5-hydroxyindoleacetic acid. Feldman JM and Lee EM. Am J Clin Nutr. October 1985

Manganese and Copper promote binding of Dopamine to Serotonin
Research has shown that iron increases the binding of dopamine and serotonin to serotonin binding proteins in calf brains. The research has shown that manganese and copper also increase the binding for dopamine only. **Manganese**, iron

and copper **promote binding** due to their ability to oxidize dopamine into dopamine-O-quinone.

Reference
Manganese and copper promote the binding of dopamine to "serotonin binding proteins" in bovine frontal cortex. C. Velez-Pardo et al. June 1995.

Seasonal Variation of Serotonin
Serotonin levels are significantly higher during winter and fall (*Seasonal variation in human brain serotonin transporter binding. N. Praschak-Rieder et al. Sept 2008*).

Essential Oils and Neurotransmitters
Linalool is used in our Remote Viewing TXP formula which appears to help quell nervous emotions and anxiety during the remote viewing sessions. And as we showed earlier, anxiety was one of the prime reasons professional athletes failed to make successful shots.

We covered in an earlier chapter that when the monoterpenes were removed from the Bergamot essential oil, that the effects ceased. Essential oil of Bergamot has been shown to release neurotransmitters in the hippocampus regions of the brains of mice (*The essential oil of bergamot enhances the levels of amino acid neurotransmitters in the hippocampus of rat:*

implication of monoterpene hydrocarbons.Morrone LA et al. Apr 2007). Another study found that limonene acted very similar to Bergamot by enhancing GABA, Dopamine and the hypothalamic-pituitary-adrenal regions of the brain[16].

This suggests that monoterpenes are playing a key role in neurotransmitter production and operation. Also we covered earlier that the effects of the essential oil Menthone, which is a monoterpene, showed moderate microtubule efficacy.

Molecules that Exhibit Quantum Effects
Hydrogen bonds, due to being the lightest element, are not the only molecule to exhibit quantum effects [17], carbon has also been shown to exhibit quantum effects [18].

Hydrogen bonds not only exhibit electrostatic effects, but enzyme catalysis, display electrostatic effects also exhibit quantum effects[19].

The reason hydrogen bonds exhibit quantum effects is due to the fact that competition occurs from intermolecular bond bending and covalent (intramolecular) bond stretching. Hydrogen bonds also exhibit the ability to "tunnel" through energy barriers, also called "Quantum Tunneling" [20] [21].

Further Reading

Quantum nature of the hydrogen bond Xin-Zheng L et al. Nov 2010

A research paper stated that when the hydrogen bond is strengthened, it stretches (*Hydrogen bonding and information transfer. Linus Pauling, 1939. www1.lsbu.ac.uk/water/hydrogen_bonding*) Hydrogen bonding has also been found to occur in tungsten salicylate[22]. Salicylic acid functions as a plant hormone and is a strong absorber of sunlight[23]. Salicylic acid is biosynthesized from the amino acid phenylalanine[23] and it also greatly helps plant photosynthesis[24] [25] [26] [27]. Himalayan salt also contains large amounts of natural hydrogen, including the rare earths and lithium.

Low Phenylalanine levels and Dopamine
A research diet placed mice on a low phenylalanine diet for a period of 4 weeks. The study found an increase in their blood tyrosine levels of about 50% above normal. The study also found increases in dopamine and myelin *(Relationship between myelin production and dopamine synthesis in the PKU mouse brain. Joseph B et al. J Neurochem. 2003 Aug;86(3):615-26)*.

Further Reading

Neuroprotective and Anti-Aging Potentials of Essential Oils from Aromatic and Medicinal Plants Muhammad Ayaz, et al. May 2017.

Where to Obtain Monoterpenes

Besides linalool, monoterpenoid phenols exist in the essential oils of origanum vulgare, pepperwort, thyme and wild bergamot. These are also high in the monoterpene Carvacrol[28].

Monoterpene Synergy

A monoterpene hydrocarbon (such as pinene) combined with linalool or limonene has been shown to exhibit additive and synergistic effects [29].

Menthone

Menthone is also a monoterpene. Menthone is used in perfumery due to its aromatic minty scent. Menthone can be found in the essential oils of peppermint, pennyroyal, mentha arvensis and pelargonium geranium[30] [31]. As a matter of fact one essential oil we use to enhance our associative remote viewing accuracy (TXP Formula) is called Rose Geranium.

Geraniol

Geraniol is a monoterpene. As just stated, the TXP formula we use to enhance our associative remote viewing sessions uses rose geranium essential oil which contains adequate amounts of geraniol. Geraniol is also found in citral and citronella[32].

Fenchone

Fenchone is a monoterpene and also a keytone. It has an odor similar to that of camphor and is a major constituent in the essential oil of fennel, containing up to 28% Fenchone[33]. As I show in my book, **The Official Guidebook of How to Make Tinctures,** Fennel is a major component in the anti-aging tincture called the St. Germain Formula. Fennel also contains the natural insecticide Anethole, which is also found in essential oils [34]. Fenchone can also be obtained from the leaves of Zanthoxylum alatum or made into an essential oil[35]. Fenchol and eudesmol can also be obtained from Eucalyptus teretecornis[35b]. Cedar oil also contains Fenchone [37] and one study using fenchone on rabbits found it enhanced the removal of mucus by 186%[37]. Also Fenchone is found in Thuja[38]. Many cedar oils are obtained from the family Cupressaceae, particularly from Thuja occidentalis and Thuja plicata[39].

Rosemary happens to contain fenchone, as well as cymene, cineole, carvacrol, carvone limonene, terpinene and thymol[40]. Many of these are strong neurocognitive protectors, protecting the brain against the effects of aging [40].

Terpenoids

Both Monoterpenes and Sesquiterpenes are what's called Terpenoids. Linalool, Linalyl Acetate, Thymol, Carvacrol, Citronellal, Geraniol, Piperitone and Menthol are all terpenoids. Higher level terpenoids (C20 and above) help form particles that create cloud condensation (*Claeys 2004, Poschl 2010*).

What is a Terpenoid?

A terpenoid creates the yellow in sunflowers, the red in tomatoes and the scent of eucalyptus [41]. Monoterpenes react strongly with light which is why they help form clouds [42]. Below is the molecular structure of terpenoid isopentenyl pyrophosphate.

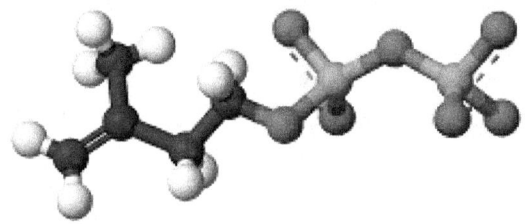

Sesquiterpenes

Sesquiterpenes exist in almost every essential oil

[43] and they are the largest group of terpenes in the animal and plant kingdom. Large quantizes are found in sagebrush which have been researched for pain control [44]. Because sesquiterpenes are larger than monoterpenes, they have more staying power (*viscous and less volatile*). Hence they are used as fixatives in the perfume industry.

Viscous oils blend well with lighter volatile oils and have a longer half-life.

What Essential oils that have a High Percentage of Sesquiterpenes?
Cedarwood, Sandalwood and Myrrh contain high amounts of sesquiterpenes[43].

Sesquiterpenes Effect on the Body
Sesquiterpenoids are being researched for the treatment of cardiovascular diseases and cancer[45]. They also show disruptive effects, with their being able to penetrate the walls of certain fungi[45]. Due to this disruptive effect, it makes sense that it would also disrupt microtubules.

The sesquiterpene Costunolide has been shown to disrupt microtubules[46] and exhibit polymerizing ability. Cedarwood also contains extremely high levels of Sesquiterpenes [47].

Sesquiterpenoids in Algae
In two regions of our associative remote viewing

sessions, conducted in Oregon and Topanga, CA, it was done in a region where algae was naturally present in high amounts. In our most recent ARV sessions in Hawaii, we have been using a bowl of algae water obtained from a nearby freshwater source. It is my hypothesis that when algae is present, when the lungs breathe in the water evaporative emissions of the algae, they may be somehow contributing to enhanced accuracy of ARV sessions. As we covered earlier, terpenes play a major role in affecting the brain's microtubules, which are found in essential oils.

A study titled Exposure to terpenes: effects on pulmonary function. and conducted by G. Hedenstierna and colleagues which was published in 1983 found that sawmill workers who worked around freshly cut wood were exposed to terpenes, which slightly altered their lung function for a short period of time with no long term negative effects. Let's explore Sesquiterpenoids in nature a little further.

Three major studies have found Sesquiterpenoids in algae[48] [49] [50]. Valerian, which we showed earlier enhances serotonin levels in the brain, also contains sesquiterpenoids[51]. What is interesting is Schulz et al. (1998) looked at eight different types of plant extracts including Piper methysticum, Valeriana officinalis, Lavandula off., Passiflora incarnata, Melissa off., Eschscbolzia californica, **Ginkgo biloba** and Hypericum perforatum and found that Valerian extract exhibited significant delta and theta brainwave activity and also decreased beta activity[52].

The Scent of Burning Incense Induces Alpha Waves

The essential oil of frankincense is composed of 75% monoterpenes, sesquiterpenols sesquiterpenes, monoterpenols and ketones. These are commonly used in sticks of incense[53] [54].

A research study found that burning incense was able to enhance fast alpha brainwave activity in bilateral posterior regions of the brain[55] [56] and when people inhaled the scent of Neroli Oil and Grapefruit oil, it was found to enhance theta activity in the brain[57].

Scents that Enhance Theta Brainwaves

Klemm et al. found that odors of birch tar, jasmine, lavender and lemon significantly enhanced theta brainwaves[58]. Also research studies found that chocolate and spearmint significantly reduced theta brainwave activity[59].

Iannilli et al. looked at the brain's electrophysiological responses to inhaling the scent of lily of the valley and strawberry odors in healthy volunteers. The study found that regions of the brain associated with processing reward was most active as they inhaled the strawberry and lily of the valley [60]. Also a study found that when a participant inhaled the scent of I. helenium root, which is a species of the **Sunflower** Plant, that it significantly enhanced alertness of the brain [59].

Research by Sowndhararajan et al. discovered that beta activities changed significantly in men compared to women when they inhaled the isomeric components terpinolene and limonene and that their absolute fast brainwave alpha activity was increased more in women compared to men when inhaling these isomers[59].

Valerian and Depression
Valerian tincture/extract has been shown to prevent depression in rats and that it increases serotonin concentrations in the brain (*Effects of Valerian on the level of 5-hydroxytryptamine, cell proliferation and neurons in cerebral hippocampus of rats with depression induced by chronic mild stress. Tang JY et al. 2008 Mar; 6(3):283-8*). Speaking from personal experience, I have found Valerian extract to be an extremely hardy tincture, having used an extract over a year old, due to keeping it in a cool place away from sunlight. When using it, it

greatly enhances the clarity and vividness of dreams.

Valeriana wallichii DC and Depression
Valeriana wallichii DC is an ayurvedic prepared formula.

A research study examined the antidepressant effects of the formula and found that it significantly increased norepinephrine and dopamine levels in mice ($p<0.05$). The study concluded that the combination demonstrated antidepressant effects by increasing norepinephrine and dopamine levels.

Reference
Antidepressant effect of Valeriana wallichii patchouli alcohol chemotype in mice: Behavioural and biochemical evidence. Sah SP et al. Feb 2011.

Could Monoterpenes be assisting Quantum Photosynthesis?

A research study found that monoterpenes existed at the upper canopy of trees in an Amazon rainforest and that the monoterpenes played an important role in photosynthesis[61]. Because plants exhibit quantum behavior during photosynthesis, perhaps there is an connection between monoterpenes and quantum phenomena. Also monoterpenes protect the process of photosynthesis during periods of abiotic stress (*Penuelas and Llusia 2002, Vickers 2009*)[61].

References. Chapter 19

(1) Highly reactive light-dependent monoterpenes in the Amazon A. B. Jardine et al. Feb 2015
(2) Effects of menthone and piperitone on growth, chlorophyll a and beta-carotene production in Dunaliella salina et al. Mina Zarei. Jan 2016
(3) Seasonal and interannual variations in whole-ecosystem isoprene and monoterpene emissions from a temperate mixed forest in Northern China Author links open overlay panelJianhuiBai et al. July 2015
(4) Geraniol and Geranial Dehydrogenases Induced in Anaerobic Monoterpene Degradation by Castellaniella defragrans Frauke Lüddeke et al. April 2012
(5) Isolation and structure/activity features of halomon-related antitumor monoterpenes from the red alga Portieria hornemannii. Fuller RW et al. Dec 1994
(6) Highly reactive light-dependent monoterpenes in the Amazon A. B. Jardine et al. Feb 2015
(7) [Influence of the composition of essential oils on their antioxidant and antiradical properties]. [Article in Russian] Misharina TA, et al. Feb 2012

(8) Essential Oils in Combination and Their Antimicrobial Properties Imaël Henri Nestor Bassolé 1,* and H. Rodolfo Juliani. Molecules 2012, 17, 3989-4006; doi:10.3390/molecules17043989

(9) Chemistry of the earthy odour of basidiomata of Cortinarius hinnuleus (Basidiomycota, Agaricales) NORBERT ARNOLD et al. Österr. Z. Pilzk. 25 (2016) – Austrian J. Mycol. 25 (2016)

(10) Effect of linalool on morphine tolerance and dependence in mice. Hosseinzadeh H et al. Sept 2012

(11) Effects of Linalool on Glutamate Release and Uptake in Mouse Cortical Synaptosome. L. F. Silva Brum et al. March 2001

(12) Linalool reverses neuropathological and behavioral impairments in old triple transgenic Alzheimer's mice. Sabogal-Guáqueta AM et al. March 2016

(13) Flavor components of monoterpenes in citrus essential oils enhance the release of monoamines from rat brain slices. Fukumoto S et al. Apr 2006

(14) Antibacterial and antifungal activity of aromatic constituents of essential oils.Dr Smaranika Pattnaik et al. Feb 1997

(15) The essential oil of bergamot enhances the levels of amino acid

neurotransmitters in the hippocampus of rat: implication of monoterpene hydrocarbons. Morrone LA et al. Apr 2007

(16) Sub-chronic effects of s-limonene on brain neurotransmitter levels and behavior of rats. Zhou W et al. Aug 2009

(17) Nuclear quantum effects of hydrogen bonds probed by tip-enhanced inelastic electron tunneling Jing Guo et al. Apr 2016

(18) Article Reference: Tim Schleif, Joel Mierez-Perez, Stefan Henkel, Melanie Ertelt, Weston Thatcher Borden, Wolfram Sander. The Cope Rearrangement of 1,5 Dimethylsemibullvalene-2(4)-d1: Experimental Evidence for Heavy Atom Tunneling. Angewandte Chemie International Edition, 2017; DOI: 10.1002/anie.201704787

(19) Electrostatic effects in enzyme catalysis: a quantum mechanics/molecular mechanics study of the nucleophilic substitution reaction in haloalkane dehalogenase. A. Soriano et al. Sept 2004

(20) Article Reference A quantum theory of hydrogen bonds. LCN Author(s): Angelos Michaelides

(21) Quantum nature of the hydrogen bond Xin-Zheng Li, et al. Feb 2011 .London Centre for Nanotechnology and Department of Chemistry, University College London, London WC1E 6BT, United Kingdom
(22) Hydrogen bonding in tungsten(VI) salicylate free acids1 Author links open overlay panelTimothy E. Baronia et al. July 1998
(23) Salicylic Acid. Wikipedia
(24) Exogenous salicylic acid improves photosynthesis and growth through increase in ascorbate-glutathione metabolism and S assimilation in mustard under salt stress. Nazar R et al. 2015
(25) Salicylic acid may indirectly influence the photosynthetic electron transport. Janda K et al. July 2012
(26) Salicylic acid alleviates adverse effects of heat stress on photosynthesis through changes in proline production and ethylene formation M Iqbal R Khan, et al. Sept 2013
(27) Effect of Salicylic Acid on Photosynthetic Pigments and Chlorophyll
(28) Percutaneous Penetration Enhancers Chemical Methods in Penetration ... edited by Nina Dragicevic, Howard I. Maibac

(29) ntimicrobial activity of six constituents of essential oil from Salvia. Sonboli A et al. April 2006
(30) Menthone. Wikipedia
(31) U.S. National Library of Medicine. Mehthone
(32) Plant-based insect repellents: a review of their efficacy, development and testing Marta Ferreira Maia and Sarah J Moore. March 2011
(33) Evaluation of the Essential Oil of Foeniculum Vulgare Mill (Fennel) Fruits Extracted by Three Different Extraction Methods by GC/MS Faiza M Hammouda et al. Jan 2014
(34) Karl-Georg Fahlbusch, Franz-Josef Hammerschmidt, Johannes Panten, Wilhelm Pickenhagen, Dietmar Schatkowski, Kurt Bauer, Dorothea Garbe, Horst Surburg "Flavors and Fragrances" in Ullmann's Encyclopedia of Industrial Chemistry, Wiley-VCH, Weinheim: 2002. Published online: 15 January 2003; doi:10.1002/14356007.a11_141.
(35) Antioxidant and Antimicrobial Properties of the Essential Oil and Extracts of Zanthoxylum alatum Grown in North-Western Himalaya. Sanjay Guleria et al. May 2013
(35B) Antimicrobial Activity of Some Essential Oils—Present Status and

Future Perspectives Sonam Chouhan et al. Aug 2017
(36) Antibiofilm and Antihyphal Activities of Cedar Leaf Essential Oil, Camphor, and Fenchone Derivatives against Candida albicans Ranjith Kumar Manoharan et al. Aug 2017
(37) Handbook of Essential Oils: Science, Technology, and Applications, Second. K. Husnu Can Baser, Gerhard Buchbaue
(38) The Chemistry of Essential Oils Made Simple: God's Love Manifest in Molecules By David Stewart
(39) Thuja occidentalis (Arbor vitae): A Review of its Pharmaceutical, Pharmacological and Clinical Properties Belal Naser et al. March 2005
(40) Resveratrol Found in Red Wine dark especially the Pinot Noir. coursehero.com
(41) Michael Specter (September 28, 2009). "A Life of Its Own". The New Yorker.
(42) Highly reactive light-dependent monoterpenes in the Amazon A. B. Jardine et al. FEB 2015
(43) Sesquiterpenes from Essential Oils and Anti-Inflammatory Activity. da Silveira e Sá Rde C, Andrade LN, de Sousa DP. Oct 2015
(44) The Use of California Sagebrush (Artemisia californica) Liniment to

Control Pain James D. Adams, Jr.. Sept 2012

(45) Sesquiterpenoids Lactones: Benefits to Plants and People Martin Chadwick, et al. June 2013

(46) A sesquiterpene lactone, costunolide, interacts with microtubule protein and inhibits the growth of MCF-7 cells. Bocca C et al. Jan 2004

(47) NTP Technical Report on the Toxicity Studies of Cedarwood Oil (Virginia) (CAS No. 8000-27-9) Administered Dermally to F344/N Rats and B6C3F1/N Mice November 2016 National Institutes of Health Public Health Service U.S. Department of Health and Human Services

(48) New aromatic sesquiterpenoids from the red alga Laurencia Okamurai yamada☆ Author links open overlay panelMinoruSuzukiEtsuroKurosawa. 1978

(49) Sesquiterpenoids in sediments of a hypersaline lagoon: A possible algal origin Author links open overlay panelV.O.Elias et al. July 1997

(50) New laurane-type sesquiterpenoids from the Chinese red alga Laurencia okamurai Yamada. Li XL et al. 2015

(51) Sesquiterpenoids and lignans from the roots of Valeriana officinalis L. Wang PC et al. Oct 2011

(52) The quantitative EEG as a screening instrument to identify sedative effects of single doses of plant extracts in comparison with diazepam. Schulz H, Jobert M, Hübner WD Phytomedicine. 1998 Dec; 5(6):449-58.

(53) Sesquiterpenoids - The Holy Fragrance Ingredients. Michael Zviely and Ming Li. May 2013

(54) pharmacol. 2011 Oct 31;138(1):212-8. doi: 10.1016/j.jep.2011.08.078. Epub 2011 Sep 12. Chemical analysis of incense smokes used in Shaxi, Southwest China: a novel methodological approach in ethnobotany. Staub PO et al. Oct 2011

(55) Iijima M., Osawa M., Nishitani N., Iwata M. Effects of incense on brain function: Evaluation using electroencephalograms and event-related potentials. Neuropsychobiology. 2009;59:80–86. doi: 10.1159/000209859

(56) Influence of Fragrances on Human Psychophysiological Activity: With Special Reference to Human Electroencephalographic Response Kandhasamy Sowndhararajan and Songmun Kim. Nov 2016

(57) Iijima M., Nio E., Nashimoto E., Iwata M. Effects of aroma on the autonomic nervous system and brain activity

under stress conditions. Auton. Neurosci. 2007;135:97–98. doi: 10.1016/j.autneu.2007.06.161

(58) Klemm W.R., Lutes S.D., Hendrix D.V., Warrenberg S. Topographical EEG maps of human responses to odors. Chem. Senses. 1992;17:347–361. doi: 10.1093/chemse/17.3.347.

(59) Influence of Fragrances on Human Psychophysiological Activity: With Special Reference to Human Electroencephalographic Response Kandhasamy Sowndhararajan and Songmun Kim. Nov 2016

(60) Source localization of event-related brain activity elicited by food and nonfood odors. Iannilli E, Sorokowska A, Zhigang Z, Hähner A, Warr J, Hummel T Neuroscience. 2015 Mar 19; 289():99-105

(61) Highly reactive light-dependent monoterpenes in the Amazon A. B. Jardine et al. Feb 2015

Chapter 20. Do Certain Essential Oils Exhibit Quantum Effects?

As we showed earlier, certain essential oils have been shown to disrupt microtubules and microtubules exhibit quantum effects. Hence the effects exhibited by the essential oil may be due to the essential oil either interacting with quantum particles or by altering the moisture content in the microtubules, enhancing their coherence. We also showed earlier that monoterpenes exist in plants, which use quantum mechanics to complete their process of photosynthesis.

It may be that the elictor effect caused by certain essential oils enhances the superposition of photons. From our research, our "TXP" formula has enhanced the ability of the mind/heart to go into coherence (HRV). Hence, enhanced coherence allows one to more clearly tap into the super positional field experienced during more viewing.

Because the smell of pleasant scents is a beneficial and rewarding experience, these essential oils may also be enhancing the release and/or production of dopamine. Also we use the TXP formula by rubbing it into our hands then inhaling it. This greatly enhances the absorption of the molecules via the lungs and through the skin.

We believe that the 2 main components in essential oils responsible for enhancing the clarity of remote viewing are the terpenes and the monoterpenes.

And as we showed in the previous chapter, essential oils alter brain chemistry and shift brain wave activity towards beta, alpha or theta, depending on the oil.

Some Essential Oils and their Molecular Composition

Rose contains Geraniol and Citronellol[1].

Rosemary contains Camphor, Cineol and Borneol[2] [3].

Sandalwood contains Santalal and Santalol[4].

Verbena contains Geraniol, Methyl-hepenone and Citral[5].

Ylang-Ylang contains Geraniol and Linalool. It also contains Acetic and Benzoic estera dn para kresol methyl ether[6].

Citronella and Citral contain Geraniol[7].

Scott Rauvers

Transfer of Information via Quantum Effects is attributed to the Coherent Resonation of Water

In our second series on remote viewing Remote Viewing, **The Complete User's Manual on Experiencing Future Consciousness**, we show in great detail how water / moisture, especially higher barometric air pressure, when ARV sessions are conducted, contribute to enhanced success of an ARV sessions. It is my hypothesis that moisture enhances the receptibility of future information possibly by enhancing the resonance occurring in the brain's microtubules.

Research by Anirban and his team showed the importance of water in the processing of information within the brain/body. They found that highly ordered water occurred inside the cylindrical cavity of what's called the microtubule[8]. Further research found that when they drained the water from the microtubule, that some of the operating properties of the microtubule ceased functioning. The study concluded that **water plays a key role in how the multiple subunits of a microtubule behave**, especially the quantum-like-effects witnessed during their operation[8].

Further Reading

Atomic water channel controlling remarkable properties of a single brain microtubule: correlating single protein to its supramolecular assembly. Sahu S et al. Sept 2013.
Discovery of quantum vibrations in 'microtubules' corroborates theory of consciousness January 16, 2014. www.phys.org. Polarizable-Vacuum Approach to General Relativity. H. E. Puthoff.

Microtubules and hyper-computation

Coherence occurring inside the water in the microtubules may could play a major role in the establishment of long-lived order in living biological systems and in higher-order brain functions, giving rise to hyper-computation and consciousness. Hence, hyper-computation in the brain may be accomplished via superluminal evanescent photons which are generated inside microtubules, by acting as quantum waveguides (*resonating cavities*). This effect is akin to the resonating effect which takes place in earth's Schuman resonance.

Summary

Water inside microtubules exhibits spontaneous super radiant quantum phase transitions favoring an energetically coherent state in which electronic clouds of water molecules oscillate in tune within a self-trapped electromagnetic field that exists within coherent domains.

This super radiance within the microtubules exhibits quantum collective behavior by the electromagnetic coherent field generated by Heart Math (HRV). It is this electromagnetic field that is able to draw in large amounts of light energy which contains information which is then interpreted by the brain during remote viewing sessions.

Heart Math and Non-Locality
Because the heart exhibits the strongest electromagnetic field out of all organs in the body, and far stronger then the brain, and even more so when it becomes coherent with the mind such as when practicing Heart Math, the electromagnetic field of energy acts as antenna, pulling in electromagnetic signals in the form of waves of light. As this light begins flowing through the heart, it moves throughout the body, including the DNA, slowing down the light's frequency, much like a power transformer has to step down raw electricity coming from the power sub-station in order to make it usable. As the water molecules within the microtubules become coherent with the water in the blood, they download the new information to the cells through superposition. The blood absorbs the

electromagnetic information imprinting it upon the water molecules in the blood. Because emotion has a powerful influence on the structure of water, a person's emotional state, while they receive this new information, governs this flow of information. Negative emotions such as hate or fear slow down or block the flow, reducing the non-local/super positional effects. The more positive and uplifting emotions such as love, joy and appreciation, bring clarity to the information flow.

Coherence and Super fluidity
When helium reaches the halfway point between boiling and absolute zero, it becomes coherent via a wave like effect that occurs via in the atoms in the helium which is similar to superconductivity. The effect is called super fluidity[9]. When helium is cooled to low temperatures, it begins demonstrating quantum effects known as "super fluidity". Super fluid also is a great conductor of heat. Physicist Nassim Haramein at the Resonance Science Foundation discovered that water plays an integral part to long-range coherence and plays a key role in the orchestration of the processing of information related with conscious awareness[10].

Super fluidity effects in Nature
An excellent way to understand super fluidity is if a spring tide is taking place at the same time as bad weather, (*due to a low pressure system*) tides will be higher. This is because lower atmospheric pressure in that region causes the water to increase its height. In the science of super fluidity, helium has the ability to climb up the walls of containers. Super fluidity can also be understood in the term of capilliary like effects where water will climb up a dry napkin, overcoming the force of gravity.

Further Reading
Solitons, Bose-Einstein condensation, and super fluidity in helium II. International Journal of Theoretical Physics. November 1987, Volume 26, Issue 11, pp 1039–1049. J. Chela-Flores and H. B. Ghassib. November 1986

Carbon Nanotubes Immersed in **Superfluid** Helium: The Impact of Quantum Confinement on Wetting and **Capillary Action**. Hauser AW1, de Lara-Castells MP2. J Phys Chem Lett. 2016 Dec 1;7(23):4929-4935. Epub 2016 Nov 18.

Further Reading
A new QED picture of water: understanding a few fascinating phenomenon. Macroscopic quantum coherence. Luigi Maxmilian Caligiuri. Jan 2015

A Quantum Computer based on Superposition

Oxford physicist David Deutsch has stated that a quantum computer may be possible using the principle of superposition. The machine would operate based on quantum principles and have a row of quantum systems that each exist in a superposition of two states. The systems would then be entangled with each other to create quantum logic gates that would then perform specific operations[11].

What is Coherence?

In nature, coherence occurs as a field of large numbers of light particles which cooperate on a collective level in a single state. An intense narrow beam of a laser is an example of coherence. In a laser's beam, all emitted photons (*light particles*) are oscillating together at a single common frequency and phase. This causes an intense light beam of a single color to emerge. At a set threshold, it reaches a critical energetic level, which then causes it to undergo a non-linear phase transition (*a jump to a higher energy level*). All photons in lasers lack individual identities which allows them to exist in the same quantum state (*Marshall, 1989*)[12]. Fluorescent and incandescent lamps emit in-coherent light due to the spreading of their photons in all directions, via a broad spectrum of frequencies.

Summary

Due to the phase transition occurring in lasers, it allows them to undergo coherence. In quantum mechanical terms, these coherent excitations are known as Bose-Einstein condensates. Bose-Einstein condensates show macroscopic and collective properties because the effects of some Bose-Einstein condensates are able to be directly observed without magnification.

Polymers and the Quantum Effect

Microtubules are polymers that are made of alpha and beta tubulin.

Polymers play a key role in the function of the quantum wave. Herbert Frohlich found that long polymer bimoleculars were able to achieve the state of quantum coherence via metabolic energy pumping. This caused non-local entanglement, which was later termed the phrase "Frohlich Condensates"[13]. In my second book of **Remote Viewing. The Complete User's Manual on Experiencing Future Consciousness,** I state that materials with a strong MOE (*Modus of Elasticity*), were good materials to enhance remote viewing. Research has shown that the MOE of tubulin is approximately 1.2GPa[14]. This MOE number is very similar to the structure of rigid plastics and Plexiglas and one of the terms for deteriorating neurons is *Neuro**plasticity***. I also state in the second book on remote viewing in great detail that Dr. Lawrence developed equipment to measure the communications taking place in plants not just at close distances, but also miles apart. He also questioned if plants may be communicating with other plant species across the galaxy. If this were true, could the method of communication used by plants be occurring via quantum non-locality?.

Vanadium
In the second book on remote viewing I also state that Tungsten has a very high MOE, being one of the best. It just so happens that Vanadium/Titanium doped with the rare earth mineral lanthanum enhances oscillations via increased vibration absorption, which boosts the absorption of far infrared radiation in the 8um to 14 um wavebands. It also enhances flexibility (MOE)[15] [16].

Vanadyl Sulfate
Vanadyl sulfate is related to Vanadium. It exist as a stable, inorganic form of vanadium, a unique trace mineral found in foods such as mushrooms and shellfish. Vanadium is thought to play a role in the formation of bones and teeth. As a nutritional supplement, vanadyl sulfate is one of the safest and most effective natural therapies for

diabetes and insulin resistance[17] and oral vanadyl sulfate has been shown to improve hepatic and peripheral insulin sensitivity with patients who have non-insulin-dependent diabetes mellitus[18].

Vanadyl Sulfate Protects the Heart
When mice 16 months of age were fed vanadyl sulfate it prevented them from developing moderate glucose intolerance. The study concluded that vanadyl sulfate was able to prevent further age-related progression of glucose intolerance and that no increase in insulin secretion occurred in the mice. The beneficial effects were due to improvement of tissue sensitivity to insulin. Vanadyl sulfate also has been shown to protect against glucose tolerance[19] [20].

References. Chapter 20

(1) Citronellol and geraniol, components of rose oil, activate peroxisome proliferator-activated receptor α and γ and suppress cyclooxygenase-2 expression. Katsukawa M et al. May 2011
(2) nvestigation of the Volatile Fraction of Rosemary Infusion Extracts Christine Tschiggerl and Franz Bucar. June 2010
(3) nvironment-related variations of the composition of the essential oils of rosemary (Rosmarinus officinalis L.) in the Balkan Penninsula. Lakušić DV et al. July 2012
(4) Synthesis and olfactoric activity of side-chain modified beta-santalol analogues. Buchbauer G et al. August 2001
(5) hemical composition and antigenotoxic properties of Lippia alba essential oils Molkary Andrea López et al. July 2011
(6) Traditional Uses, Phytochemistry, and Bioactivities of Cananga odorata (Ylang-Ylang) Loh Teng Hern Tan et al. July 2015
(7) Plant-based insect repellents: a review of their efficacy, development and testing Marta Ferreira Maia1,2 and Sarah J Moore. March 2011

(8) Atomic water channel controlling remarkable properties of a single brain microtubule: correlating single protein to its supramolecular assembly. Sahu S et al. Sept 2013
(9) New scientist magazine. Vol 23 page 772 - K. Mendelssohn of Oxford
(10) Resonance Science Foundation.. William Brown. Confirmation of Quantum Resonance in Brain Microtubules. Biomolecules exhibit quantum mechanical behavior
(11) Stanford Encyclopedia of Philosophy. Jun 16, 2015. Amit Hagar and Michael Cuffaro
(12) A Quantum Biomechanical Basis for Near-Death Life Reviews. Thomas E. BeckJanet E. Colli. March 2003
(13) Biological Coherence and Response to External Stimuli. Fröhlich, Herbert. 1988
(14) Flexural rigidity of microtubules and actin filaments measured from thermal fluctuations in shape. Gittes F et al. Feb 1993
(15) Effect of lanthanum doping on the far-infrared emission property of vanadium–titanium slag ceramic Kewei Zhang et al. Feb 2017
(16) Current oscillations in vanadium dioxide: Evidence for electrically

triggered percolation avalanches Tom Driscoll, et al. Sept 2012

(17) Effects of vanadyl sulfate on carbohydrate and lipid metabolism in patients with non-insulin-dependent diabetes mellitus. Boden G et al. Sept 1996.

(18) Oral vanadyl sulfate improves hepatic and peripheral insulin sensitivity in patients with non-insulin-dependent diabetes mellitus. Cohen N et al. June 1995

(19) Comparative study on the preventing effects of oral vanadyl sulfate and dietary restriction on the age-related glucose intolerance in rats. Michela Novelli et al. Aging Clinical and Experimental Research October 2005, Volume 17, Issue 5, pp 351–357.

(20) Bhuiyan SMD, Shioda N, Fukunaga k. Targeting protein kinase B/Akt signaling with vanadium compounds for cardioprotection. Cardiovasc and Renal 2008;12:1217-27.

Chapter 21. Does Consciousness operate at a Measurable Frequency?

Current research debates whether coherence originates in cortical networks or the thalamus. However 'thalamo-cortical **40Hz**' stands as the promising candidate for the neural-level substrate for consciousness[1].

Further Reading
Synchronization of Fast (30-40 Hz) Spontaneous Oscillations in Intrathalamic and Thalamocortical Networks Mircea Steriadee et al. The Journal of Neuroscience, April 15, 1996.

A thalamic reticular networking model of consciousness. Byoung-Kyong Min et al. Ma 2010.

Geomagnetic Storms and 40Hz
What is most interesting is geomagnetic storms that occur on earth do so in the 40hz range as experiments on rats with Persinger have shown [2]. Hence enhanced 40Hz frequencies could be causing a disturbance in brainwaves during above average geomagnetic activity.

High Temperature quantum mechanical vibrations have been detected in the brain's neurons. The frequency exists at approximately 1 million cycles per second. This frequency may be responsible for causing wave interference creating electrical oscillations that are related to conscious awareness existing in the 40Hz range referred as "beat frequencies". High frequency oscillations (*known as 'coherent 40Hz'*) mediate conscious experience[3] [4].

Perhaps the 40 cycles per second of - quantum coherence is attributed to this self-collapse phase. Mediators in deep meditative states have reported "flickerings of consciousness" occurring while they meditate[5] [6] [7]. These flickerings have been measured and occur in the coherent 40-Hz oscillations range [5] [6] [7].

Further Reading

Human oscillatory brain activity near 40 Hz coexists with cognitive temporal binding. Joliot M, Ribary U, Llinás R. Proc Natl Acad Sci USA. 1994;91:11748–11751.

Quantum computation in brain microtubules? The Penrose-Hameroff "Orch OR" model of consciousness. I. Quantum computation and consciousness

Stimulus-Dependent ? (30 – 50 Hz) Oscillations in Simple and Complex Fast Rhythmic Bursting Cells in Primary Visual Cortex Jessica A. Cardin. June 2005.

Stimulus-dependent neuronal oscillations in cat visual cortex: Inter-columnar interaction as determined by cross-correlation analysis. Engel AK, König P, Gray C, Singer W. European Journal of Neuroscience. 1990;2:588–606

Temporal coding in the visual cortex: New vistas on integration in the nervous system. Engel AK, König P, Kreiter AK, Schillen TB, Singer W. Trends in Neuroscience. 1992a;15:218–226

Towards a neurobiological theory of consciousness Francis Crick and Christo] Koch. 1990. Seminiars in the Neurosciences.

Stimulus-specific neuronal oscillations in orientation columns of cat visual cortex. Gray CM, Singer W. Proc Natl Acad Sci U S A. 1989 Mar; 86(5):1698-702.

Summary
Discrete conscious events may be taking place at approximately 40 times a second. It takes between 100 and 10,000 neurons for a single conscious event [8] hence this frequency may be pulsating through the neurons of the brain at this frequency. The frequency between 13 and 39 cycles per second is beta and the frequency between 40 and 80 cycles per second is known as gamma.

Further Reading

A Giant Neuron Has Been Found Wrapped Around The Entire Circumference of The Brain. This could be where consciousness forms. Bec Crew 28 Feb 2017. Science Alert

The Ajna Light

The Ajna light takes advantage of what's known as hypnagogia. Hypnagogia exists as the transitional state as the mind passes from wakefulness to sleep. The Ajna light uses LED's interfaced with your computer to induce the effect. You can learn more about the Ajna light at www.ajnalightmeditation.com.

During early testing of the Ajna Light, the flicker of the LED's was tested over a wide range of frequencies. Their research coincided with published papers showing the following:

4Hz - 1 flickering light effect, 3 flickering lights effect; strong, 5 flickering lights; very strong

10Hz - 5 flickering lights; hallucinations, more color, an overall 3D effect; a physical release experienced

13Hz - 5 flickering lights: green color felt stronger, overall nice feeling

19Hz - 5 flickering lights: very relaxing, 2 Dimensional, more superficial

24Hz - 5 flickering lights: really loved it; more laughing, spreading out in 2 dimensions

31Hz - 5 flickering lights: felt much more relaxed; running for 80 seconds relaxing, physiological effects of breathing slowing very quickly, non invasive; felt a release

33Hz 5 flickering lights: the vagus nerve fading to parasympathetic; calm, pleasure and ecstasy experienced when the frequency was given for a long time; a gentle state, too fast for the brain to process, so felt as if different frequencies were happening; enjoyment and felt visceral release.

GABA and Consciousness
Research has found that GABA located in the basal forebrain (PV neurons) causes brain waves that are linked to certain states of consciousness. It is the coherence in the brain that allows for consciousness to organize perception and perform analysis of data from the world around us.

Researchers decided to see what would happen if they turned the brain's PV neurons off and on using laser light. The researchers found that when the PV neurons were activated using laser light, that the animal's cortex exhibited more gamma activity (*Cortically projecting basal forebrain parvalbumin neurons regulate cortical gamma band oscillations Tae Kima et al. May 2015*) Because both laser light and LED's emit coherent light, it could be why flickering LED's enhance gamma waves. This has been proven in a study where a strip of flickering LED lights were shown to mice for an hour, (flashing at 40 hertz). The study found it caused increased gamma brain waves. This in turn reduced their beta-amyloid levels by half in their visual cortex of the mice that were in the early stages of Alzheimer's. The study found that 24 hours later, their amyloid levels returned to normal in the region of the brain that processes information from the eyes. A follow-up study using mice with higher levels of amyloid buildup that were shown flickering LED's for an hour for several days showed that their levels of free-floating amyloid and amyloid plaques decreased (*Gamma frequency entrainment attenuates amyloid load and modifies microglia. Iaccarino HF. Dec 2016*).

Hence, the coherent photons may be boosting gamma wave brainpower by stimulating photons in the body's nervous system, allowing the body to heal itself

Can Meditation Enhance Superposition?
I had shown in the second edition of remote viewing that mediators were more intuitive on average compared to non-mediators. Research

has also shown that meditation induces hypothetical quantum dipole oscillations[9]. This in turn regulates protein changes via quantum computation leading to effects such as delocalization, electron tunneling and super positioning properties that occur within the microtubules. The study found that the effects experienced were similar to those experienced via meditative and healing practices such as praying and chanting. Hence further studies may show that prayer has a quantum effect associated with it. This may explain where the insights and information during remote viewing sessions are coming from. It may also mean that there exists as "gateway" from which information flows, utilizing electron tunneling and superposition. The process of quantum superposition was proposed early on by Penrose who showed it to be possible via quantum gravity 'objective reduction' calculations (*Penrose 1996*)[10].

Summary

The microtubules in the brain behave as strong oscillating systems that filter out and amplify signals. This in turn generates "*conscious recurring moments*" that terminate at the "*collapse of the wave function*" in our 3D space-time geometry (*Hameroff and Penrose 2014*)[11].

Quantum Collapse and the Brain's Microtubules

The probability of looking at a target during remote viewing is akin to finding a particle following wave mechanical principles. These particles are smeared across time and space and the probability of finding it occurs when it is observed. Hence quantum events are non localized potential possibilities.

The remote viewer collapses the quantum wave function, producing an event. Hence the act of observation deflates the infinite potential quantum wave functions forcing it to manifest in just one specific fashion.

Penrose (1994) states quantum systems "*self-collapse*" by growing after reaching a critical mass of energy[12]. Quantum coherence begins building in the brain's microtubules until a threshold of energy is reached. Next comes the resultant "*self-collapse*" (OR), creating the perceived "*instantaneous now*" event thus creating a smooth experience of consciousness and our perception of the flow of time.

How structures transition from quantum mechanics to classical physics is still under intense discussion. It may be however that the transition occurs as a gradual phase occurring at around 200 Angström (1 Angström equals the size of an helium atom).

Physicist Giuseppe Arcidiacono and Biochemist Salvatore Arcidiacono stated objects at the quantum scale are forced to choose between causes which occur from the past (*as diverging waves*) and events occurring in the future (*as converging waves*). Hence results of these two choices are not able to be determined in advance. As a consequence, quantum objects exist in states of uncertainty and chaos (*Arcidiacono, 1991*) until they are observed[13] [14].

Further **Reading**
Dynamics of thalamo-cortical network oscillations and human perception. Ribary U. 2005.

Chapter 21. References

(1) Consciousness. Adam Zeman. Dept of Clinical Neurosciences. Brain (2001)
(2) Suppression of experimental allergic encephalomyelitis is specific to the frequency and intensity of nocturnally applied, intermittent magnetic fields in rats. Cook LL and M.A. Persinger. Oct 2000
(3) The natural history of consciousness, and the question of whether plants are conscious, in relation to the Hameroff-Penrose quantum-physical 'Orch OR' theory of universal consciousness Peter W Barlow. July 2015
(4) Confirmation of Quantum Resonance in Brain Microtubules. Biomolecules exhibit quantum mechanical behavior. William Brown. Resonance Science Foundation
(5) A phenomenology of meditation-induced light experiences: traditional buddhist and neurobiological perspectives Jared R. Lindahl, et al. Jan 2014
(6) Attentional processes and meditation. Hodgins HS1, Adair KC.. Dec 2010
(7) Elliott M. (1998). Synchronous information presented in 40-Hz flicker enhances visual feature binding.

(8) Consciousness, Microtubules and The Quantum World Interview with Stuart Hameroff, MD, in Alternative Therapies (May 1997 3(3):70-79 by Bonnie Horgan).

(9) Quantum Resonance & Consciousness. Contzen Pereira. September 2015

(10) Quantum computation in brain microtubules? The Penrose–Hameroff 'Orch OR' model of consciousness By Stuart Hameroff

(11) Consciousness in the universe: a review of the 'Orch OR' theory. Hameroff S1, Penrose R. Mar 2014

(12) Orchestrated reduction of quantum coherence in brain microtubules: A model for consciousness Stuart Hameroff, Roger Penrose

(13) Causality, retrocausality and consciousness Antonella Vannini. Syntropy 2006, 3, pag. 227-234 ISSN 1825-7968

(14) THE FUNDAMENTAL EQUATIONS OF POINT, FLUID AND WAVE DYNAMICS IN THE DE SITTER-FANTAPPIÉ-ARCIDIACONO PROJECTIVE RELATIVITY THEORY Leonardo Chiatti

Chapter 22. Types of Meditation and its effect on Brainwave Activity

There are many research studies showing that people who meditate are more intuitive on average compared to non-mediators. I show numerous examples in my second remote viewing book **Remote Viewing. The Complete User's Manual on Experiencing Future Consciousness**. Hence the brainwave frequencies exhibited by mediators may act as a way to control / limit the signal to noise ratio that occurs during remote viewing as well as give one more control over their emotions, especially anxiety which can negativity impact remote viewing performance.

Types of Meditation and Brainwave Patterns
Different types of meditation exhibit differing forms of electroencephalographic (EEG) activity. Let's examine some research studies in greater detail confirming this.

A research study published in 2017 compared practitioners of three different types of meditation. 1 - **Himalayan Yoga**. 2 - **Vipassana** 3 - **Isha Shoonya**. The study also included a control group. The study found that all the mediators exhibited stronger parieto-occipital 60hz to 110hz gamma amplitude oscillations compared to the control group. The study also found that stronger 7hz to 11hz alpha activity took place in the group practicing the Vipassana meditation compared to all other groups. The group practicing the Himalayan Yoga Meditation exhibited lower 10hz to 11hz activity. The study concluded that meditation practice causes distinct changes in the EEG gamma frequency range which are consistent to a variety of meditation practices[1].

The 10hz Frequency
10Hz is related to Alpha brainwaves and 10hz at low frequencies also occurs during certain phases of sleep[2]. Magnetic effects have been shown to take place at 10hz. This effect occurs when a sensation of flickering light in the eye takes place when it is in an environment with ELF magnetic fields greater than 10 mT and with frequencies over 10 Hz (*Magnetophosphenes: a quantitative analysis of thresholds. P. Lövsund, et al. May 1980).*

Research suggests the magnetic field interaction leading to magnetophosphenes occurs in the eye's retina (*Tenforde, 1990*). Takahashi et al. (1986) examined DNA synthesis in Chinese hamster cells (V79) after they were exposed to Helmholtz coils via pulsed magnetic fields[3]. DNA synthesis became enhanced about 13% at 10 Hz ($p<0.01$) and 30% at 100 Hz (p<0.001) with no significant effects shown at other frequencies[3].

McLeod et al. (1987) showed protein biosynthesis in neonatal bovine fibroblasts was reduced by low frequency sine wave electrical fields. Data showed the optimal frequency range for the effect was at 10 Hz[3]. Also **10hz high frequency** in the brain is in alpha when the eyes are closed and at **10hz low frequency** when the brain is asleep[4] [5]. 10Hz has also been shown to reduce the effects of epilepsy and reduce sleep disturbances [6].

Research by Wever showed that when volunteers were subjected to "*natural radiation fields*" or to 10 Hz at 2.5 V/m, that it caused changes in their body-temperature and activity-rest cycles when they were shielded from earth's natural frequency[7]. People who meditate have been shown to exhibit brainwave frequencies of between 4Hz and 13 Hz [8].

What are Gamma Brain Waves?
Gamma brain waves have been measured as the fastest brainwave frequency that have the smallest amplitude. People experiencing gamma brain waves commonly report such sensations as "*feelings of blessings*" which is commonly reported by monks or experienced mediators.

Flickering Light and Brainwave Activity
I had noticed on a few occasions while remote viewing when my LED flashlight batteries were failing that a flickering would start to occur. This flickering seemed to help the remote viewing. Research has shown that when mice were exposed to flickering LED lights (on-off at 40 times per second) that the flickering stimulated gamma brainwaves and that it may be a promising candidate for Alzheimer's treatment [9]. Also another research study found that 10 Hz electroencephalographic (EEG) alpha rhythms stimulated memory in older people [10].

How to Generate 10Hz and 40Hz Gamma
10hz can be generated by creating a tone of 335 Hz in the right ear and generating a second tone of 345 Hz in the left ear. This causes a subjectively perceived binaural beat to occur at 10 Hz. Instead of hearing two separate tones, only one tone occurs that fluctuates in frequency or loudness.

Gamma frequencies (40 Hz) can be made by creating a tone of 320 Hz in the left ear and 360 Hz in the right ear. To enhance the effect white noise can be generated between 20 Hz and 10 kHz band filtered to both ears to enhance clarity (*Oster, 1973*)[11]. One type of software (besides listening to 10hz or 40hz online YouTube videos) that can do this for you is Brainwave Generator (www.bwgen.com). (*Oster, 1973*).

Also studies have shown that gamma wave brainwave activity has been shown to take place in the brains of those exhibiting seizures [12]. Hence taking herbs that control seizures while listening to 40Hz gamma may enhance the effects, although research is necessary to confirm this. GABA has also been shown to increase alpha brainwaves and enhance the immune system [13]. Another research study found that GABA played a major role in gamma brain wave oscillations with healthy levels of GABA enhancing gamma oscillations [14].

Gamma waves are also associated with extremely high levels of cognitive functioning and peak concentration which is why increased gamma waves have been shown to occur in meditators [15]. 40Hz Gamma Wave Binaural Tone MP3's can be found on online video file sharing sites and I go into far greater detail about how 40Hz Gamma Waves can be used to enhance remote viewing in my first remote viewing book **Wormhole Theories, Sunspot Activity and Remote Viewing Stocks**.

40-80 Hz activity in the brain has also been observed by von der Marlsburg and Schneider, 1986; Gray and Singer, 1989 and Crick and Koch, 1990.[16]. Gamma waves may also play important roles in helping the brain identify visual cues in both subliminal and conscious stimuli[17] [18] [19].

Methods that Amplify 40Hz Gamma

Adding weak noise (random fluctuations) enhances 40hz gamma[20] and Spatiotemporal patterns that occur in the neurons of the brain are related to the noise that occurs[21].

Further Reading

Noise in the nervous system A. Aldo Faisal et al. April 2008

Genetic Analysis of Circadian Responses to Low Frequency Electromagnetic Fields in Drosophila melanogasterGiorgio Fedele. et al. Dec 2014

Alpha and Gamma waves and Creativity

A research study showed subjecting volunteers to alpha (10 Hz) and gamma (40 Hz) enhanced creativity. The participants listened to the binaural beats for three minutes before performing a series of tasks. The study showed that the binaural beats at both frequencies enhanced performance in the divergent, but not convergent thinking tasks. The study also found that people who had a low eye blink rate benefited from listening to the alpha binaural-beat stimulation and those who had high eye blink rates were unaffected or impaired by the alpha and gamma binaural-beat stimulation. White noise in the background was also included in the stimulus which amplified the binaural-beats[22].

What is the Eye Blink Rate?
The eye blink rate is a non-invasive indirect method of measuring dopamine function in the brain. Higher eye blink rates mean higher dopamine functioning[23].

Gamma frequencies have long been postulated to be related to consciousness via neuronal synchronization and the periodic structure of waves occurring as gamma delta beat frequencies[24]. This behavior is very similar to the alternating interference that occurs during double slit experiments[25]. Gamma oscillations occur at between 30 Hz and 80 Hz[26].

Nicotine Enhances Right Brain Functioning
A research study found that low levels of nicotine activate the left side of the brain, whereas large doses create more activity on the right side of the brain[27]. The left brain rules logic and the right brain rules intuition.

Phenylacetaldehyde Enhances Photon Emission
A research study found that adding

phenylacetaldehyde to bacteria, especially gram-negative bacteria, enhanced the emission of their photons. Gram negative bacteria have a thinner wall and extra external lipophilic membrane which is why it possibly enhanced their light emission. The spectrum of the light observed was in the 500 nm range. This is similar to that of when free amino acids, protein (*bovine serum albumin*) or isopropylamine interact with phenylacetaldehyde (*Photon emission by bacteria challenged with phenylacetaldehyde. A possible distinction between gram-positive and gram-negative bacteria. Surpili MJ et al. March 1993;57(3):564-9*).

What is Phenylacetaldehyde?
Phenylacetaldehyde is used in the synthesis of polymers for controlling the rate of polymerization and fragrances. It occurs extensively in nature as it is biosynthetically extracted from the amino acid phenylalanine. Natural sources of Phenylacetaldehyde can be found in buckwheat, chocolate, flowers, and communication pheromones that occur in various insects.

Phenylacetaldehyde has a rose, honey-like sweet with a green and grassy aroma. It is commonly added to narcissi, hyacinth or rose fragrances. Phenylacetaldehyde is also found in some flavored cigarettes and beverages.

Phenylacetaldehyde has also been used to make phenylalanine (*via the Strecker reaction*) in order to make aspartame sweetener.

Hemispheric Balancing

It just may be so that individuals that can completely focus their minds on a single task and relax their body causing their brain waves to enter the alpha rhythm (*or for other people the theta rhythm*) all without falling asleep, (*reducing the frequency of the brainwaves*) that they can perform certain mental tasks at that specific brainwave level which would otherwise not be possible during other brain wave states.

While in alpha or theta, the individual performing an associative remote viewing session is able to project his or her awareness or consciousness into the future. How is this possible? Due to the fact that at the alpha or theta brainwave levels the left brain hemisphere, including those of the right brain, are synchronized. It is this synchronization that allows one to tap into the tremendous knowledge and power hidden within the subconscious mind. This same synchronization effect also occurs at 40hz gamma. Hence, training ourselves to think at the alpha or theta level without falling asleep offers tremendous untapped potential. The alpha rhythm also can boost the body's immune system keeping us healthy. It also allows one to remove stress and boost mental ability. Some of the signs that the body has reached the alpha state include the following:

- Heart rate and breathing slow down
- The person becomes completely passive physically and mentally
- A person's body temperature may rise or even perspire
- Eyes may start moving rapidly (REM). This is the same REM that takes place while one is dreaming
- The person's galvanic skin resistance (GSR) increases
- The person's body may start swaying pleasantly and slowly

Stochastic Resonance

Stochastic resonance exists as a signal that is too weak to be detected by standard sensors. We can think of it as a quantum type effect. The signal is commonly boosted by adding white noise to the signal, due to white noise containing a wide spectrum of frequencies. This is because frequencies in white noise correspond to the original signal's frequencies that end up resonating with one other. Hence the original signal becomes amplified while not amplifying the rest of the white noise. This causes an enhancement of the signal-to-noise ratio that makes the original signal much more prominent. This also means the added white noise allows the signal to become detectable by the sensor[28]. This is why white noise is used in some binaural beat frequencies to amplify the original frequency.

Noise Amplifies Electrical Signals in the Brain
Natural sources of electrical noise occur in our brain and play important roles in cognitive functioning. This noise comes from random firings of neurons and electrochemical reactions which may play a part in stochastic resonance. The effect termed 'stochastic resonance' exists in a wide range of living organisms as well as systems.

Shoichi Kai and Toshio Mori of the University of Kyushu in Japan found that stochastic resonance in information processing exists in the human central nervous system (*T Mori and S Kai 2002 Phys. Rev Lett. 88 218101*). Most complex systems that have weak periodic signals are able to be strengthened by noise. The process takes place when random peaks in the noisy signal coincide with regular peaks in the periodic signal. It is most effective when the noisy signal exists at a specific amplitude that is relative to the periodic signal. This effect already exists in our body in the sense of touch and the control of blood pressure in our brain, which is why flotation tanks eliminate bodily sensation, possibly reducing internal noise.

During Kai and Mori's research study, they shone light signals into eyes of 5 students while at the same time measuring their brain waves. Noisy signals were shone into the student's left eyelids and periodic signals shone into their right eyelids while the students were resting. During this time, Kai and Mori measured the intensity of the student's alpha brain waves. As they did this, they found a sharp peak occurring at 5 Hz (*frequency of the periodic signal*). However after increasing the strength of the noise signal that was relative to the periodic signal, they found that a 'harmonic' peak emerged in the student's alpha brainwaves at 10 Hz. They also observed that when they increased the noise signal, the peak first intensified then diminished. Kai and Mori stated the harmonic peak may be a stochastic resonance occurring in the visual cortex of the human brain. This is because it reaches a maximum at a specific signal to noise ratio. They also stated that the effect took place solely in the student's brain instead of their eyes because they positioned a light proof screen between the student's eyes so as to avoid signals interfering outside their head. Kai and Mori state this effect may play important roles in complex brain functions such as cognition and perception (*T Mori and S Kai 2002 Phys. Rev. Lett. 88 218101*).

10 Hz Current Induces Alpha Brainwave Rhythmus

A research study by the University of North Carolina (UNC) School of Medicine showed that low doses of electric current at 10-hertz enhances alpha brain wave activity. The study also found that it enhanced
creativity by 7.4% in adults. The study also suggests that people with depression show impaired alpha
oscillations (*Functional Role of Frontal Alpha Oscillations in Creativity. Caroline Lustenbergera et al et al. April 2015*).

Seasonal Variation of Sesquiterpenes in the Essential Oil (Lamiaceae)

A research study looking at Ocimum gratissimum L. (Lamiaceae) to treat the central nervous system distilled it in essential oil form each season of the year. The study looked at the main component eugenol. Its activity on the body's central nervous system was evaluated as well as its anticonvulsant activity against seizures. The study found that the essential oil distilled during Spring protected animals against seizures induced by electroshock and that during all 4 seasons 1,8-cineole and eugenol were the most abundant compounds with Spring generating the highest level of sesquiterpenes. The study concluded that the variation of the compounds explain differences observed in biological activity occurring in essential oils that are distilled at different seasons of the year.

Reference
Effects of seasonal variation on the central nervous system activity of Ocimum gratissimum L. essential oil. CM Freire et al. J Ethnopharmacol. 2006 Apr 21;105(1-2):161-6. Epub 2005 Nov 21.

References. Chapter 22.

(1) Increased Gamma Brainwave Amplitude Compared to Control in Three Different Meditation Traditions Claire Braboszcz, et al. Jan 2017
(2) Influence of Electric, Magnetic, and Electromagnetic Fields on the Circadian System: Current Stage of Knowledge Bogdan Lewczuk
(3) Evaluation of the Potential Review Draft Carcinogenicity of (Do Not Electromagnetic Cite or Quote) Fields. United States Environmental Protection Agency. October 1990
(4) Effects of 6-10 Hz ELF on Brain Waves David S. Walonick Minneapolis, MN, May 1990
(5) The 10 Hz Frequency: A Fulcrum For Transitional Brain States. Garcia-Rill E et al. 2016
(6) Low-frequency electroacupuncture suppresses focal epilepsy and improves epilepsy-induced sleep disruptions. Yi PL et al. July 2015
(7) Biologic Effects of SElectric and Magnetic Fields Associated with Proposed Project Seafarer. Committee on Biosphere Effects of Extremely- Low -Frequency Radiatio. Assembly of Life Sciences. National* Academy of Sciences. 1977

(8) Buddha's Brain: Neuroplasticity and Meditation Richard J. Davidson, et al. Jan 2008

(9) Iaccarino, H.F., et al. Gamma frequency entrainment attenuates amyloid load and modifies microglia. Nature. 2016 Dec 7;540(632):230-235.

(10) 10 Hz flicker improves recognition memory in older people Jonathan Williams,

(11) The impact of binaural beats on creativity Susan A. Reedijk, Anne Bolders, and Bernhard Hommel. Nov 2013

(12) Induced visual illusions and gamma oscillations in human primary visual cortex. Adjamian P1

(13) Relaxation and immunity enhancement effects of gamma-aminobutyric acid (GABA) administration in humans. Abdou AM et al. 2006

(14) GABA level, gamma oscillation, and working memory performance in schizophrenia Chi-Ming A. Chen, et al. Mar 2014

(15) Buddha's Brain: Neuroplasticity and Meditation Richard J. Davidson, et al. Jan 2008

(16) Toward a Science of Consciousness: The First Tucson Discussions and

Debates edited by Stuart R. Hameroff, Alfred W. Kaszniak, Alwyn Scott

(17) Hughes JR. (1964). Responses from the visual cortex of unanesthetized monkeys. pp. 99–153. In: Pfeiffer CC, Smythies JR, (Eds), International review of neurobiology vol. 7, Academic Press, New York OCLC 43986646

(18) Gregoriou GG, Gotts SJ, Zhou H, Desimone R (Mar 2009). "High-frequency, long-range coupling between prefrontal and visual cortex during attention". Science. 324 (5931): 1207–1210.
Bibcode:2009Sci...324.1207G. PMC 2849291 Freely accessible. PMID 19478185.
doi:10.1126/science.1171402.

(19) Melloni L, Molina C, Pena M, Torres D, Singer W, Rodriguez E (Mar 2007). "Synchronization of neural activity across cortical areas correlates with conscious perception". J Neurosci. 27 (11): 2858–65. PMID 17360907. doi:10.1523/JNEUROSCI.4623-06.2007

(20) Stochastic Resonance Modulates Neural Synchronization within and between Cortical Sources Lawrence M. Ward, et al. Dec 2010

(21) Key role of coupling, delay, and noise in resting brain fluctuations Gustavo Decoa et al. Feb 2009

(22) Auditory Beat Stimulation and its Effects on Cognition and Mood States Leila Chaieb, et al. May 2015
(23) Spontaneous eye blink rate as predictor of dopamine-related cognitive function-A review. Jongkees BJ1, Colzato LS. Dec 2016
(24) Oscillatory Correlates of Visual Consciousness Stefano Gallotto et al. July 2017
(25) Consciousness and the double-slit interference pattern: Six experiments Dean Radin et al. January 2012
(26) Gamma and beta frequency oscillations in response to novel auditory stimuli: A comparison of human electroencephalogram (EEG) data with in vitro models Corinna Haenschel et al. Dec 1999
(27) Smoking, nicotine dose and the lateralisation of electrocortical activity. Norton R1, Brown K, Howard R. 1992
(28) Ward LM, Doesburg SM, Kitajo K, MacLean SE, Roggeveen AB (Dec 2006). "Neural synchrony in stochastic resonance, attention, and consciousness". Can J Exp Psychol. 60 (4): 319–26. PMID 17285879.doi:10.1037/cjep2006029.

Chapter 23. Can Photons Travel Backwards Through Time?

Photon Emissions from Living Organisms
Photon emissions have been observed emitting from organs, bacteria, cells and entire organisms for decades [1]. The emitted frequency of these emissions occurs at an extremely **low frequency** range and they also exhibit Compton scattering [2]. Other research has shown that photon emissions occur in the presence of changes in weak (nanoTesla) geomagnetic fields [3]. Microtubules also emit single photons of light and are common seen behaving as microscopic pulse-lasers within living cells (*Popp, Li, Nagl, and Klima, 1983*)[4].

Geomagnetic activity has been shown to enhance bioluminescence in bacteria (*Berzhanskaya et al., 1996*)[5]. Research has also found that **photons are emitted from the left hemisphere of the brain**, but not from the right side of the brain and that these brain photons may be potentially interacting with global geomagnetic activity[6]. Photon emissions from living systems that are ultra weak have also been found to be highly coherent[7].

Thomas Edison, Luminescence and Silver Sterling Mine

Ogdensburg, New Jersey in Sussex County, sits on the Crossroads of the American Revolution. It was the site of the first industrial city (Paterson) and also home to Thomas Edison, the great inventor. Mines in this region have produced over 357 types of minerals with many being found in the region of Sterling Hill Mine. 357 types of these minerals account for about 10 percent of minerals known to science and 35 of the minerals are not found anywhere else and remarkably **91 of these minerals fluoresce**.

The Sterling Hill region is rich in **zinc** and iron deposits with most ore being composed of franklinite. Franklinite consists of zinc, **manganese** oxide and iron. It also contains zincite (*also called zinc oxide*) and willemite (*also called zinc silicate*).

Thomas Edison purchased an iron mine in Sparta during 1890 which he named the New Jersey and Pennsylvania Concentrating Works. He invested about $2 million of his own money and raised $1 million from investors. During 2005 an astronomical observatory was built on the museum grounds, housing several telescopes.

Zincite
Also known as zinc oxide it can be produced

synthetically. Zinc oxide is a wide-bandgap semiconductor in the II-VI semiconductor group.

Manganite
In our second book on remote viewing, we show extensively how the mineral manganese enhances remote viewing. Manganese is found in abundance in pineapple and brown rice. This may be due to manganese's ability to maintain coherence during Heart Math (HRV).

A research study titled: Penetration of the electromagnetic field through plate-shaped bulk polycrystalline sintered lead and Y–Ba doped manganite conducted by A Rinkevich and colleagues in 2002 found that mixing manganite with lead, barium and yittrium had an effect on the electromagnetic field. Another study found that Yttrium manganite shows ferromagnetic properties *(Mechanochemical synthesis of yttrium manganiteAuthor links open overlay panel M. Počuča-Nešić et al. March 2013).*

A counterpart of the element manganese in mineral form is called manganite. Manganite is very similar to iron. Iron is used in industry and manganese is used in alloys one of which is to harden steel. Manganite is commonly found in high pressure environments, most notably the ocean floor. **Nsutite** which is a natural form Manganite is used as the cathodic material in common zinc-carbon dry-cell **batteries**. Manganite is also found in medicines and batteries. Manganite is composed of manganese oxide-hydroxide and quality specimens are found at Ilfeld in the Harz Mountains of Germany. Manganite is commonly found in **calcite** and barite veins (traversing porphyry). During prehistoric times it was used as a **pigment** and a fire starter by Neanderthalers. Manganite will reduce the combustion temperature of wood from 350 degrees Celsius to 250 degrees Celsius.

Manganite is also found with **apatite**, clay minerals, biotite, quartz and feldspars. Manganite is also used for purifying water, for plant fertilizers as a catalyst, as a livestock feed additive, and as a colorant for bricks.

Todorokite

Todorokite is commonly found above ground and is associated with Manganite and is found with calcite. It has a very high refractive index (approx 1.74). At the Sterling Hill mine region in New Jersey, todorokite has been found in small collections of manganese oxides in mines on the surface.

Birnessite

Birnessite is a very soft mineral and is an oxide mineral of manganese. It is formed by precipitation in oceans, lakes and groundwater. Birnessite contains a large amount of manganese and is composed of **sheets of water molecules**. Recent research is showing Birnessite may be used to harvest sunlight in order to split water into oxygen and hydrogen (*Birnessite. Wikipedia*). Pictured below is a specimen of Birnessite.

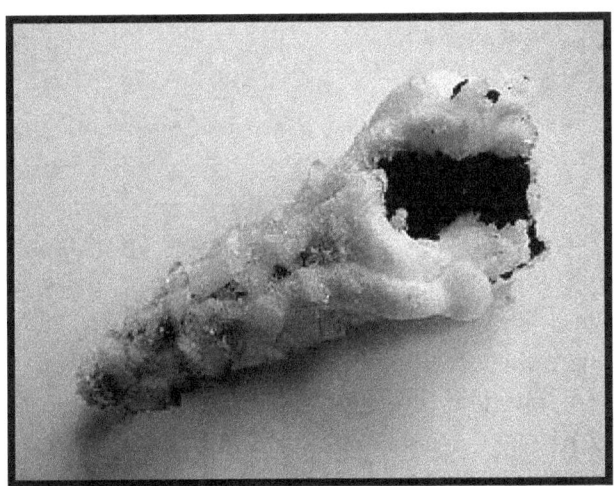

Further Reading

Electromagnetic Field Effect On Luminescent Bacteria. IEEE Trans. Magn. 31, 4274–4275. Berzhanskaya, L.Y., Beloplotova, O.Y., Berzhansky,

V.N., 1995.

The Time Travelling Photon Experiment

Now let's return to photons and time travel. The most recent mathematical models describing space and time allow for the possibility for time to turn backwards. This allows for the existence of closed time-like curves (*also called CTC*) which allow one to loop back to the past. Research by scientists at the University of Queensland, Australia used photons to simulate quantum particles travelling through the fabric of time. They discovered that one photon was able to pass through a wormhole and interact with its older self[8]. In another experiment conducted in 2007, researchers in France pumped photons into a device and demonstrated that their actions retroactively changed something which had already happened previously[9] [10] [11].

Below is a quote from the experiment

*"If we attempt to attribute an objective meaning to the quantum state, curious paradoxes appear: quantum effects mimic instantaneous action-at-a-distance as well as **influence of future actions** on past events, even after the events have been irrevocably recorded".*

Summary
Photons can influence past events. Hence what is happening in the present may be a result of interactions that happened in the past, proving that time flows as a loop.
Romijn (2002) [12] proposed that virtual photons (*constituent units of the electromagnetic field*) are carrier units of consciousness and others suggest zero-energy tachyons (*Hari, 2008*)[12], or hypothetical particles (*Eccles, 1992*)[12], both of which travel faster than the speed of light, are assumed to imbue biological substrates the quality of consciousness. Hence, photons may be the carrier for sending (*asking questions*) receiving information (*image retrieval*) during associative remote viewing sessions.

Quantum Superposition and Travel to the Past
Quantum superposition allows time travel to the past. The explanation states that as a photon goes back in time to switch off the machine that

sent it, until it does so, it will at **all times be in a superposition of states** (*having multiple or directly conflicting states occurring within the same moment*).

Now let's next explore how remote viewing may be connecting the viewer with alternate universes and timelines.

References. Chapter 23

1. Imaging of light emission from the expression of luciferases in living cells and organisms. Lee F. Greer. 2002
2. Compton Scattering and the Emission of Low Frequency Photons. AA(Swiss Federal Institute of Technology, Zurich, Switzerland). 1947
3. Inverse relationship between photon flux densities and nanotesla magnetic fields over cell aggregates: Quantitative evidence for energetic conservation Michael A. Persinger, et al. May 2015
4. A Quantum Biomechanical Basis for Near-Death Life Reviews. Thomas E Beck et al. March 2003
5. Photon emissions from human brain and cell culture exposed to distally rotating magnetic fields shared by separate light-stimulated brains and cells Blake T. Dotta et al. Masrch 2011
6. Differential Spontaneous Photon Emissions from Cerebral Hemispheres of Fixed Human Brains: Asymmetric Coupling to Geomagnetic Activity and Potentials for Examining Post-

Mortem Intrinsic Photon Information Justin N. Costa et al. June 2016
7. Biophoton emission. New evidence for coherence and DNA as source. F. A. Popp et al. March 1984
8. Experimental simulation of closed timelike curves. Martin Ringbauer et al. May 2014. Centre for Engineered Quantum Systems, School of Mathematics and Physics, University of Queensland, Brisbane, QLD 4072, Australia Centre for Quantum Computation and Communication Technology, School of Mathematics and Physics, University of Queensland, Brisbane, QLD 4072, Australia
9. Experimental delayed-choice entanglement swapping Xiao-song Ma et al. March 2012. Nature Physics 8, 479–484 (2012)
10. Delayed choice for entanglement swapping ASHER PERES Department of Physics, Technion—Israel Institute of Technology, 32 000 Haifa, Israel
11. ARTICLE REF - Delayed-choice Experiment in Cavity QED. Rameez-ul-Islam. National Institute of Lasers and Optronics, Islamabad

12. Electromagnetic fields as structure-function zeitgebers in biological systems: environmental orchestrations of morphogenesis and consciousness Nicolas Rouleau1,2 and Blake T. Dotta. Nov 2014

Chapter 24. Remote Viewing and Alternate Timelines

There is an excellent online-video made by Remote Viewer Courtney Brown showing how remote viewing interacts with parallel universes. The video is called **Courtney Brown on Multiple Universes**. It is highly recommended for anyone wanting a clear picture of how remote viewing interacts with alternate timelines.

Parallel Worlds and the Biophysical Field
Ervin Laszlo (1995) stated memory may be stored in a holographic memory field that is collective in nature and that this field exists outside the boundaries of the physical body[1] [2]. This fits neatly with the experiences by clairvoyants who state they access information contained in "bio-energetic fields" that surround their client. This field may also be responsible for Jung's concept of "*the collective unconscious*" which exists as a vast pool of archetypal images.

In astronomy, the interstellar medium (ISM) is the matter and radiation that exists in the space between the star systems in a galaxy. This matter includes gas in ionic, atomic, and molecular form, as well as dust and cosmic rays. It fills interstellar space and blends smoothly into the surrounding intergalactic space. The ISM is a region where quantum effects might be observed in bulk.

Improve your Remote Viewing Accuracy Techniques using Quantum Microtubules

Research by cell developmental biologists discovered the morphogenetic fields that control cell activity changes calcium levels in the cells via field effects. It is these morphogenetic fields that enable the biophysical body to regulate cellular chemistry and also govern interactions taking place in the cell's nucleus[3].

During some states of consciousness that take place during remote viewing memories from the brain and biophysical body are possibly being transferred via biophysical fields which in turn affect the neurons. This is interrelated with the control of calcium concentrations in the brain.

Changes of calcium can also take place via EM interaction. This occurs when a neuron is depolarized, which causes a huge influx of calcium ($Ca2+$) through its membrane. This depolarization effect and its associated ions produce a low intensity magnetic field (*Electromagnetic fields as structure-function zeitgebers in biological systems: environmental orchestrations of morphogenesis and consciousness. Nicolas Rouleau and Blake T. Dotta. November 2014. Front Integr Neurosci. 2014; 8: 84. doi: 10.3389/fnint.2014.00084*).

Summary

Remote viewing exists as the result of a biophysical field that travels throughout multiple timelines that is contiguous with our own, yet makes itself known to us in the quantum realm. While remote viewing, the biophysical field of the person travels to the target site through the quantum field. This quantum field is smeared into many possible states. When the biofield of the remote viewer arrives at the target site that is being remote viewed, the remote viewer's biophysical body may sometimes have difficulty observing the target much like a dreamer trying to realize she is dreaming. This may explain why at times bilocation occurs, as if the remote viewer actually exists at that distant location.

Perhaps the parallel reality is also unwittingly experienced in out-of-body-experiences and during lucid dreaming. Emotion may be the prime driving force that links parallel universes.

Further Reading
Possible disruption of remote viewing by complex weak magnetic fields around the stimulus site and the possibility of accessing real phase space: a pilot study. Koren SA and, Persinger MA. Dec 2002

Conclusions
Remote viewing sessions allow one to view multiple timelines occurring within our own universe. The very process of associative remote viewing causes changes in our present timeline,

where the new timeline becomes a new sequential path of experiences existing on a hyper-surface. Hence, remote viewing the future causes one to establish a new sequence of future events in the present universe.

To put it eloquently, when the remote viewing session begins, many possible timelines come into play, but after the session, we remember only a **single time-line**.

The question remains however, are we perceiving parallel universes close to our own during remote viewing, compared to the many time lines existing within multiple universes?

References. Chapter 24

1. Organism and Psyche in a Participatory Universe* Dr. Mae-Wan Ho.
2. A Quantum Biomechanical Basis for Near-Death Life Reviews. Thomas E. Beck and Janet E. Colli. March 2003
3. Bioelectric mechanisms in regeneration: unique aspects and future perspectives Michael Levin, Ph.D. May 2009

Chapter 25. Neutrinos and Parallel Universes

Hydrogen and Alternate Universes
Ranga-Ram Chary, astronomer at NASA at the California Institute of Technology, used Planck data to search for evidence of alternate universes. He found three regions in the sky where there is 4,500 times the amount of hydrogen compared to other regions [1] [2].

Parallel Universes have been examined in great detail by Fred Alan Wolf [3].

Further Reading
Quantum phenomena modeled by interactions between many classical worlds. Hall MJW, Deckert D-A, Wiseman HM. Phys. Rev. X. 2014;4:041013.

The Planck Space Telescope (*pictured below*)

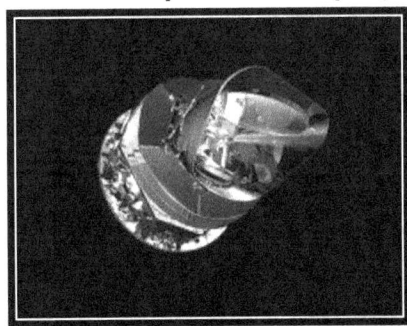

Further Reading
Spectral Variations of the Sky: Constraints on Alternate Universes. R. Chary.. Dec 2015

Neutrinos are particularly suited to straddle between worlds due to their ability to change flavors[4], their extremely small mass, their close resemblance to electrons, and most of all their abundance as they flow throughout our earth and our bodies.

Physicists have verified that quantum particles can occupy superposition by using neutrinos separated by 450 miles, the farthest such test to date [5] [6] [7]. A group of physicists from MIT shot a beam consisting of one "*flavor*" of neutrino, which is a tiny subatomic particle that rarely interacts with other matter. Hundreds of miles away they detected a different flavor of neutrino. This indicated the neutrinos were traveling at quantum superposition via a mid-flight mixture of flavors[5] [6] [7].

This study adds credibility to the long-held theory that states subatomic particles can exist in superposition while hurtling through space-time. Neutrinos created in the early universe are also being explored as a way to create a unified theory of quantum mechanics and gravity, including string theory and loop quantum gravity.

Recommended Reading
Model of a multiverse providing the dark energy of our universe. Eckhard Rebhan (Sep 2017)

Exploring the Universe with Very High Energy Neutrinos. A. Kappes, for the Ice Cube Collaboration (Jan 2015)

Cyclic multiverses. Konrad Marosek, Mariusz P. Dabrowski, Adam Balcerzak (Submitted on 14 Sep 2015 (v1), last revised 7 Jul 2016 (this version, v2)

A tale of two modes: Neutrino free-streaming in the early universe. Lachlan Lancaster, Francis-Yan Cyr-Racine, Lloyd Knox, Zhen Pan (Submitted on 21 Apr 2017 (v1), last revised 5 Jul 2017.

Do Neutrinos Behave like Quantum Waves? Quantum mechanics states particles at times behave like waves, and vice versa. Neutrinos exhibit effects very similar to quantum behavior in that they "*oscillate*"[8]. This oscillation is caused by the flavor of the neutrino[8].
Neutrinos may be a misunderstood form of superluminal particles called tachyons[9]. Research has stated tachyons may be compatible with Lorentz invariance. Lorentz-violating neutrino oscillations have been proven to exist in quantum gravity[10][11].

Further Reading
Jeong, E.J., "Neutrinos must be tachyons," arXiv:hep-ph/9704311 (1997)

Apparent Lorentz violation with superluminal Majorana-tachyonic neutrinos at OPERA?F Tamburini and M Laveder. Feb 2012. The Royal Swedish Academy of Sciences

Exploration of possible quantum gravity effects with neutrinos II: Lorentz violation in neutrino propagation Alexander Sakharov et al. 2009

Pair Creation Constrains Superluminal Neutrino Propagation Andrew G. Cohen and Sheldon L. Glashow. 27 October 2011

Electromagnetic fields as structure-function zeitgebers in biological systems: environmental orchestrations of morphogenesis and consciousness Nicolas Rouleau1,2 and Blake T. Dotta. Nov 2014

Tachyons
Tachyons have been researched by well-known physicists such as Feinberg, Sudarshan and Recami[12]. There is a very well thought-out research paper suggesting that mental units called "*psychons*" by Eccles may be tachyons (*hypothetical particles*) [13]. Hence tachyon theories may be applicable to brain physics.

Eccles proposed there existed an association between dendrons and what he called psychons. Dendrons are dendrite bundles that act as basic anatomical units of the neocortex for reception[14]. The paper states a zero-energy tachyon may behave as a trigger for exocytosis (*a quantum tunneling process*) and that the energy is not occurring at a single presynaptic terminal but at all terminals in the dendron. This theory is consistent with tachyons, which states they occur as non-local phenomena and become absorbed non-locally and instantaneously via detectors behaving in a cooperative and coherent way.

Further **Reading**
Eccles's Psychons Could Be Zero-Energy Tachyons Syamala D. Hari. June 2008. Neuro Quantology |Vol 6 |Issue 2|Page 152-160.

Neutrinos Travel Faster than Light via Super-Luminosity
The neutrino detector OPERA has detected superluminal neutrinos [15].

Why Aren't Parallel Universes "bumping" into one Another?

Our universe may be just one in an infinite number of other universes all linked by quantum waves. If this is true, wouldn't these universes be bumping into each other from time to time?, especially at the very start of our universe, where collisions that occurred may have bumped into other universes.

Research has shown that there is a mysterious "*cold spot*" in the observable universe. Some space experts believe this cold region is the result of our universe having collided with another universe during its early stages and that this cold region is the occurrence from our universe having grown like a bubble from a vacuum. The cold spot measures 1.8 billion light years in diameter and is about 0.00015 cooler than its surroundings[16] [17]. Hameroff and Penrose state that the reason the mind does not see multiple universes (*outside of remote viewing*) is due to the quantum coherence occurring in the brain's microtubules[18] [19]. Hence, the effects of mild de-coherence caused by essential oils and other substances on the brain's microtubules remove this veil.

If we exist in separate "bubble" universes and our reality is based upon the quantum collapse wave function, multiple universes could occupy the same space and pass right through one another invisibly due to their different frequencies, much like swarms of insects travelling in opposite directions pass through one another without colliding. This would also mean that events affecting our universe may also be exhibiting "*ripple effects*" in nearby universes. In a sense of the word "*parallel realities*" implies that realities already coexist, right next to each other. In other words, there is no actual time; we simply move from one already-existent universe to another already-existent universe, and we are doing so billions of times every second. In this way, we generate the illusion of movement, space, time, motion, change and so on. Nothing actually changes structurally speaking. Everything already exists; everything happens all at once; everything exists all at once.

There are endless timelines, and each timeline consists of billions of parallel realities per second. Endless timelines with endless stories, each with billions of parallel realities happening each second moved through by consciousness.

Right now you are in a particular frequency and you are not recognizing all the other alternate timeline realities that are also available.

Parallel Universes and Healing
If emotions are the driving force between parallel worlds, exercises that heal emotions may be one way to improve health and well-being.

All it takes is a change of heart, a change of intention, a change of mindset and a change of belief. When this change is executed and seen all the way through, we "*shift timelines.*" Negative emotions no longer attract negative experiences. You have actually changed your timeline. Oddly enough, not only does this mean you have changed your future experience (*what you will be experiencing*), it also means that the body-mind vibrational spectrum that you are now inhabiting and are conscious of, has a different past. You now notice a parallel picture that corresponds to the vibratory state you have now shifted into.

So, now you have access to the future and

the past simultaneously. You have access to this new future self, this new parallel timeline that you chose. Being in that energy, you will naturally have a different type of memory and a different type of future.

Changing the Past to Change the Future
One way to specifically do this is to actually change your past. Here is one simple method that you can learn to apply. Take an emotionally disturbing event you remember from your past, perhaps one that you don't really enjoy; one that you don't prefer. It may be one that you wish wasn't yours, even though you can see how it has benefited you, how it has helped you to become who you are. You can take that memory that you have a bit of, a nasty after-taste from, and change it, quite literally. Simply **overwhelm** that past memory with your new chosen frequency vibration energy.

Let's say the new vibratory state you prefer over your memory is love, joy, support and respect. Take those frequencies (*or any one of them*) and overwhelm that moment in your past with the particular frequency that you chose and see what happens. Begin "*energetically massaging*" your image of the past with a different vibratory state. Overwhelm it. Insert it. Superimpose your new vibratory state onto your image, onto the memory from the past that you don't really enjoy. Exude the energy of love and light and the vibration of aliveness and respect and support and of being loved. The more you superimpose this into the past, the more you start to see immediate change in the present. If it does not make sense to your present memory of your timeline, simply allow that to be OK, and keep superimposing your chosen vibratory state onto that past image. Watch it change and allow this to be a permission symbol for you to feel that you have actually changed, in the present, to a different parallel timeline that is more aligned; that is of a freer, more transparent nature. More transparent to what? More transparent to your Heart's true desires; more transparent to who you truly want to manifest through your body-mind-individuation experience. As current Timelines Shift, this also shifts the coordinate location and energetic contents of the past and future timelines existing on the earth plane, all at the same time.

This is not a delusion, as such. This is actually you shifting into a parallel timeline, where you had a slightly different (*or a significantly different*) experience of a very similar moment. Now you are in a parallel reality which is of a higher vibratory state. A more evolved, more expanded version of the timeline of your

experience and alternate self. As you keep changing that past experience, you are actually, in the present, moving yourself into a different timeline. You are now shifting into a different timeline more dramatically than you would if you were shifting through parallel realities on *automatic pilot."*
So try this out and see what happens. Some things that may happen:

The same person who perhaps abused you, suddenly starts hugging you sincerely with no other meaning than to simply hug you and love you and support you unconditionally.

The boss who fired you, now suddenly gives you a big commission, or a reward, or a promotion.

References. Chapter 25

1. Looking Within Our Universe For Something Beyond. February 22, 2016. Caltech.edu
2. Spectral Variations of the Sky: Constraints on Alternate Universes R. Chary.. Dec 2015
3. Parallel Universes have been examined in great detail by Fred Alan Wolf. ISBN-10: 0671696017
4. A quantum-information theoretic analysis of three-flavor neutrino oscillations Quantum entanglement, nonlocal and nonclassical features of neutrinos Subhashish Banerjee, et al. Oct 2015
5. Article Ref - MIT scientists find weird quantum effects, even over hundreds of miles. Jennifer Chu. July 2016. MIT News. Mit.edu
6. Article Ref #2 - Physicists Detect Neutrinos in Superposition Hurtling Through Spacetime. Kelsey Houston Edwards. July 2016. PBS.org
7. Violation of the Leggett-Garg Inequality in Neutrino Oscillations J. A. Formaggio, D. I. Kaiser, M. M. Murskyj, and T. E. Weiss Phys. Rev. Lett. 117, 050402 – Published 26 July 2016

8. Atmospheric neutrinos and discovery of neutrino oscillations Takaaki Kajita. Apr 2010
9. Jeong, E.J., "Neutrinos must be tachyons," arXiv:hep-ph/9704311 (1997)
10. Apparent Lorentz violation with superluminal Majorana-tachyonic neutrinos at OPERA? F Tamburini and M Laveder. Feb 2012. The Royal Swedish Academy of Sciences
11. Exploration of possible quantum gravity effects with neutrinos II: Lorentz violation in neutrino propagation Alexander Sakharov et al. 2009
12. Comments on Musha's theorem that an evanescent photon in the microtubule is a superluminal particle. Hari SD.. July 2014
13. Electromagnetic fields as structure-function zeitgebers in biological systems: environmental orchestrations of morphogenesis and consciousness Nicolas Rouleau1,2 and Blake T. Dotta. Nov 2014
14. Evolution of consciousness. Eccles JC Proc Natl Acad Sci U S A. 1992 Aug 15; 89(16):7320-4.

15. Constraints and tests of the OPERA superluminal neutrinos Xiao-Jun Bi et al. Sept 2011
16. Article Ref – Wikipedia. CMB cold spot.
17. Technical Ref - Local Voids as the Origin of Large-Angle Cosmic Microwave Background Anomalies. I. Kaiki Taro Inoue and Joseph Silk. 2008
18. Consciousness in the universe: A review of the 'Orch OR' theory StuartHameroff and RogerPenroseb. 2013
19. Orchestrated Objective Reduction of Quantum Coherence in Brain Microtubules: The "Orch OR" Model for Consciousness. Stuart Hameroff & Roger Penrose. 1996.

Chapter 26. Microtubules and The Quantum Brain

Indeed if information received during remote viewing sessions occurs as a quantum process through non-locality/superposition, there must exist scientific evidence of this occurring.

As we covered earlier, Hameroff and Penrose state the brain operates in the quantum range. It may be that some dynamics of brain functioning occur via delocalized quantum wave function operation. This would account for the amazing feats of some people such as incredibly short reaction times of some athletes or for geniuses to solve problems via intuitive insights[1].

The amount of activity occurring in microtubules at the nanosecond scale allows for a potentially huge increase in the brain's computational capacity. Biological cells contain approximately 107 tubulins (*Yu & Bass 1994*)[2], allowing for nanosecond switching in microtubules to take place at approximately 1016 operations per second (per neuron). This huge capacity may account for the behaviors of the single celled organisms Paramecium, which elegantly avoid obstacles, swim and find food and mates all without the aid of a nervous system or even synapses. This intuitive ability may be due to "*radar like*" echo effects taking place in their microtubules with the large computational capacity processing the information received from the echoes.

Summary
The universe is communicating with itself, much like the neurons in our brain communicate with one another. Each neuron in our brain consists of a network of microtubules. At certain times our mind may be tapping into this universal galactic communications network during remote viewing.

How many Neurons does the brain contain?
There are approximately 100 billion neurons in a human brain[3]. It is theorized that within these neuron networks exist precognitive circuits that have tuned themselves to function as adaptive responses for remote events. In the future we may witness a merging of artificial intelligence with artificial precognition technologies. Perhaps the large neurons gathered from whales or other large animals will be arranged in such a way as to create a precognition signal that can be recorded electronically before a future event occurs. This would be a huge benefit for aircraft, crime detection or other technologies that involve safety and security.

Further Reading

Quantum computation in brain microtubules? The Penrose–Hamero? 'Orch OR' model of consciousness. Stuart Hameroff Departments of Anesthesiology and Psychology, The University of Arizona, Tucson, AZ 85724, USA.

Why Meditation may strengthen the connection with the Liquid Crystalline structures within our body

Living organisms, including a large part of the human body, exist in a liquid crystalline state (*Ho, 1999; 1998b*)[4].

Liquid crystals encompass a wide range of fluidity, from semi-solid proteins to solid crystal to gel-like cellular fluids. Compared to calcium phosphate crystals in bones which are solid, collagen in human bone is semi-solid and is thus referred to as a liquid crystal. Biological organisms such as ourselves are composed mostly of liquid crystalline materials including cell membranes, connective tissue and other tissues that fill up the spaces between the organs.

Liquid crystals, used in calculator displays, possess ideal properties that allow them to communicate at the intercellular level. Hence a coherent liquid crystalline human body allows for intercommunication that is non-local and instantaneous (*Ho, 1998a*)[5].

Are Microtubules Quantum Computers?
Tuszynski & Brown state that microtubules may be quantum computers[6]. A quantum computer described by Lloyd (1993) is described as having arrays of weakly coupled quantum systems with sequences of electromagnetic pulses that cause transitions in quantum states within a crystal lattice[6]. The existence of specific microtubule lattice geometry (*such as helical patterns that mimic the Fibonacci series*) and massive parallelism may also facilitate quantum error correction in the future design of quantum computers.

Further Reading
Quantum computation in brain microtubules? The Penrose–Hameroff 'Orch OR' model of consciousness By Stuart Hameroff

Facts about Microtubules

- Neuronal microtubules are specifically necessary for consciousness
- Quantum coherence exists in microtubules
- Microtubule-based cilia/centriole structures are quantum optical devices

A detailed interior of a Microtubule
Each microtubule contains a hollow cylindrical

tube made of tubulin proteins. The outer core measures 25n. There are millions of these proteins along the microtubule of the brain's neuron and they are connected in such a way that create the neuron firing process[7].

Each microtubule contains individual peanut-shaped dimmers known as tubulins that are electrical dipoles that measure between 4 and 8 nanometers. They are rigid and contain a hexagonal lattice, each with its own internal coaxial hole. This hole is what allows liquids based on simple molecules and lipids to circulate. This circulation causes changes in dielectric functions and linear responses. This in turn changes the individual tubulin's many body states[7]. It is interesting to note that most antidepressants will target lipid rafts [8].

Matsuyama and Jarvik (1989) have linked Alzheimer's disease to microtubule dysfunction[9]. This would make sense because Alzheimer's is characterized as a condition where the moisture in the brain "*dries up*" and microtubules contain within them resonating water.

Research has shown water inside microtubules contains conditions for the formation of macroscopic coherent quantum states[10]. This allows for water molecules to oscillate in phase with an electromagnetic field. It is the coherent interaction that occurs between the water's electric dipoles and the existing electromagnetic field that generates stable and ordered structures in macroscopic spatial regions. This coherence causes the photons to go into "*self-induced transparency*" which penetrate the optical medium and propagate it. This could lead to remarkable potential such as coherent optical supercomputers.

Further **Reading**
Possible Existence of Superluminal Photons Inside Microtubules and the Resulting Explanation for Brain Mechanism Takaaki Musha1, 2, Luigi Maxmilian Caligiur. October 2015

Quantum Vacuum Dynamics, Coherence, Superluminal Photons and Hypercomputation in Brain Microtubules Luigi Maxmilian Caligiuri and Takaaki Musha. Applied Numerical Mathematics and Scientific Computation

Quantum Coherence In Microtubules: A Neural Basis For Emergent Consciousness? Stuart R. Hamerof. Journal of Consciousness Studies, 1, No.1, Summer 1994, pp. 91ñ118

Quantum optical coherence in cytoskeletal microtubules: implications for brain function Mari Jibu et al. 1994

Quantum Mechanics In Cell Microtubules: Wild Imagination Or Realistic Possibility ?N.E. Mavromatosa and D.V. Nanopoulosb. August 18-22 1997. Structural Biology

Nanotubes and microtubules as quantum information carriers Alan G. Michette et al. October 2004

Is quantum brain dynamics involved in some neuropsychiatric disorders? Author links open overlay panel. E. Pessa et al. May 2000

Electronic QED coherence in brain microtubules LUIGI MAXMILIAN CALIGIURI. Foundation Of Physics Research Center. WSEAS TRANSACTIONS On BIOLOGY And BIOMEDICINE

Consciousness and Neurons
Papers published in physics of life reviews state that consciousness comes from deep fine scale activities occurring inside the brain's neurons[11] [12] [13]. Hence, neurological, cognitive and mental conditions may arise from microtubule vibrations that interact with the quantum field. Quantum coherence may be occurring via interactions involving ions which pass through ion channels inside the neuronal membrane (*the quantum Icing model*).

Microtubules in plants
Microtubules in plants create right handed arrays in the plant's roots and are also responsible for the clockwise **rotation**/direction (*both left and right*) witnessed in their growth pattern[14] [15] [16]. They are also responsible for overcoming earth's gravity [17].

Piezoelectric Properties of Microtubules
Microtubules also behave as "*electrets*" consisting of oriented dipoles that are predicted to have piezoelectric properties (*Athenstaedt, 1974; Mascarenhas, 1974*). They are also reputed to have ferroelectric properties (*Tuszynski et al, 1995*)[19]. I show in my second book of remote viewing **Remote Viewing. The Complete User's Manual on Experiencing Future Consciousness** that piezoelectricity plays a key role in successful associative remote viewing sessions. This is due to the fact that before an earthquake, piezoelectricity increases[20], acting as a precursor to an earthquake. Hence using piezoelectricity during ARV sessions, helps detect future signals. What is also interesting is the material grapheme can be used to generate piezoelectricity [21] and grapheme has been used to modulate linearly polarized infrared light[22].

Microtubule Structures found in Basalt and Pumice Stone

Basalt and pumice stone both display similar structure characteristics to microtubules in that they both exhibit small circular cavities that absorb and trap water moisture.

Microtubule   Basalt   Pumice Stone

Microtubule structures have been found in rocks on the sea floor over 3.5 billion years old. They are similar in structure to microtubules created by organisms found in today's sea floor pillow basalt rocks[23]. Another interesting point is the microtubules in our brain measure 60 micros in length and the size of the rock microtubules mentioned in the study measure 50 micrometers in length[24]. As a side note, we have found that holding a piece of pumice stone in our hands seems to reduce the time it takes for Heart Math coherence to take place. It may be that the cavities in the pumice stone resonate with the EMF field emitted form the heart, creating an enhanced EMF field emitted by the heart.

A History of Microtubules

Microtubule shaped structures have been around longer than man and they were utilized by organisms early in earth's history.

The size of the organisms that existed at the beginning of the Cambrian explosion are very similar in size to what we would predict as being necessary for a waking conscious experience. In summary, they existed as tiny sea urchins and small worms[25].

The spines of sea urchins closely resemble the complexes of the microtubules in our brain[26]. Could it be that these were earth's first conscious entities and that one day in earth's past history quantum coherence reached a threshold in the nervous systems of worms or the spines of sea urchins to cause "*self-collapse*", suddenly causing conscious experience to take place?

Microtubules have been researched for their properties in spine development [27] [28]. It is also interesting to note that the food dye coloring called brilliant blue G (BBG) has been used to heal spinal cords in mice and also for healing optic nerve injuries [29].

Further Reading

Quantum optical coherence in cytoskeletal microtubules: Implications for brain function.

References. Chapter 26

1. Linoleic acid: Is this the key that unlocks the quantum brain? Insights linking broken symmetries in molecular biology, mood disorders and personalistic emergentism. Massimo Cocchi and Gustav Bernroider. April 2017
2. Remarks on the Number of Tubulin Dimers Per Neuron and Implications for Hameroff-Penrose Orch OR Danko Georgiev. NeuroQuantology | December 2009 | Vol 7 | Issue 4 | Page 677-679
3. Equal numbers of neuronal and nonneuronal cells make the human brain an isometrically scaled-up primate brain. Azevedo FA et al. April 2009
4. Quantum Coherent Liquid Crystalline Organism Dr. Mae-Wan Ho Invited lecture at European Quantum Energy Medicine Conference, Copenhagen, 19th September, 2008
5. A Quantum Biomechanical Basis for Near-Death Life Reviews. Thomas E. BeckJanet E. Colli. March 2003
6. Quantum computation in brain microtubules? The Penrose–

Hameroff 'Orch OR' model of consciousness By Stuart Hameroff
7. Desai A. & Mitchison TJ (1997). "Microtubule polymerization dynamics.". Annu Rev Cell Dev Biol. 13: 83–117. PMID 9442869. doi:10.1146/annurev.cellbio.13.1.83
8. Antidepressants Accumulate in Lipid Rafts Independent of Monoamine Transporters to Modulate Redistribution of the G Protein, Samuel J. Erb, et al. July 2016
9. Hypothesis: microtubules, a key to Alzheimer disease. S S Matsuyama and L F Jarvik. Oct 1989
10. Possible Existence of Superluminal Photons Inside Microtubules and the Resulting Explanation for Brain Mechanism Takaaki Musha1, 2, Luigi Maxmilian Caligiur. October 2015
11. Article Ref - Discovery of quantum vibrations in 'microtubules' inside brain neurons supports controversial theory of consciousness. Science Daily. January 2014
12. Tech Ref - Stuart Hameroff and Roger Penrose. Consciousness in the universe: A review of the 'Orch

OR' theory. Physics of Life Reviews, 2013 DOI: 10.1016/j.plrev.2013.08.002
13. Tech Ref #2 - Stuart Hameroff, Roger Penrose. Consciousness in the universe. Physics of Life Reviews, 2013; DOI: 10.1016/j.plrev.2013.08.002
14. Helical Microtubule Arrays and Spiral Growth Clive Lloyd and Jordi Chan.)ct 2002
15. Microtubules in Plants Takashi Hashimoto. Apr 2015
16. Helical microtubule arrays in a collection of twisting tubulin mutants of Arabidopsis thaliana Takashi Ishida, et al. May 2007
17. Cortical microtubules are responsible for gravity resistance in plants Takayuki Hoson, et al. June 2010
18. Information Processing in Microtubules: Biomolecular Automata and Nanocomputers. Stuart R. HameroffSteen Rasmussen.
19. Ferroelectric Behavior in Microtubule Dipole Lattices: Implications for Information Processing, Signaling and Assembly/Disassembly. Jack Adam Tuszynski et al. June 1995

20. recursory changes in seismic velocity for the spectrum of earthquake failure modes M.M. Scuderi, et al. Aug 2016
21. Engineered Piezoelectricity in Graphene Mitchell T. Ong and Evan J. Reed. Dec 2011.
22. Efficient modulation of orthogonally polarized infrared light using graphene metamaterials Yudong Cui and Chao Zeng. March 2017
23. Signs of Life Found in Ancient Lava By Richard A. KerrApr. 22, 2004 , 12:00 AM. www.sciencemag.org
24. Microtubules in basalt glass from Hawaii Scientific Drilling Project #2 phase 1 core and Hilina slope, Hawaii: evidence of the occurrence and behavior of endolithic microorganisms. Walton AW. Aug 2008
25. Growth, interaction and positioning of microtubule asters in extremely large vertebrate embryo cells T.J. Mitchison et al. Aug 2012
26. Consciousness, Microtubules and The Quantum World. Interview with Stuart Hameroff, MD, in Alternative Therapies (May 1997 3(3):70-79 by Bonnie Horgan).

27. Microtubules in Dendritic Spine Development and Plasticity Jiaping Gu and James Q. Zheng. Dec 2009
28. Microtubule stabilization reduces scarring and enables axon regeneration after spinal cord injury Farida Hellal, et al. Jan 2011
29. Brilliant blue G treatment facilitates regeneration after optic nerve injury in the adult rat. Ridderström M1, Ohlsson M. Dec 2014

Chapter 27. Microtubule and Essential Oils

The Action of Essential Oils at the Microscopic Scale

Essential oils exhibit hydrophobicity. Hydrophobicity is the ability for an essential oil to partition with lipids present in cell membranes of mitochondria and bacteria. As it does so, it renders the membrane more permeable. ie: making them more flexible[1] [2] [3].

The ability of hydrocarbons to interact with a cell's membrane allows for easy penetration of carvacrol into the cell[4] [5]. Pei et al. stated the synergistic effects of carvacrol with eugenol and thymol with eugenol may be due to the fact that thymol and carvacrol cause disintegration in the outer membrane of E. coli[6]. This allows for eugenol to enter the cytoplasm and then combine with the proteins. The synergistic effect of cinnamaldehyde/eugenol may be due to the interaction with different enzymes or proteins[6].

Geraniol
Geraniol enhances resistance of Gram-negative bacterial species (*Enterobacteraerogenes, E. coli, P. aeruginosa*)[7].

Overall **Gram-positive** bacteria are more susceptible to essential oils. This is due to the fact that Gram-negative bacteria contain a dense rigid outer membrane, rich in lipopolysaccharide[8].

Further Reading

Geraniol Restores Antibiotic Activities against Multidrug-Resistant Isolates from Gram-Negative Species. Vannina Lorenzi, et al. March 2009

Dietary Geraniol by Oral or Enema Administration Strongly Reduces Dysbiosis and Systemic Inflammation in Dextran Sulfate Sodium-Treated Mice Luigia De Fazio et al. March 2016

The Terpenes

The most widely represented class of hydrocarbons in essential oils is the terpenes. Terpenes are one of the major constituents for a plant's defensive system, especially airborne terpenes [9]. The protection properties of Terpenes include drought, heat and biotic stresses and pathogen and herbivore attacks. We could say Terpenes act like antibodies, akin to the way our body's immune system fights unwanted invaders. What is most interesting is the scent known as **Geosmin**, which occurs just before it rains, is a terpene [10] and they also are found in Cyanobacteria[11].

Borneo

Borneol is found in rosemary and acts as a natural

insect repellent and is commonly used in traditional Chinese medicine[12]. What is most interesting is the anti-aging herb, Chrysanthemum which is also good for the eyes, contains approximately 12.5% Borneol and 11% **camphor** [12]. The original name given to Borneo was Borneo camphor. It was named in the year 1842 by French chemist Charles Frédéric Gerhardt[13]. Borneol is able to be produced by the reduction of camphor. Many essential oils also contain Borneol[14].

Chrysanthemum has also been shown to be good for the heart by lowering hypertension and blood pressure (*Chrysanthemum morifolium extract improves hypertension-induced cardiac hypertrophy in rats by reduction of blood pressure and inhibition of myocardial hypoxia inducible factor-1alpha expression. T. Gao et al. Dec 2016. Pharm Biol. 2016 Dec;54(12):2895-2900. Epub 2016 Jun 7*).

Further Reading
Screening Study of Leaf Terpene Concentration of 75 Borneo Rainforest Plant Species: Relationships with Leaf Elemental Concentrations and Morphology Jordi Sardans et al. (2015)

Essential Oil Synergy
This section can be useful when making your own TXP formulas for enhancing coherence via the brain's microtubules.

Combinations of monoterpenes alcohols with phenolics produces synergistic effects. Examples include a combination of phenolics, such as carvacrol with thymol and both components with eugenol[15] [16]. These show strong synergistic effects against E. coli strains.

- Carvacrol and Linalool show synergism
- Eugenol and Linalool show synergism
- Menthol and Geraniol show synergism
- Pinene and Limonene show synergism
- Rosemary and Oregano show synergism
- Chloramphenicol and Thymol show synergism
- Rosemary and Camphor exhibit major synergy [17]
- Eugenol and Cinnamaldehyde, synergize with fluconazole to **fight fungi** infections[18]
- **Cinnamon and lemon** synergize with the anti fungal mediation Amikacin[19]

Geraniol

- Rose Geraniuim essential oil contains 29% Citronella and 12% Geraniol [21]. Geraniol can be found in Lemon Thyme essential oil (66.59%.) and Savory essential oil (50.4%)[22]. Geraniol synergizes with

Chloramphenicol[22]. Chloramphenicol is an antibiotic used to treat eye infections. Moon Carrot synergizes with chloramphenicol[22]. Moon Carrot contains up to 67% sesquiterpene hydrocarbons. Geraniol has also been shown to enhance the effectiveness of certain antibiotics. Geraniol has also been found in immortelle (*Helichrysum italicum*)[22]. The leaves of Helichrysum italicum exhibit a strong curry like smell.

- Cinnamon and Clove show synergism (*Antibacterial activity of cinnamaldehyde and clove oil: effect on selected foodborne pathogens in model food systems and watermelon juiceS. Siddiqua,. et al. 2015 Sep*).

Summary

It may be that certain essential oils, besides enhancing flexibility of the brain's microtubules, also enhance the flexibility of the arteries and improve the performance of the heart [24]. This is due to the fact that remote viewing during favorable solar weather conditions are conditions where the strength of the heart is stronger [25]. It is also interesting to note that blood pressure is lower at nighttime and higher during the daytime [26].

Seasonal Variation of Blood Pressure

Blood pressure is higher during winter compared to summer with the greatest effect taking place in older people. Blood pressure is also affected by the daily air temperature.

Reference

Seasonal variation in arterial blood pressure. P J Brennan et al. Oct 1982. Br Med J (Clin Res Ed). 1982 Oct 2; 285(6346): 919–923.PMCID: PMC1499985.

Essential Oils that Lower Blood Pressure

Ylang-Ylang[27]

Lavender [28]

Marjoram lowers sympathetic nervous system activity and also stimulates the parasympathetic nervous system [29].

Neroli creates comfort and emotional soothing. It is also effective in cardiac palpitations resulting from shock or fear[30].

Linalool reduces blood pressure [30B].

Citronella has been shown to significantly decrease blood pressure and reduce heart rate after inhalation[30c].

The Composition of Ylang Ylang Essential Oil
Benzyl acetate (4), linalool (8), and benzyl benzoate (3) were the three major compounds identified from the essential oils of Ylang Ylang with the percentage of 18.2%, 14.1%, and 12.3%, respectively.

Reference
Essential oils with insecticidal activity against larvae of Aedes aegypti (Diptera: Culicidae). Vera SS, Zambrano DF, Méndez-Sanchez SC, Rodríguez-Sanabria F, Stashenko EE, Duque Luna JE. Parasitol Res. 2014 Jul; 113(7):2647-54.

Another research study revealed the main components in ylang-ylang essential consisted of farnesyl acetate, benzyl acetate, geranyl acetate, cinnamyl acetate, methylanisole, methyl benzoate, benzyl benzoate, **linalool, geraniol** and benzyl salicylate. Linalool was the main component (28%) and caused the strong floral scent of ylang-ylang. Hydrocarbons in ylang-ylang consisted mainly of sesquiterpenes and monoterpenes with germacrene D and β-caryophyllene showing 63% of **hydrocarbon** content. The sesquiterpenes γ-Muurolene and farnesyl acetate were also found in ylang-ylang essential oil.

The study also found that the higher grade of Ylang Ylang Essential Oil commonly called "*Ylang Ylang Extra*", or "*Ylang Ylang II*" contains high amounts of linalool, p-cresyl methyl ether (p-methylanisole), benzyl acetate, methyl benzoate and geranyl acetate. The other grades contain more sesquiterpene hydrocarbons such as germacrene D, β-caryophyllene and α-farnesene.

Reference
The pharmacology of benzyl alcohol and its esters. The effect of benzyl alcohol, benzyl acetate and benzyl benzoate when given by mouth upon the blood pressure, pulse and alimentary canal Charles M. Gruber, Ph.D., M.D. October 1923

Further reading
Effects of Ylang-Ylang aroma on blood pressure and heart rate in healthy men. Da-Jung Jung, et al. April 2013

Rose Oil has been found to provide a relaxed and calming effect, suggested as a way to relieve depression (*Relaxing effect of rose oil on humans.T. Hongratanaworakit. Nat Prod Commun. 2009 Feb;4(2):291-6*).

Barometric Air Pressure and Blood Pressure
Because our most successful ARV sessions would

always take place when the barometric air pressure had reached its peak, this may be due to its effect on blood pressure. A research study found that when the weather was cooler, there was an increase in blood pressure *(Effects of Simulated Heat Waves with Strong Sudden Cooling Weather on ApoE Knockout MiceShuyu Zhang,1 Zhengzhong Kuang,2 and Xiakun Zhang3, Int J Environ Res Public Health. 2015 Jun; 12(6): 5743–5757.Published online 2015 May 26. doi: 10.3390/ijerph120605743PMCID: PMC4483669)* and the accuracy of our associative remote viewing always peaked during early spring when temperatures were cool, but not freezing.

Another study found that when barometric air pressure was lower, people had more migraines and that rapid changes in barometric pressure caused an increase in blood pressure[32].

Seasonal Variation and Blood Pressure
A study found that during spring and winter there were differences in blood pressure[33].

Further Reading
Effects of aromatherapy on blood pressure, heart rate variability, and catecholamines in the pre-hypertension middle aged women. Jung YJ. Department of Nursing, Catholic University, Seoul, Korea, 2007.

The effects of the inhalation method using essential oils on blood pressure and stress responses of clients with essential hypertension. Hwang JH. . Taehan Kanho Hakhoe chi. 2006;36(7):1123-1134

Eugenol and linalool: Comparison of their antibacterial and antifungal activities Rehab Mahmoud Abd El-Baky and Zeinab Shawky Hashem. October, 2016

Bergamot and Blood Pressure
A study showed that mixing ylang ylang, lavender and bergamot lowered blood pressure [34] and five scientific studies reviewing Bergamot (*Peng et al., 2009; Seo, 2009; Chang and Shen, 2011; Liu et al., 2013; Ni et al., 2013*) found that Bergamot reduced blood pressure, reduced heart rate stress responses [35].

Synergistic Ratios
Thymol and Cinnamaldehyde exhibit synergy at the 1:1 ratio. Eugenol and Cinnamaldehyde exhibit synergy at lower levels of eugenol[36].

Summary
It may be that starting off a remote viewing session when blood pressure is slightly lower or being in an environment that that allows for lower blood pressure is advantageous. This is

because the mild stress put on the body's system during a remote viewing session may cause a gradual rise in blood pressure, hence starting off at slightly lower blood pressure may reduce this stress. Further research is necessary to see if this hypothesis holds true.

References. Chapter 27

1. The Cell: A Molecular Approach. 2nd edition. Cooper GM. Sunderland (MA): Sinauer Associates; 2000.
2. Antimicrobial Properties of Plant Essential Oils against Human Pathogens and Their Mode of Action: An Updated Review Mallappa Kumara Swamy, et al. Dec 2016
3. Effect of Essential Oils on Pathogenic Bacteria FilomenaNazzaro, et al. Nov 2013
4. Effect of Essential Oils on Pathogenic Bacteria FilomenaNazzaro, et al. Nov 2013
5. Essential Oils in Food Preservation: Mode of Action, Synergies, and Interactions with Food Matrix Components Morten Hyldgaard, et al. Jan 2012
6. Evaluation of Combined Antibacterial Effects of Eugenol, Cinnamaldehyde, Thymol, and Carvacrol against E. coli with an Improved Method. Ruisong Pei et al. Sept 2009
7. Dietary Geraniol by Oral or Enema Administration Strongly Reduces Dysbiosis and Systemic Inflammation in Dextran Sulfate Sodium-Treated Mice Luigia De Fazio et al. March 2016
8. Cell wall structure of Gram-negative bacteria, Gram-positive bacteria, mycobacteria and fungi.. Nature Revies. Nature Reviews Microbiology 13, 620–630 (2015)
9. Plant terpenes: defense responses, phylogenetic analysis, regulation and clinical

applications Bharat Singh and Ram A. Sharma. April 2014
10. Terpene synthases are widely distributed in bacteria Yuuki Yamada et al. October 2014
11. Terpenoids and Their Biosynthesis in Cyanobacteria BagmiPattanaik and Pia Lindberg. Jan 2015
12. Volatiles of Chrysanthemum zawadskii var. latilobum K Kyung-Mi Chang and Gun-Hee Kim. Sept 2012
13. Charles Frédéric Gerhardt. Wikipedia
14. Plants containing borneol Archived 2015-09-23 at the Wayback Machine. (Dr. Duke's Phytochemical and Ethnobotanical Databases)]
15. Essential Oils in Combination and Their Antimicrobial Properties Imaël Henri Nestor Bassolé and H. Rodolfo Juliant. Molecules 2012, 17, 3989-4006; doi:10.3390/molecules17043989
16. Stratum corneum absorption and retention of linalool and terpinen-4-ol applied as gel or oily solution in humans. Cal K, Krzyzaniak M. June 2006
17. Penetration-enhancement underlies synergy of plant essential oil terpenoids as insecticides in the cabbage looper, Trichoplusiani Jun-Hyung Taka,1 and Murray B. Isman1
18. Antibiofilm activity of certain phytocompounds and their synergy with

fluconazole against Candida albicans biofilms. Khan MS. Ahmad. March 2012
19. Increasing antibiotic activity against a multidrug-resistant Acinetobacterspp by essential oils of Citrus limon and Cinnamomumzeylanicum. Guerra FQ, Mendes JM, Sousa JP, Morais-Braga MF, Santos BH, MeloCoutinho HD, Lima Ede O Nat Prod Res. 2012; 26(23):2235-8.
20. Essential Oils and Their Components as Modulators of Antibiotic Activity against Gram-Negative Bacteria PetrutaAelenei, et al. July 2016
21. Rose geranium essential oil as a source of new and safe anti-inflammatory drugs. Boukhatem MN et al. Oct 2013
22. Essential Oils and Their Components as Modulators of Antibiotic Activity against Gram-Negative Bacteria PetrutaAelenei, et al. July 2016
23. Influence of cinnamon and clove essential oils on the D- and z-values of Escherichia coli O157:H7 in apple cider. Knight KP1, McKellar RC. Sept 2007
24. Geraniol blocks calcium and potassium channels in the mammalian myocardium: useful effects to treat arrhythmias. deMenezes-Filho JE. et al. Dec 2014
25. Influence of local geomagnetic storms on arterial blood pressure. Dimitrova S et al. Sept 2004

26. Clinical relevance of nighttime blood pressure and of daytime blood pressure variability. Palatini P et al. Sept 1992
27. Effects of Ylang-Ylang aroma on blood pressure and heart rate in healthy men Da-Jung Jung, et al. April 2013
28. Essential Oil Inhalation on Blood Pressure and Salivary Cortisol Levels in Prehypertensive and Hypertensive Subjects In-Hee Kim et al. November 2012
29. Essential Oil Inhalation on Blood Pressure and Salivary Cortisol Levels in Prehypertensive and Hypertensive Subjects In-Hee Kim et al. Nov 2012
30. Essential Oil Inhalation on Blood Pressure and Salivary Cortisol Levels in Prehypertensive and Hypertensive Subjects In-Hee Kim, et al. Nov 2012
30B - Antioxidant activity of linalool in patients with carpal tunnel syndrome Geun-HyeSeol et al. July 2015
30C - The Harmonizing Effects Of Citronella Oil On Mood States And Brain Activities. WinaiSayowan Et Al. April 2012
31. The effect of ambient temperature and barometric pressure on ambulatory blood pressure variability. Jehn M et al. Nov 2002
32. Examination of fluctuations in atmospheric pressure related to migraine Hirohisa Okuma, et al. Dec 2015
33. Evaluation of the impact of atmospheric pressure in different seasons on blood

pressure in patients with arterial hypertension. Kamiński M et al. 2016
34. [The effects of the inhalation method using essential oils on blood pressure and stress responses of clients with essential hypertension]. Hwang JH1. December 2006
35. Citrus bergamia essential oil: from basic research to clinical applicationMichele Navarra et al. March 2015
36. Penetration-enhancement underlies synergy of plant essential oil terpenoids as insecticides in the cabbage looper, Trichoplusia ni Jun-Hyung Taka, and Murray B. Isman. Feb 2017

Chapter 28. Essential Oils and their Effects on Brainwave Activity

Linalool and Brainwave Patterns
Inhalation of Linalool has been found to produce a decrease in beta waves after work, compared to before work, probably as a relaxation effect[1]. Alpha brainwaves are associated with calmness and Beta brainwaves occur when the brain is alert.

Another study found that inhaling the scent of **Jasmine** enhanced **beta wave** activity in the brain[2]. Also inhaling Angelica Gigas root essential oil was found to cause significant changes in the absolute low beta activity of the brain [3].

Research has also shown that men did not show any significant change when inhaling geosmin (*the scent smelled before rain comes*), however changes did occur when men inhaled **methylisoborneol**, which is a substance similar to geosmin. Methylisoborneol caused **significant increases in alpha brainwave activity**, especially in the frontal region. Women however were affected by both geosmin and methylisoborneol after inhalation. They showed decreases in their high beta and low beta brainwave activity. Hence, some women may have a better sense of smell compared to men. The meta-study concluded that the scents of geosmin and methylisoborneol produced major alternations in brainwave activity according to gender[4].

Research by Haehner showed that women were able to identify lime, orange and lemon when they inhaled them and that they exhibited behavioral changes compared to men. The study also found that women were more sensitive to odor molecules compared to men (*Gender Differences in Electroencephalographic Activity in Response to the Earthy Odorants Geosmin and 2-MethylisoborneolMinju Kim et al. August 2017*).

Another study found that women were able to identify and detect different tastes and odors compared to men [4]. Hence, could this be another reason women are more intuitive then men?, showing a link between intuition and scents?

What is Methylisoborneol?
Methylisoborneol is a derivate of the substance Borneol and it is found in large quantizes in blue-green **algae** living in saline lakes in South Western Manitoba, Canada [5]. The smell of Methylisoborneol consists of a unique earthy/musty odor and it is sometimes responsible for the earthy smell of some drinking water [5]. It can be found in **turmeric** essential oil from Pakistan [7] and **tea tree** essential oil [7]. It is also present in cork taint in winemaking[8].

Both Methylisoborneol and Geosmin are present in algae blooms[9]. Some types of algae, especially blue-green types, produce both Methylisoborneol and Geosmin[10]. Both Methylisoborneol and Geosmin are responsible for the musty earthy scent smelled when one is in the presence of an algae bloom.

If cyanobacteria dies, it will release Methylisoborneol. Cyanobacteria can be purchased as a food supplement. Two types of these supplements are Aphanizomenon flos-aquae and Spirulina (Arthrospira Platensis).

What is Geosmin?
Geosmin is produced by large amounts of soil bacteria when it rains after a dry spell of weather.. It is also found in some **algae**. Geosmin is also experienced as an off-flavor component in some rural drinking water supplies such as Phoenix Arizona.

One of the most interesting discoveries of working with associative remote viewing is the discovery that certain **scents that have strong odors**, especially those in the parts per million range seem to enhance remote viewing. The algae water that we have nearby during our remote viewing sessions contains substances similar to Methylisoborneol which we believe is enhancing our remote viewing sessions. As we showed previously, Methylisoborneol enhances alpha brainwave activity.

Algae dissolves Radiation
Another discovery which I outline in much greater detail in the second edition of remote viewing **Remote Viewing. The Complete User's Manual on Experiencing Future Consciousness**, was the substance Tritium which is a mild beta emitter (*a form of radiation*). Tritium enhances remote viewing we believe due to its luminescent properties. It is also interesting to note that radiation lowers blood pressure [11]. Tritium radiation is so harmless it is sold as key chains online. As long as the Tritium remains sealed, much like the radioactivity in a smoke detector, it is harmless. Only if you breathe in the vapors is it toxic. It is interesting to note that Algae has been proposed as a solution to radioactive waste[12].

MSG
MSG is also called monosodium glutamate and also exists **as a very strong scent**. MSG is used as a flavor enhancer for meats. It is also found in cheese and tomatoes. Glutamates are responsible for brain neurotransmitters [13]. Studies have also shown that monosodium glutamate enhances GABA levels[14].

References. Chapter 28

1. Odor distinctiveness between enantiomers of linalool: difference in perception and responses elicited by sensory test and forehead surface potential wave measurement. Sugawara Y et al. Feb 2000
2. Effects of Inhaled Rosemary Oil on Subjective Feelings and Activities of the Nervous System Winai Sayorwan, et al. Dec 2012
3. Effect of essential oil and supercritical carbon dioxide extract from the root of Angelica gigas on human EEG activity. Sowndhararajan K et al. Aug 2017
4. Gender Differences in Electroencephalographic Activity in Response to the Earthly Odorants Geosmin and 2-Methylisoborneol. Minju Kim et al. August 2017
5. Geosmin and 2-Methylisoborneol from Cyanobacteria in Three Water Supply Systems George Izaguirre et al. Marsh 1982
6. Composition and Antimicrobial Activity of the Essential Oil from Leaves of Curcuma longa L. Kasur Variety Z. Parveen, et al. Feb 2013
7. The role of flavor and fragrance chemicals in TRPA1 (transient receptor potential cation channel, member A1) activity associated with allergies Satoru Mihara and Takayuki Shibamoto. March 2015

8. Cork Taint of Wines: Role of the Filamentous Fungi Isolated from Cork in the Formation of 2,4,6-Trichloroanisole by O Methylation of 2,4,6-Trichlorophenol María Luisa Álvarez-Rodríguez, et al. Dec 2002
9. HARMFUL ALGAL BLOOMS: Musty Warnings of Toxicity Kris S. Freeman. Nov 2010
10. Geosmin and 2-Methylisoborneol from Cyanobacteria in Three Water Supply Systems GEORGE IZAGUIRRE,* CORDELIA J. HWANG, STUART W. KRASNER, AND MICHAEL J. McGUIRE Water Quality Branch, The Metropolitan Water District of Southern California, La Verne, California 91750 Received 16 July 1981/Accepted 10 November 1981
11. Fall in blood pressure during radiation therapy. Larsson LE, et al.June 1976
12. Algal Bioremediation of Waste Waters from Land-Based Aquaculture Using Ulva: Selecting Target Species and Strains Rebecca J. Lawton et al. Oc 2013
13. Glutamate as a neurotransmitter in the brain: review of physiology and pathology. Meldrum BS. April 2000
14. The two-step biotransformation of monosodium glutamate to GABA by Lactobacillus brevis growing and resting cells. Zhang Y et al. JUne 2012

Chapter 29. The Thalamus Region of the Brain and Remote Viewing

Our physical bodies are surrounded by a mantle of natural electromagnetic energy which our eyes sense. 90% of this electromagnetic energy becomes filtered out in the thalamus region of our brain[1]. A Russian researcher working in collaboration with Indian researcher Shah, and using Kirlian technology, took pictures of the electromagnetic energy field around the body and showed that disease appeared first in the body's aura which then became transferred to the physical body [1]. The process took between 6 to 8 months prior to the appearance of the disease in the physical body[1].

The thalamus region deals with attention and the thalamus is controlled by our brain's limbic system. The limbic system governs our decision-making, our emotions and our memory. Information that is filtered through this part of the brain forms a visual mental model which we experience as reality.

Recommended Reading
A Critical Analysis of Chromotherapy and Its Scientific Evolution. Samina T. Yousuf Azeemi* and S. Mohsin Raza

Research has shown that certain regions in the brain's left lateral parietal cortex and thalamus, behaved differently when participants thought about the future and the past, compared to thinking about the present[2]. The cortex region of the brain is used for visualization.

Let's next look at the 5 main regions of the brain responsible for the processing of information.

- The limbic system generates information processing
- The **hypothalamus** processes **memory**
- The thalamus governs our attention
- The reticulate gyrus governs information processing
- The amygdala, contained in the limbic system, governs emotions and pleasure

Summary
The multiple avenues of communication between the biophysical body and the brain during remote viewing exist in a hierarchical order.

References. Chapter 29

1. A Critical Analysis of Chromotherapy and Its Scientific Evolution Samina T. Yousuf Azeemi* and S. Mohsin Raza. Dec 2005
2. Consciousness of subjective time in the brain Lars Nyberg. et al. June 2010. Mental time travel.

Chapter 30. Tungsten as a Photon Light Emitter

I mention in great detail in the second book on remote viewing Remote Viewing. The Complete User's Manual on Experiencing Future Consciousness, that the very durable metal Tungsten, was discovered by Dr. Kozyrev to exhibit strong effects on space/time[1]. Also our research found using Tungsten metal or minerals that contained Tungsten greatly enhanced the accuracy of our associative remote viewing sessions of the Dow Jones.

Transition Metal Dichalcogenide Research has shown that Transition Metal Dichalcogenide (TMD's) can emit quantum light via single photons of light[2][3] and is also used for transistors [4].

What is a TMD? TMD is short for Transition Metal Dichalcogenide, which is a semiconductor that can be layered a few atoms thick. TMD's are commonly combined with graphene and hexagonal boron nitride to create **van der Waals** heterostructures.

The following illustration is **Transition Metal Dichalcogenide**.

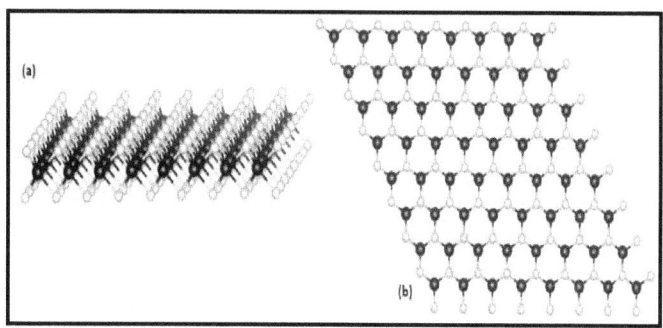

Tungsten disulfide and its closely related companion metal **Molybdenum** happen to be **TMDs**. These materials can easily be purchased online. Tungsten diselenide and Tungsten disulphide have been used to build quantum emitters[5].

Molybdenum in foods
Foods high in molybdenum include mung beans, lentils and peas, cereals and leafy vegetables.

Lunar Phase and Mung Beans
A research study found that the First Quarter moon after a full moon supermoon and near the new moon that the length of mung bean sprouts was affected. Also the potassium levels were also affected (*Macronutrient K Variation in Mung Bean*

Improve your Remote Viewing Accuracy Techniques using Quantum Microtubules

Sprouts with Lunar Phases. Kanan Deep and Raj Mittal. Nuclear Science Labratories, Physics Departent. Punjabi University, India. March 2014). This study shows that supermoon full moons have effects on mung bean roots. A supermoon occurs when the moon is closest to earth. It is also known as a perigee moon.

This makes sense because research has shown that magnetic variations are greater when the moon is in perigee (*The Moon's influence on the Earth's magnetism. Chapman, S.. SAO/NASA Astrophysics Data System (ADS)).*

Further Reading
Interaction of the moon with the earth's magnetosphere. Otto Schneider. March 1967

What is Tungsten Disulfide? Tungsten Disulfide contains a hexagonal structure similar in appearance to molybdenum disulfide and occurs naturally in the mineral **Tungstenite**.

Molybdenite crystals contain the same hexagonal structure as tungstenite, the only difference being tungstenite has a higher specific gravity. The atoms of Tungstenite are bonded to six selenium ligands with a pyramidal geometry. Tungsten Disulfide is also used to create a catalyst reaction when hydro-treating crude oil. Because of their ability to readily **absorb large amounts of light**, they are used in electrochemical solar cells, giving them the ability to pass 95% of the received light. Tungsten Disulfide's unique structure also allows it to create LED's of any color. The following picture is the molecular structure of Tungsten Disulfide.

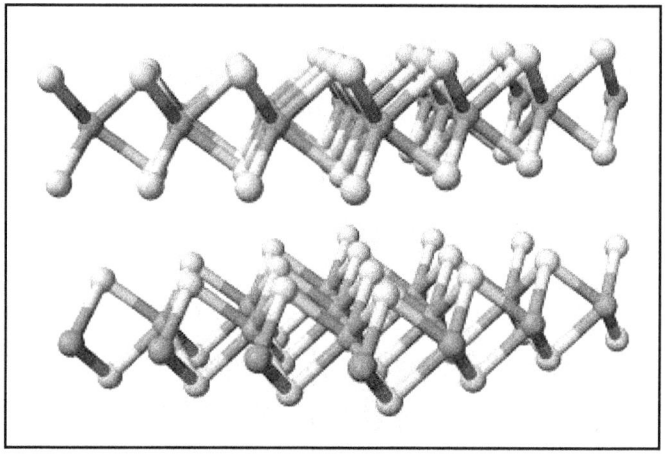

Further Reading
Electron transfer and coupling in graphene–tungsten disulfide van der Waals heterostructures. Jiaqi He et al. Nov 2014

Microtubules and the van der Waals force
Microtubules consist of a soft-matter structure with high regularity and a cylindrical symmetry. They exhibit this softness due to forces that underlie their polymer formation[6]. This structure is brought about by van der Waals forces and

weak hydrogen bonds[6]. The stronger covalent forces are due to the formation of polypeptide chains and specific amino acids[6].

This is a major finding, because van der Waals forces are a primary component of the weak force, which is utilized by the substance Tritium. I outline this connection in greater detail in my second remote viewing book.

The strength of the weak force and *"weak materials"* lie in their ability to absorb and hold high energies without consuming themselves, much like a catalyst reaction. These can be strong magnetic fields, high frequency or UV light.

One example is the rare elements, which are weak enough to cut with a knife, are used in low voltage light bulbs, which when stimulated by high frequency UV light, create a fluorescent glow that amplifies the light. Another example is paramagnetic materials will absorb a strong magnetic field, yet not retain the magnetic field. This allows for a dipole magnetic moment to occur. Tritium is a low gamma ray emitter, yet of high frequency.

One interesting thing I would like to point out is the hard mineral such as zircon is abundant in common beach sand. Zircon is a hard mineral used in sandpapers. Beach sand in general is soft and has a velvety feel. The soft (weak) rare minerals such as lanthanum are found in granite.

Granite in general is one of the hardest materials used for building very durable, long lasting structures. These are perfect illustrations of how the weak materials/forces are responsible for the overall strength of a substance.

Zircon occurs in most granites (*Chemical characteristics of zircon from A-type granites. K. Breiter. 2014*) and Zircon is a common accessory to trace mineral constituent of most granite and felsic igneous rocks.

Composition of Rare Earth in Common Minerals

Apatite is rich in the rare earth Yttrium[7]. Granite contains the rare earths Sphene, Apatite and K. Feldspar[7]. Schist and Zircon are rich in Ytterbium[7]. Allanite is rich in Lanthanum[7]. Xenotime is rich in Yttrium[7]. Basalt contains Europium (positive anomaly) and Rhyolite contains Europium (negative anomaly)[7]. Ilmenite contains Lanthanum, Gadolinium and Ytterbium[7]. Basalt has also been found to contain an abundance of lanthanide and yttrium, which is similar in composition to meteorites[8]. Europium is a very reactive rare earth element which is used to dope plastics to make lasers and it is also a good absorber of neutrons[9].

References. Chapter 30

1. **THE TORSION FIELD AND THE AURA** by Claude Swanson, Ph.D
2. Quantum Optics with Transition Metal Dichalcogenides. Quantum Photonics Group. Department of Physics Institute for Quantum Electronics
3. Large-scale quantum-emitter arrays in atomically thin semiconductors Carmen Palacios-Berraquero, et al. Feb 2017
4. Recent Advances in Electronic and Optoelectronic Devices Based on Two-Dimensional Transition Metal Dichalcogenides Mingxiao Ye, et al. June 2017
5. Large-scale quantum-emitter arrays in atomically thin semiconductors. Palacios-Berraquero C et al. May 2017
6. Nonlinear dynamics of C–terminal tails in cellular microtubules. Dalibor L Sekulić et al. July 2016
7. https://www.ncbi.nlm.niEcosystem Composition Controls the Fate of Rare Earth Elements during Incipient Soil Genesis Dragos G. Zaharescu, et al. Feb 2017
8. Some aspects of the geochemistry of yttrium and the lanthanides Geochimica et Cosmochimica Acta By: Michael Fleischer. .usgs.gov

9. https://education.jlab.org/i. Jefferson Labs. The Element Europium

Chapter 31. Microtubules and the Schuman Resonance

We show in book #2 of our 3 part series on remote viewing that precognition may be the result of an "*echo*" type effect caused by the Schuman resonance. Because the brain's microtubules are hollow and instruments used to measure the Schuman resonance are also hollow or tube like shaped, it would make sense that the sensations felt during precognition/remote viewing would have a radar-type echo effect associated with them.

A number of Biophysics studies states that living systems on earth are tuned into the background frequency of earth via the Schumann resonance. Perhaps the water in our bodies is what allows the microtubules to go into resonance with earth's Schuman resonance.

Proof that the Schuman resonance may be influencing microtubules is that long stainless steel cylinders are used as an antenna to pick up and measure earth's Schuman resonance frequencies. Microtubules in the body are hollow and Earth's Ionosphere contains within it a hollow cavity that goes into resonance from lightening strikes, which creates the Schuman Resonance. The following illustration shows how the various frequencies of the Schuman resonance *'bounce'* around earth's ionosphere.

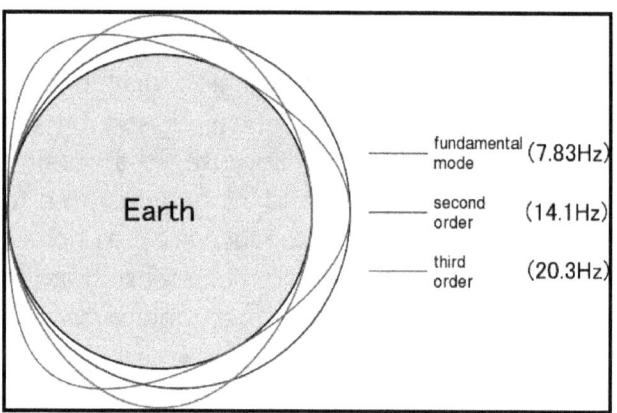

US Patent #20150102675 titled: Power Receiver for Extracting Power from Electric Field Energy in the Earth, shows that a hollow cylinder surrounding the ground shaft is used to tap into earth's Schumann resonances.

It may well be that the Schuman resonance excites the tubulin regions in microtubules, enhancing the clarity of remote viewing sessions during the right times, as we have always obtained our most accurate remote viewing sessions when the Schuman resonance is at more intense levels (*favorable solar weather conditions*)
. It is also interesting to note that an entire science fiction television series named "The

Time Tunnel" is about a tunnel shaped device that people enter into to travel through time.

The Schumann Resonance Affects the Parahippocampal gyrus

A study found the Schumann resonance has been shown to affect the brain's Parahippocampal gyrus[1]. The study concluded that certain conditions may cause effects in information processing that occur for brief periods of time. The study also concluded that both natural and technology-based variables affecting the Schumann resonance may affect human brain activity. These include dream-related memory consolidation and modifications of cognition [1]. Other studies have found that the parahippocampal region is also affected by geomagnetic activity and that the frequencies of 7.81Hz and approximately 20Hz, (which are the first and third harmonics of the **Schumann resonance) affected brainwave activity**[2]. Other studies found that theta brainwaves (4–7 Hz) in the right prefrontal sensor were correlated with the steady-state magnetic field of the earth[2].

Further Reading
Schumann Resonance Frequencies Found within Quantitative Electroencephalographic Activity: Implications for Earth-Brain Interactions.

References. Chapter 31

1. Human Quantitative Electroencephalographic and Schumann Resonance Exhibit Real-Time Coherence of Spectral Power Densities: Implications for Interactive Information Processing. Michael A. Persinger and Kevin S. Saroka.2015.
2. Similar Spectral Power Densities Within the Schumann Resonance and a Large Population of Quantitative Electroencephalographic Profiles: Supportive Evidence for Koenig and Pobachenko Kevin S. Saroka, David E. Vares, Michael A. Persinger. Jan 2016

Scott Rauvers

Chapter 32. How Tobacco, Photosynthesis and Manganese all relate to one another

Manganese and Photosynthesis
Manganese plays an important role in plant photosynthesis (*On the Role of Manganese in Photosynthesis Kinetics. Euglena gracilis et al. Jan 1971), (Light-Dependent Oxygen Metabolism Of Chloroplast Preparations Stimulation By Manganous lons. M. Helen et al. Department Of Botany, University Of Minnesota, Minneapolis*).
When Manganese is combined with lead sulfide it results in enhanced sensitivity in quantum dot solar cells (*Exploring the effect of manganese in lead sulfide quantum dot sensitized solar cell to enhance the photovoltaic performance Dinah Punnoose, et al.*) and Manganese is found in tobacco smoke (*Toxic Elements in Tobacco and in Cigarette Smoke: Inflammation and Sensitization R.S. Pappas, Ph.D. July 2011*) .

RH4032
RH4032 is commonly found in tobacco and also acts as a microtubule destabilizer and powerful anti-microtubule agent[1]. RH4032 also binds strongly to the roots of tobacco plants[1]. RH4032 is a Benzamide that is highly soluble in water. The

Benzamide molecule is shown in the following picture.

Elicitors

These are used by plants as a defense mechanism[2]. One of the first elictors was discovered by Gianinazzi and Kassans (1974) who found that polyacrylic acid, which absorbs and retains water, caused it to swell to many times its original volume[3]. Hence if our nervous system shows pre-sentient effects, especially during times of higher water moisture, perhaps plants are exhibiting similar presentiment effects using elictors. Plants are able to detect and respond to gravity, changes in temperatures and light.

Acetylsalicylic Acid

Acetylsalicylic Acid (Aspirin) has been found to increase a tobacco plant's resistance against TMV and cause an accumulation of PR proteins (*White 1979*)[4]. Salicylic acid comes from the inner bark of the willow tree and functions as a plant hormone. Salicylic acid can also be biosynthesized from the amino acid phenylalanine. Birch oil and Oil of Wintergreen contain large amounts of **Methyl Salicylate**[5]. Pictured is the molecular symbol for Salicylic Acid.

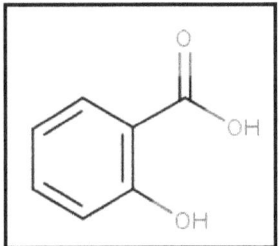

Phenylpropanoids

Phenylpropanoids are made by plants from Phenylalanine[6]. Dopamine production in the body comes from the amino acids phenylalanine and tyrosine[7]. Phenylalanine is the precursor to tyrosine. Phenylpropanoids give plants protection from UV light and defend against pathogens and berbivores. They also help facilitate pollen and act as scent generators for insects to pollinate the plant[8].

Reduced Iron Enhances Manganese Retention

I show in my second book on remote viewing **Remote Viewing. The Complete User's Manual on Experiencing Future Consciousness** that having lower levels of iron in the body enhances manganese levels and that manganese levels are a key part of successful remote viewing.

Research is showing that alterations in dopamine biology may enhance manganese neurotoxicity and that excessive long term levels of above average manganese can cause neurochemical disturbances. Other research has shown that brains that have decreased levels of GABA occur as a result of manganese and a research study involving mice found that an increase in their GABA levels occurred when they were exposed to manganese and had iron deficiency. The study concluded that manganese enhances GABA concentrations.

Reference
Manganese exposure alters extracellular GABA, GABA receptor and transporter protein and mRNA levels in the developing rat brain. Joel G. Anderson et al. Neurotoxicology. Nov 2008. Hence taking GABA boosting foods and foods high in Manganese may synergize with one another.

References. Chapter 32

1. Covalent Binding of the Benzamide RH-4032 to Tubulin in Suspension-Cultured Tobacco Cells and Its Application in a Cell-Based Competitive-Binding Assay David H. Young* and Veronica T. Lewandowski. Sept 2000
2. Role of Elicitors in Inducing Resistance in Plants against Pathogen Infection: A Review Meenakshi Thakur and Baldev Singh Sohal. Dec 2012
3. Water absorption, retention and the swelling characteristics of cassava starch grafted with polyacrylic acid. Witono JR et al. March 2014
4. The Effects of Salicylic Acid and Tobacco Mosaic Virus Infection on the Alternative Oxidase of Tobacco' Adrian M. Lennon. et al. Plant Physiol. (1997) 11 5: 783-791
5. SALICYLIC ACID. Toxnet. Toxicology Data Network
6. Phenylpropanoid. Science Direct. www.sciencedirect.com/
7. Dopamine precursors and brain function in phenylalanine hydroxylase deficiency. Lou HC
8. Flavonoids: biosynthesis, biological functions, and biotechnological applications María L. Falcone Ferreyra, et al. Sept 2012

Chapter 33. The TXP Formula

The TXP formula was derived from mixing various combinations of essential oils with one another and then using the EM Wave 2 device to find out which combination of essential oils caused the most rapid and longest lasting ability for the mind / heart to go into coherence (HRV). This formula is rubbed into the hands before practicing heart math to create mind/body coherence. A fresh batch is always made before each new ARV session.

- ➢ **2 oz water**
- ➢ **4 drops linalool essential oil**
- ➢ **10 drops geraniol (available from rose geranium essential oil)**
- ➢ **8 drops tea tree essential oil**
- ➢ **8 drops lemon myrtle essential oil**

Chapter 34. Favorable Environments and Solar Weather Conditions for Successful Associative Remote Viewing Sessions

Speaking from years of associative remote viewing experience, I have found the very best conditions that create successful associative remote viewing sessions are the ones that are shielded from electromagnetic radiation. Hence in Topanga, CA the sessions were done in a rural ravine shielded from general electromagnetic interference and in Hawaii I received the best results while in an old WW2 underground bunker shielded from all forms of electromagnetic radio signals.

Now let's take a more in-depth look at environments that are conductive to successful remote viewing sessions.

Geomagnetic Activity Levels
Accuracy of remote viewing is enhanced when earth's **geomagnetic energy** levels (**Middle Latitude Fredericksburg K-indices**) are between 7 and 11, especially when the geomagnetic activity has peaked and is descending and is not forecast to 'jump' to higher levels from the date of the remote session out until the target date. It is during this 'sweet spot' that we believe the brain may have an enhanced ability to

comprehend more information. Pictured is the '*sweet spot*' of geomagnetic energy.

Date	A	Middle Latitude Fredericksburg K-indices							
2017 03 23	9	4	3	2	2	1	2	1	1
2017 03 24	7	2	3	2	0	2	2	2	1
2017 03 25	3	0	2	0	0	2	2	1	1
2017 03 26	3	1	0	0	1	1	1	1	2
2017 03 27	34	2	3	5	5	5	4	5	4
2017 03 28	-1	5	5	3	3	3	2	3	-1

You can real time earth geomagnetic levels at the address below.

http://legacy-www.swpc.noaa.gov/ftpdir/indices/DGD.txt

As just mentioned, there must not be a sudden rise or excessive KP values all through until the target date, which is a maximum of 4 days out. You can get a visual graph of this by visiting the address below:

http://www.spaceweather.gc.ca/forecast-prevision/long/sflt-1-eng.php

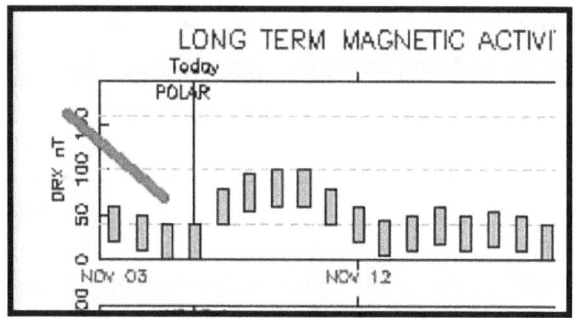

Additional favorable conditions

Solar wind speed at 350 or less with best results as it just begins declining into the 350 range. Wang Sheely Solar Wind Speed Forecast http://legacy-www.swpc.noaa.gov/ws/

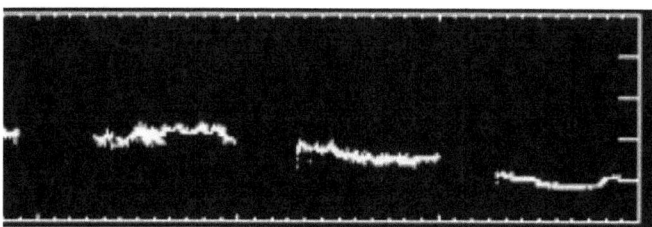

Barometric air pressure having peaked and heading towards low barometric air pressure.

10.7 cm radio flux rising / steady. This parameter seems to enhance the viewing of

numerical numbers during remote viewing.
http://legacy-www.swpc.noaa.gov/ftpdir/indices/quar_DSD.txt

Intuitive and emotional **biorhythm peaking**, especially when they both peak together at the same time. Numerous biorhythm charts can be found online.

Between 72 hours before or after a **first quarter or full moon** when more water moisture is present in the air. This seems to be one of the main parameters necessary for a successful associative remote viewing session.

Earth's Magnometer

One of the most telling signs that conditions are prime for a successful associative remote viewing session is earth's magnometer will have just come out of a period of major disturbance. This will occur most often as earth's geomagnetic activity is settling down after a peak in activity.

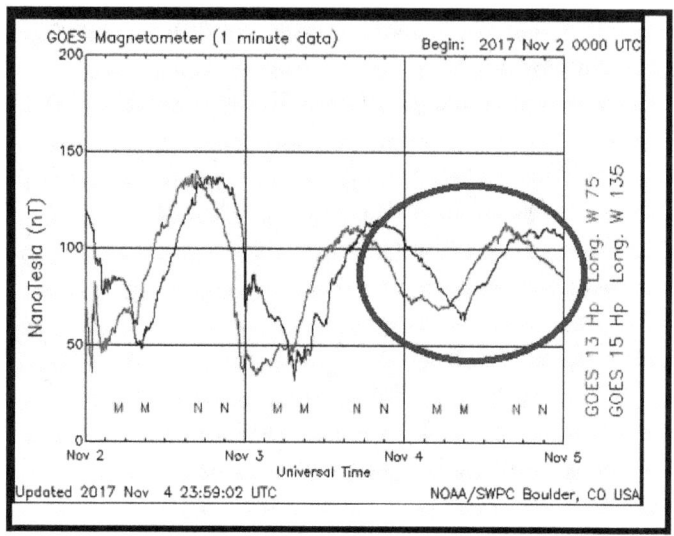

Real Time Magnometer Activity

http://www.swpc.noaa.gov/products/goes-magnetometer

Historical Solar Weather Data
http://legacy-www.swpc.noaa.gov/ftpdir/warehouse/

If it is hard to obtain heart math coherence before the ARV session, this is a good sign that one or more of the solar weather properties is off-key. One should recheck solar weather values after the ARV session and abandon the results if the properties have been found to be unfavorable. Dreams after an ARV session are a very good indicator of how successful the ARV session was.

Dark, foreboding dreams are a sign that the ARV session may have been a failure. Dreams that leave you feeling energized, light and are of good feelings are a good clue that the ARV session was a success.

Summary
This creative / intuitive energy flows though the universe and into the earth. As we go into heart math coherence (HRV) we go into coherence with our earth. While in coherence, this energy enters our heart where our brain deciphers the information.

Taking foods, herbs or inhaling essential oils good for the heart and it's arteries are most beneficial during this time. Especially lithium. Lithium has been shown to enhance the magnitude of delta and theta brainwaves and lithium is found at high levels in the minerals spodumene, petalite and amblygonite and lepidolite.

References
Lithium use in special populations. E. Mohandas and V. Rajmohan. July 2007

Lithium Resources of North America. James J. Norton And Dorothy Mckenney Schlegela Contribution To Economic Geology. Geological Survey Bulletin, 1027-

Brain oscillations in bipolar disorder and lithium-induced changes. Murat Ilhan Atagün. 2016. Neuropsychiatr Dis Treat. 2016; 12: 589–601. Published online 2016 Mar 7. doi: 10.2147/NDT.S100597. PMCID: PMC4788370

Favorable Gravity Bouguer Environments for Remote Viewing
After conducting associative remote viewing sessions in 3 locations over a period of 3 years and researching the gravity waves occurring in the region, the best regions for remote viewing are regions where there is **minimal to no gravity disturbances or anomalies**. You can research the Bouguer Gravity intensity in your region by visiting the address below.

https://mrdata.usgs.gov/general/map.html

The Tao

Chinese Taoism believes in an essence known as "*chi*" which exists as an interfering coherent energy that is photonic in nature and flows from the sun and stars. Because chi plays a role in intuition, regions and environments conductive to healthy chi are also regions where remote viewing accuracy is enhanced.

As we move out of the Age of Pisces, which was a Yang element and into the Age of Aquarius, which is a Yin element, we may find our intuitive abilities not as strong, which is why learning remote viewing as we enter the Age of Aquarius is a valuable tool. Yang properties are favorable environments for remote viewing. Let's take a look at environments which are conductive to the flow of Yang energy.

The Properties of Yang -

Energy flows downwards. High Pressure Environments (such as barometric high pressure, the bottom of a swimming pool, moon overhead etc.). Straight (less circular). Centripetal (moving inwards). Compact, compressed and high density. High Altitude. Colors are red, orange and yellow. Small size. Warm. Hard and Rough. Inland, away from large bodies of water. Syntropic or anti-entropic. Dry. Flows fast. Stars. Deep tones and low notes. Mild acidic.

Minerals -
Compacted Crystalline. Alkaline reserve minerals (somewhat heavier elements). Solid. Overt and extroverted. Piezoelectric. Narrow and Compact.

The Seven System -
Moves inwards. Right Spin. Moves downwards toward earth going from head to feet. Good solid wobble and impact.

Body Properties -
More red blood cells. More linear. High blood sugar. Tense, contracted. Rapid Pulse. Standing or Sitting (vertical body). Younger. Slim, linear, apple shaped. All pituary and pineal hormones. Testosterone, Aldosterone. Higher Blood Pressure. Farming, Gardening. Deep Sleep. Some Sun.

Nutritional Properties -
Fast oxidation. High Na/K. Low Ca/Mg. Parasympathetic. Step Down. Love, Joy and Contentment. Pure, Uplifted and Logical. Morality. Free Markets. Red Lamp Sauna, Ozone Machine, Foot and Hand reflexology. Zinc dominant.

Food Properties -
Fish, Eggs, Whole Grains, Salt and cooked vegetables. Roots that grow below the ground.

Butter, cream, vegetable oils (except coconut and palm oils).

Supplements
Minerals, Glandulars.

Voice
Lower Pitch.

Chakras
Akash - Thyroid - Throat - Space - Hearing - Ears - Ego/Pride

Earth - Rectal - Adrenal - Solid - Smell - Nose - Elimination - Attachment

Clear - Grey - Mind - Pituary Gland - Crown - Red (experience) - Psychic Awareness

Yellow - Fire - Eyes/Sight - Solar Plexus - Dark Blue (experience) - Digestion

Chapter 35. The Brain as a Hologram and the Field of Zero-Point Energy

D. Pollen and M. Trachtenberg expressed holographic brain theory to explain photographic memories in certain people. They state such individuals have more vivid memories because they are able to access very large regions of their memory holograms[1].

Hameroff suggests microtubule cavities, or centrioles, act as waveguides for evanescent photons that are responsible for quantum signal processing[2] [3]. How might this information processing occur? Frohlich (1970), Penrose (1956) and Onsager (1956) state that the tubulins in the microtubules exhibit excitations that exist in the gigahertz range[4] [5].

The current theory is that microtubular structures act as coherent fiber bundle sets that store holographic images. This is much like a fiber-optic holographic system. Research is starting to show that superluminal photons that occur inside microtubules due to evanescent waves provide the necessary energy to record and retrieve a quantum coherent holographic memory that is holographic in nature[4] [5]. Georgiev states consciousness is the result of quantum computations that occur via applied laser-like pulses inside quantum gates that exist within the brain's cortex (*Possible Existence of Superluminal Photons Inside Microtubules and the Resulting Explanation for Brain Mechanism Takaaki Musha and, Luigi Maxmilian Caligiuri. Advanced Science-Technology Research Organization, Yokohama, Japan. August 10, 2015*). The electrical activity going on in the brain's nerve cells are far too slow and variable for processing the large amount of information present within it. Hence a quantum holographic effect occurring in the brain's microtubules which convey frequencies in the terahertz through gigahertz, megahertz and kilohertz frequencies is the only explanation for the vast storage and retrieval of information of the brain[6].

Further **Reading**
Possible Existence of Superluminal Photons Inside Microtubules and the Resulting Explanation for Brain Mechanism Takaaki Musha and, Luigi Maxmilian Caligiuri. Advanced Science-Technology Research Organization, Yokohama, Japan. August 10, 2015.

Additional **Reading**
Talbot, M. The Holographic Universe; Harper Perennial: New York, NY, USA, 1991; pp. 23–24.

Pribram, K.H.; Nuwer, M.; Baron, R.J. The holographic hypothesis in memory structure in brain function and perception. In Contemporary Development in Mathematical Psychology;

Atkinson, R.C., Krantz, S.H., Luce, R.C., Suppes, P., Eds.; W.H.Freeman & Co.: New York, NY, USA, 1974; pp. 416–467.

Hameroff, S.R. Information processing in microtubules. J. Theor. Biol. 1982, 98, 549–561.

Georgiev, D.D. Bose-Einstein condensation of tunneling photons in the brain cortex as a mechanism of conscious action, 2004. Available online: http://cogprints/3539/1/tunneling.pdf

Georgiev, D.D. Quantum computation in the neuronal microtubules: Quantum gates, ordered water and superradiance, 2004. Available online: http://arxiv.org/abs/quantph/0211080

Smith, T. Quantum Consciousness. Water, Light speed, and Microtubules. Available online: http://www.valdostamuseum.org/hamsmith/QuanCon2.html

Veselago, V.G. The electrodynamics of substances with simultaneously negative values of e and ? . Soviet Physics Uspekhi 1968, 10, 509–514.
Ung, B. Metamaterials: A Metareview. Available

online:
http://www.polymtl.ca/phys/doc/art_2_2.pdf

Veselago, V.; Braginsky, L.; Shklover, V.; Hafner, C. Negative refractive index material. J. Comput. Theor.Nanosci. 2006, 3, 1–30.

Jibu, M.; Hagan,S.; Hameroff, S.R.; Pribram, K.H.; Yasue, K. Quantum optical coherence in cytoskeletal microtubules: Implications for brain function. BioSystems 1994, 32, 195–209.

Satinover, J. The Quantum Brain; John Wiley & Sons, Inc.: New York, NY, USA, 2001.

Albrecht-Buehler, G. Rudimentary form of cellular "vision". Proc. Natl. Acad. Sci. USA 1992, 89, 8288–8292.

Hameroff, S.R. A new theory of the origin of cancer: quantum coherent entanglement, centrioles, mitosis, and differentiation. BioSystems 2004, 77, 119–136.

Jibu, M.; Pribram, K.H.; Yasue, K. From conscious experience to memory storage and retrieval: The role of quantum brain dynamics and boson condensation of evanescent photons. Int. J. Mod. Phys. B 1996, 10, 1753–1754.

Jibu, M.; Yasue, K. What is mind?—Quantum field theory of evanescent photons in brain as quantum theory of consciousness. Informatica 1997, 21, 471–490.

Jibu, M.; Yasue, K.; Hagan, S. Evanescent (tunneling) photon and cellular vision. BioSystems 1997, 42, 65–73.

Recami, E. A bird's-eye view of the experimental status-of-the-art for superluminal motions. Found. Phys. 2001, 31, 1119–1135.

Recami, E. Superluminal tunneling through successive barriers: Does QM predict infinite group velocities? J. Modern Opt. 2004, 51, 913–923.

Musha, T. Possibility of high performance quantum computation by superluminal evanescent photons in living systems. BioSystems 2009, 96, 242–245.

Ziolkowski, R.W. Superluminal transmission of information through an electromagnetic material. Phys. Rev. E 2001, 63, doi: 10.1103/PhysRevE.63.046604.

Claus, R.O. Noncontact Measurement of High Temperature Using Optical Fiber Sensors; Final Report, NAG-1-831 (NASA-CR-186975), Virginia Polytechnic Institute and State University: Blacksburg, VA, USA, 1990; pp. 40–50.

References. Chapter 35

1. Holographic View of the Brain Memory Mechanism Based on Evanescent Superluminal Photons Takaaki Musha. Information 2012, 3, 344-350; doi:10.3390/info3030344
2. Comments on Musha's theorem that an evanescent photon in the microtubule is a superluminal particle. Hari SD.. July 2014
3. Possibility of high performance quantum computation by superluminal evanescent photons in living systems. Musha T. June 2009
4. CONSCIOUSNESS, FREE WILL AND QUANTUM BRAIN BIOLOGY – THE 'ORCH OR' THEORY1a Stuart Hameroff
5. Quantum Computation in Brain Microtubules? The Penrose-Hameroff 'Orch OR' Model of Consciousness [and Discussion] Author(s): Stuart Hameroff and P. Marcer. Aug. 15, 1998
6. Holographic View of the Brain Memory Mechanism Based on Evanescent Superluminal Photons Takaaki Musha Advanced Science-Technology Research Organization. August 2012

Chapter 36. The Zero Point Field and Memory

The zero-point field is commonly known as the quantum vacuum. The vast memory storage capabilities present outside the physical body may be due to the zero-point field. The electromagnetic zero-point field's existence has been demonstrated by Steve Lamoreaux (1997)[1].

The central nervous system and brain are not the locations of memory itself but may exist as an organic processes that interacts directly with the zero-point energy field at varying quantum levels. Hence the zero-point may provide the answer as to why unlimited amounts of information are able to be retrieved and written via quantum mechanical processes.

The zero-point field is comprised of non-local continuous quantum energy fluctuations that allow for instantaneous communication. The body's DNA and microtubules communications systems provide the means for a person's memories and interaction of the zero-point field serves to explain reviews of events from a perspective of persons with whom they had previously interacted with. The zero-point field is not just the source of all matter and energy but is the continuum that non-local communications may be occurring across unlimited distances.

Summary

DNA, the brain and microtubules are able to communicate non-locally and with a virtually unlimited memory storage capacity due to quantum behavior. At the molecular level, the complexity of microtubules is incomprehensible. However at the quantum level, their process of communication is relatively simple.

Waking Conscious states similar to Microtubules

1- Marshall (1989) states the bose Einstein condensation exists as a coherent quantum state and that it occurs among neural proteins (*penrose 1987, Bohm and Hiley 1993, Jibu and Yasue 1995*),

2 - Preconscious to conscious transitions (*Stapp 1993*),

3 - Collapse of quantum wave function in presynsptic axon terminals (Beck and Eccles 1992).

Are Microtubules Interacting with the Quantum Foam?

Microtubules are not only very, very tiny structures, but also show quantum fluctuations (*Orchestrated Objective Reduction of Quantum Coherence in Brain Microtubules: The "Orch OR" Model for ConsciousnessStuart Hameroff & Roger Penrose.*

1996), (Elsevier. *"Discovery of quantum vibrations in 'microtubules' inside brain neurons supports controversial theory of consciousness." Science Daily. ScienceDaily,* January 2014). This small size makes it a promising candidate for it interacting with the quantum foam. Microtubules measure 60 microns long and the quantum foam is estimated to be 10-3510-35 meters in length. To summarize, the microtubule's dimension is approximately 1027,1027 times the length of the quantum foam. So we can scale this up to get a better picture of its dimensions. If a single section of quantum foam was 1 inch in length, then a microtubule would be the size of the observable universe. Perhaps the interaction is taking place via a type of resonance/oscillation.

The Planck Scale and Quantum Consciousness
The Planck scale is roughly measured at 10-33 centimeters. It is at this level space-time geometry goes from smooth to "*granular*".

It is at the Planck scale that quantum particles (*also called virtual photons*) are continuously popping in and out of our existence. This is often called the "*quantum foam*". At this level also exists what's known as the zero point energy field. The question is, is the quantum foam alive? Does it have its own consciousness? and does it contain coherent information?

Because microtubules may be interacting with the quantum foam, they may be doing so as mini Planck black holes. Haramein[2] states that these exist as white hole / black hole systems.

The Diosi-Penrose theories state this exists as a bubble geometry in space-time (*On Diósi-Penrose criterion of gravity-induced quantum collapse. Shan Gao Unit for HPS & Centre for Time, SOPHI, University of Sydney*). Haramein states the quantum effects are occurring via white hole / black hole effects and that the oscillations are responsible for the collapse of the wave function proposed by Hameroff and Penrose. Haramein also proposes that the structure manifold of space time is fractal in nature and not flat and smooth as shown in the standard models.

Reference
The Origin Of Spin: A Consideration Of Torque And Coriolis Forces In Einstein's Field Equations And Grand Unification Theory N. Haramein¶ and E.A. Rauscher. The Resonance Project Foundation, haramein@theresonanceproject.org .Tecnic Research Laboratory, 3500 S. Tomahawk Rd., Bldg. 188, Apache Junction, AZ 85219 USA

Further Reading
Detection of Space-time Fluctuations by a Model Matter Interferometer. Ian C. Percival and Walter T. Strunz 1997

Do Birds Utilize The Quantum Realm?
Many of us have heard or read about situations where animals, including birds, were able to flee an approaching climate related disaster before it happened, such as an approaching tsunami or other major disaster. Besides a bird's ability to detect changes in **infrasound**[3] perhaps birds also can see changes taking place in the Quantum fields in nature. Infrasonic waves propagating from mountain ridges and oceanic coasts along earth's surface over thousands of kilometers[3] may be easily visible to birds.

Birds may sense Earth's magnetic field via Quantum Fields
Thorsten Ritz at University of California, Irvine demonstrated how magnetic disturbances that detect transitions between types of quantum-mechanical atomic spin states may disrupt the internal compass of the European robin (*Erithacus rubecula*).

Thorsten stated birds have an inner sensor system that consists of spin states which flip in response to changes in Earth's magnetic field.

These changes produce signals that the bird's brain detects[4] [5].

What does Quantum Mechanics Mean? One of the most simple explanations to explain quantum mechanics was developed by Dr. Tom Bell who stated that a photon or subatomic particle that splits in two is linked forever with its counterpart[6].

Quantum Mechanics and Bird Navigation Early research proposed that some type of apparatus existed in a bird's eye that initiated a chemical response during navigation. Birds seem to be able to maintain delicate spin states for extraordinarily long times of up to 100 microseconds. Research showed that the internal magnetic compass utilized by robins is disrupted by extremely small levels of magnetic *'noise'*. When the noise (*a tiny oscillating magnetic field*) was introduced into the bird's environment, it disabled the Robin's sense of direction. Once the noise was removed, the bird's compass returned to normal functioning.

This simple experiment shows that magnetic fields are causing an alteration in the body chemistry of birds and perhaps humans as well. This sounds a lot like how high levels of earth's geomagnetic activity disrupt the accuracy of remote viewing sessions [7] [8], perhaps by altering serotonin levels.

Chickens will orientate themselves via Earth's magnetic field[8] [9] and this effect may also be taking place in pigeons, zebra finches and honeyeaters[8] [9]. As we showed early in our second book on remote viewing that manganese (*found in pineapple and brown rice*) enhances remote viewing studies, have also shown that when mice were exposed to a magnetic field, that the levels of manganese in their bodies **INCREASED**[10].

Activity in the body has been shown to coincide two days after a new and full moon. This has been found to be due to acetylcholine and serotonin *(cardioacceleratory activity)*. Both these chemical transmitters exhibit a circadian rhythm. Work by Rounds (*Wichita State Universtiy, Kansas*) suggested that this effect takes place due to a "spill over" type effect occurring in the nervous system due to it varying intercellular mediators which affect the substances in the blood. Most likely acetylcholine, serotonin and noradrenaline are most likely the substances responsible as they are affected by the new and full moon. Hence the moon may be influencing cellular processes through an intermediary action such as the nucleotides cGMP and cAMP.

Reference

Reference: New Scientist Jun 12, 1975. Page 599

Spontaneous deliveries have been shown to take place during full moons (*Incidence of lunar position in the distribution of deliveries. A statistical analysis.* G. Ghiandoni et al. March 1997)

Further Reading

The Quantum Robin. Navigation News Pages 15 through 17. (The_Quantum_Robin.pdf)

Quantum effects in biology: Thorsten Ritz. Department of Physics and Astronomy, University of California, Irvine. 2011.

Birds can Sense Polarized Light
Marshall Stoneham of University College in London and his team stated birds use a technique that detects light polarization. This is commonly called Haidinger's brush and is a strange effect that superimposes a faint, yellow bow-tie shape on our eye's field of vision. Haidinger's brush is thought to come from the way blue light-absorbing lutein molecules arrange themselves in concentric circles in our eyes. This is interesting because in our second book on associative remote viewing I show that Haidinger's brush is one way for the eyes to see polarized light. Haidinger's brush is seen much

more easily at sunset when the light of the sun is polarized. Our ARV sessions had the best results when conducted at the 1/2 moon (first quarter) heading towards the full moon. The first quarter moon is a time where the light of the moon is at **maximum polarization**. Hence a strong environment of polarized light enhances the results of remote viewing. Now let's get back to Stoneham's study.

Stoneham's math showed that a magnetic field is able to produce a similar distortion in the visual field of birds and that the orientation of the field would change as the magnetic field changed. This effect only exists if quantum states were able to last long enough to affect a bird's light sensing molecules. Hence the vision of some birds consists of a heads up display, much like those used on the windscreens of luxury cars[11] [12].

Physicists from the University of Geneva showed how human eyes are able to detect a large Bell inequality violation (*Bell inequalities are shown to be violated by quantum mechanical predictions*).. This proves the existence of quantum entanglement[13].

Further Reading
Retinal and post-retinal contributions to the quantum efficiency of the human eye revealed by

electrical neuroimaging Gibran Manasseh, et al. Nov 2013

Evidence Of Macroscopic Quantum Phenomena And Conscious Reality Selection Cynthia Sue Larson. Cosmos and History: The Journal of Natural and Social Philosophy, vol. 10, no. 1, 2014

Quantum Entanglement: Can We "See" the Implicate Order? Philosophical Speculations. Michele Caponigro et al. Sept 2010

Towards Quantum Experiments with Human Eyes as Detectors Based on Cloning via Stimulated Emission Pavel Sekatski. Physical Review Letters Week Ending 11 September 2009

The polarization sense in human vision Author links open overlay panel. AlbertLe Floch et al. Sept 2010

References. Chapter 36

1. Casimir effect. Wikipedia.
2. Quantum Gravity and the Holographic Mass Nassim Haramein. February 2013
3. Pigeon homing from unfamiliar areas An alternative to olfactory navigation is not in sight Hans G Wallraff. April 2014
4. Aeticle Ref - ISA GROSSMAN SCIENCE 01.27.1104:20 PM IN THE BLINK OF BIRD'S EYE, A MODEL FOR QUANTUM NAVIGATION
5. Techinical Ref - Sustained Quantum Coherence and Entanglement in the Avian Compass Erik M. Gauger, Elisabeth Rieper, John J. L. Morton, Simon C. Benjamin, and Vlatko Vedral Phys. Rev. Lett. 106, 040503 – Published 25 January 2011
6. Bell's Experiment in Quantum Mechanics and Classical Physics Tom Rother. Aug 2013
7. Article Ref - IN THE BLINK OF BIRD'S EYE, A MODEL FOR QUANTUM NAVIGATION. LISA GROSSMAN SCIENCE. Jan 2011.
8. Tech Ref - Quantum effects in biology: Bird navigation. 22nd Solvay conference on Chemistry.

Thorsten Ritz. Department of Physics and Astronomy, University of California, Irvine. 2011
9. Quantum effects in biology: Bird navigation. 22nd Solvay conference on Chemistry. Thorsten Ritz. Department of Physics and Astronomy, University of California, Irvine. 2011
10. The influence of a magnetic field on manganese transport into rat brain. Vojtisek M et al. Dec 1996
11. A new model for magnetoreception A. Marshall Stoneham,. March 2012.
12. New Scientist, 27 November 2010, p 42. (arxiv.org/abs/1003.2628).
13. Towards Quantum Experiments with Human Eyes as Detectors Based on Cloning via Stimulated Emission Pavel Sekatski et al. Sept 2009

Further Reading
The Casimir effect as a possible source of cosmic energy. Physics Letters A,223, 163-166. Southwick, S. M., Morgan, C. A., Nicolaou, A. L., and Charney, D. S. (1997).

Chapter 37. Variations of Water Moisture Caused by Moon Phases

Scientific Studies of Moon Phase and Water Moisture

Research by the Solar Institute showed that remote viewing accuracy is greatly enhanced during certain moon phases. The best moon phases are up to 72 hours before the moon's first quarter and up to 72 hours before the full moon. We also show in greater detail in our second book on associate remote viewing that the period of the moon's first quarter also happens to be the moon phase that exhibits the strongest polarized light.

The illustration on the following page shows the best times for ARV according to moon phase.

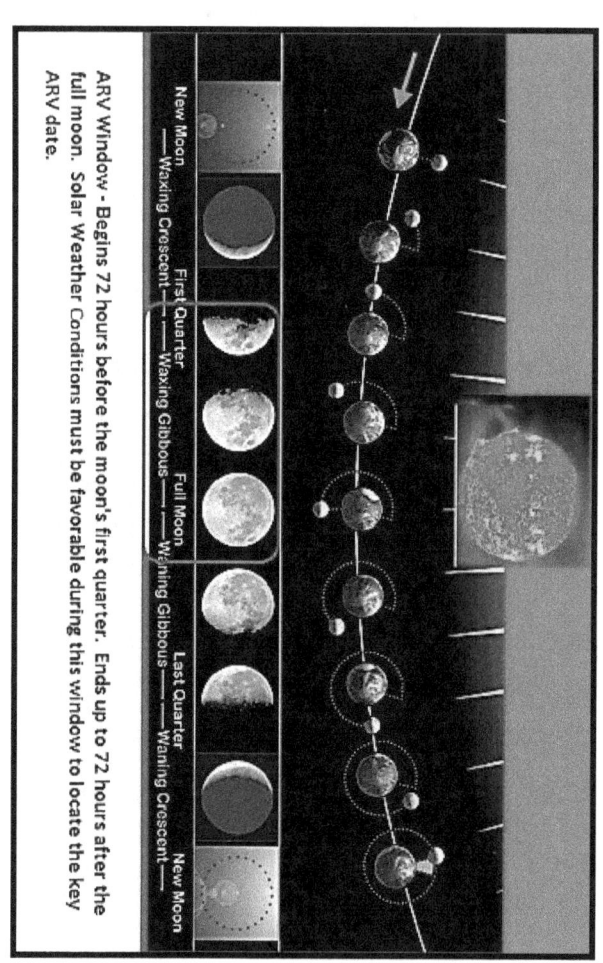

ARV Window - Begins 72 hours before the moon's first quarter. Ends up to 72 hours after the full moon. Solar Weather Conditions must be favorable during this window to locate the key ARV date.

Now let's take a look at some detailed scientific studies regarding moon phase, air pressure and water moisture.

Research on the amount of water seeds absorb according to the moon phase has been carried out by Innamorati and Signorini (1980) and Spruyt (1987) in addition to the seasonal variation [1] [2]. Research has found that a circaseptan lunar rhythm exists with how much water is absorbed by seeds.

Brown and Chow 1973 found that the absorption of water by bean seeds showed a constant pattern beginning at the moon's first quarter and lasting until the last quarter moon with the strongest absorption of water occurring just after the first quarter, full and last quarter moon[3] (*shown below*).

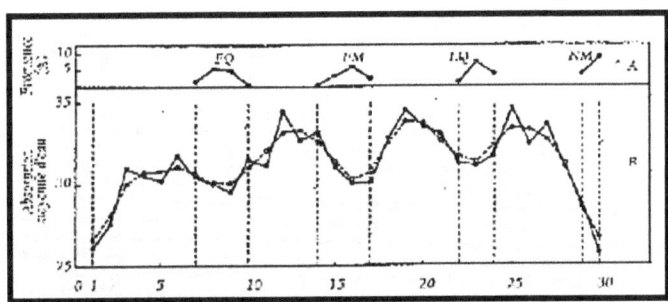

Also seeds have been shown to germinate faster when planted 3 days before a full moon (*shown in the following image*)[4].

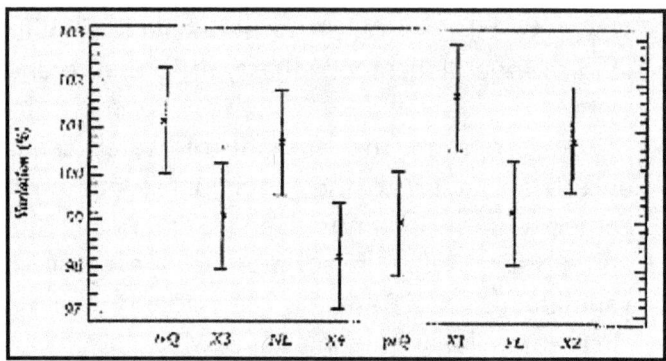

Pinto Beans have also been found to absorb more water during full moons[5].

Further Reading

Indigenous Knowledge Informing Management of Tropical Forests: The Link between Rhythms in Plant Secondary Chemistry and Lunar Cycles. Vogt, K. A., Beard, K. H., Hammann, S., O'Hara Palmiotto J., Vogt, D. J., Scatena, F. N., Hecht, B. P. (2002): Ambio Vol. 31 No. 6, Sept 2002: 485- 490.

Research conducted by Semmens (1923) found that the polarized moonlight quickened seed germination[6]. Baly and Semmens found that polarized light had an effect on starch hydrolysis in that it increased the rate of decomposition[7]. This effect was also studied by other researchers and not found to occur[7].

Further Reading
Lunar Rhythms In Forestry Traditions – Lunar-Correlated Phenomena In Tree Biology And Wood Properties. Ernst Zürcher. Jan 1999

Polarized Light and Effects on Microorganism
I go into greater detail about how polarized light enhances remote viewing in my second edition on remote viewing, in which the moon's polarized light is at maximum at the moon's first quarter and almost non-existent during the full moon. It is also interesting to note here that Pinus nigra (the Austrian pine or black pine), **rotates the rays of polarized light to the right** (dextrogyre)[8]. Pine is a terpene and our TXP formula uses Tea Tree Oil, which contains terpene hydrocarbons. Also our TXP formula utilizes limonene (from Basil Essential Oil). Limonene is a terpene hydrocarbon and it also rotates the rays of polarized light to the right. Basil, Lavender, Bay Leaves, Goldenrod (a bright yellow flower) and Mugwort all contain Linalool[9]. The molecular structure of Linalool is shown in the following image on the next page.

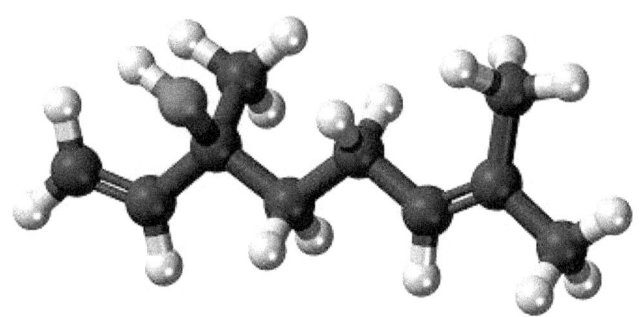

A list of Dextrogyre Substances
Manilla Elemi is strongly dextrogyre. Carvol and Carvene are strongly dextrogyre with Carvene being very strong. Safrene is dextrogyre. Corriander, Sweet Fennel, E. Globulus (*also called Eucalyptus globulus, southern blue-gum or the Tasmanian bluegum*), and the leaves of Myrtus Chequen are dextrogyre. Also Sweet Orange Oil (a citrus) rotates the rays of polarized light to the right[10]. Oil of Myrtle is Dextrogyre[11] and we use Lemon Myrtle in our TXP formula. Eucalyptus oil and Fenchol are also strongly dextrogyre[12]. Coniine is dextrogyre (*American Druggist, Volumes 17-18*). Coniine is a poisonous substance found in poison hemlock (*Conium maculatum*). Coniine is also found in the yellow pitcher plant (*Sarracenia flava*) as well as fool's parsley.

Research by Macht (1925) found that polarized light affected ferments and pharmacologically active cultures and Macht and

Hill (1925), Bhatnager and Lal (1926) and Macht (1926) found that seedlings grew quicker under polarized light. Semmens found long durations of polarized light actually damaged the plants.

Research conducted by Dastur and Gunjikar (1934) found that elliptical polarized light caused the rate of photosynthesis to increase[13]. Basil essential oil contains the following substances: citral, geraniol, nerol, linalool and fenchone *(Characterization of Geraniol Synthase from the Peltate Glands of Sweet Basil1Yoko Iijima,. Jan 2004).*

Further **Reading**
The Effect of Circularly Polarized Light on the Photosynthesis and Chlorophyll of Synthesis of Certain Marine Algae. G.C. MCLEOD. Wood's Hole Oceanographic Institution, Woods Hole, Mass.

Circularly Polarized Light with Sense and Wavelengths To Regulate Azobenzene Supramolecular Chirality in Optofluidic Medium. L. Wang et al. Sept 2017

Theta Brainwaves and Rotating Polarized Light
Could it be that certain essential oils/terpenes that **rotate polarized light to the right** enhance the mind / heart resonance that takes place during an ARV session? and might this also be enhancing theta brainwaves?

Clozapine
The substance **Clozapine**, which exhibits dextrogyre properties, (*rotates polarized light to the right*) has been shown to enhance theta brainwaves.

Reference
Effects of Psychotropic Drugs on Quantitative EEG among Patients with Schizophrenia-spectrum Disorders. June Hyun et al. Aug 2011. ; 9(2): 78–85. doi: 10.9758/cpn.2011.9.2.78. Clin Psychopharmacol Neurosci. PMCID: PMC3569080

Clozapine is used to treat suicidal behavior in patients with schizophrenia. People who are depressed usually have low levels of dopamine. Also we use Ginkgo Biloba extract in our ARV sessions to enhance the brain's alertness, especially due to the fact that ARV sessions are conducted around midnight when the mind could use more awareness.
    A research study looked at the clinical effects of ginkgo biloba used as an adjunct to clozapine. The study found that Ginkgo was effective in decreasing the patient's negative symptoms, however not their psychopathology symptoms. The study concluded that Gingko was useful for enhancing the effects of clozapine in patients who showed resistance to standard schizophrenia medications.

Reference

A placebo-controlled study of extract of ginkgo biloba added to clozapine in patients with treatment-resistant schizophrenia. A. Doruk et al. July 2008. Int Clin Psychopharmacol. (4):223-7. doi:10.1097/YIC.0b013e3282fcff2f.

Fenchyl

Fenchyl, an isomer of borneol, produces a powerful **camphor**-like scent and is used in **strawberry** and other berry flavors. Fenchyl can be found in Stachys tibetica essential oil, which is a folk medicine remedy used in India, Ladakh and Tibet for psychiatric disorders. Fenchyl is also found in apple juice (*The role of flavor and fragrance chemicals in TRPA1. Satoru Mihara and Takayuki Shibamoto. March 2015*).

Other Essential Oils used to heal psychological disorders include: bergamot, pine, eucalyptus, caraway, lavender, lemon, geranium, juniper, lemongrass, orange, peppermint, mint, rosemary, thyme, ylang-ylang, sage and tea tree.

Summary

Polarized light which is strongest at the first quarter (half moon) moon, combined with increased water moisture in the air, enhances associative remote viewing sessions due to the water moisture enhancing resonance in the brain's microtubules.

Now let's take a look at research studies looking at how much rain falls according to the phase of the moon.

Rainfall according to Phase of Moon Meteorological observations conducted in Germany and Paris found that the maximum number of rainy days occurred from the moon's first quarter and the full moon. Rain was less likely to fall from the last quarter to new moon[14] [15] [16] [17]. The following picture shows a chart from their research study.

	Rainy days.	Rain in inches.
On the day of new moon	4·365	3·98
On the day following	4·365	3·99
On the day of the first octant	3·968	2·48
On the day following	3·571	4·68
On the day of the first quarter	3·175	1·88
On the day following	3·175	3·92
On the day of the second octant	3·968	4·67
On the day following	3·175	6·58
On the day of full moon	3·571	1·63
On the day following	3·175	1·86
On the day of the third octant	3·571	1·88
On the day following	1·984	2·60
On the day of the last quarter	2·778	2·34
On the day following	3·175	3·45
On the day of the fourth octant	3·175	2·53
On the day following	2·778	2·10

Observations over a 16 year period at Augsburg Swabia, Bavaria, Germany found the following details shown below[18]

Epochs.	Number of clear days in 16 years.	Number of overcast days in 16 years.	Quantity of rain in 16 years in inches.
New moon..........	31	61	26·551
First quarter........	38	57	24·597
Second octant.......	25	65	26·728
Full moon..........	26	61	24·686
Last quarter........	41	53	19·536

Summary

Peaking high pressure (*dry air and a minimal chance of rain*) drops to lower air pressure (*increased chance of rain*). This cycle occurs on average most often from the period of new moons to first quarter moons. Hence the window of ARV accuracy begins up to 72 hours before the moon's first quarter.

Perigee and Apogee Moon and Rainfall
Another study looking at moon phases found that it rained most often during the moon's first quarter[19].

	PHASES OF THE MOON.					
	New Moon.	First quarter.	Full Moon.	Last quarter.	Perigee.	Apogee.
No. of rainy days coincident with the days of the Moon's phases.	77 days.	82 days.	79 days.	60 days.	93 days.	78 days.

The same study also looked at the moon's distance to the earth and found that during perigee moons rainfall was much more likely to occur[19]. Our research has also found that our ARV sessions are more accurate during perigee moons, especially during super moon full moons, which is the time the full moon is closest to earth.

We shall soon show that barometric air pressure is higher when the moon is in apogee and air pressure is lower when the moon is in perigee. Hence, lower barometric air pressure increases the chance of rain more likely to occur.

One research study found that planting potato, bean, winter rye, carrot, radish or mustard during a perigee moon exhibited a significantly positive reaction, compared to other periods of sowing[20].

Weather and Trauma

A research study found that when **the weather was good**, there was a significant increase in cases of trauma[21], and as we have shown in this book, trauma/violence/intense emotion enhance the clarity of remote viewing. Our research has shown that favorable solar weather conditions enhance remote viewing. These are also conditions where the **weather is usually good**. Favorable solar weather conditions include lower solar wind speeds (350 or less), a stable 10.7cm radio flux and lower geomagnetic activity. Below is a typical cloud formation that occurs most of the time when the sun's 10.7cm radio flux is increasing.

One of the more telling signs that solar weather is at favorable periods for a successful remote viewing session, is that the clouds in the sky will have a "ripple" or "wavy" type effect. These clouds kind of look like earth from outer space when cropped into a circle (*parallel earth*?).

Time Travel, the Sun and Science Fiction
There have been numerous science fiction time

travel movies and television episodes where the sun is present. Below are just two examples from the science fiction series Star Trek.

Star Trek Motion Picture **The Voyage Home**. The spaceship uses the sun as a sling shot to go back in time to retrieve a whale. What is interesting is whales have some of the largest neurons. Neurons make up a major part of the brain's microtubules.

Star Trek Episode #78 **All Our Yesterdays**. In this episode, an entire civilization has travelled back in time to escape their planet's sun going supernova. The crew arrive on the planet just hours before the sun goes supernova and travel back in time.

Air Pressure and Moon Phase
If indeed more water moisture is in the air during these moon phases, then there must exist a variation in air pressure as well, because low air pressure = increased chance of rain.
    The University of Washington conducted a study looking at the link between lunar forces and rainfall. The study re-discovered the effect, shown in previous research studies, that oscillations in air pressure are linked to the phases of the moon (*first detected in 1847*)[22].

Also temperature changes are related to moon phase which was discovered in 1932. An earlier study by UW confirmed air pressure varies with the phases of the moon[23]. This study found that when the moon is overhead or underfoot (*at the opposite side ie; below the earth from where one stands*) that air pressure is higher. When the moon is directly overhead, the gravity of the moon causes Earth's atmosphere to bulge upwards. This causes earth's air pressure to rise. The higher air pressure increases the temperature which allows the warmer air to hold more moisture.

Barometric Air Pressure and Moon Distance
A research study conducted in October 1808 looked at the distance of the moon to the earth and sought to see if it had an effect on earth's barometric air pressure. Perigee moons are when the moon is at its maximum closeness to earth. The study found that during Apogee moons, earth's barometric air pressure was higher. The study also looked at the phase of the moon and found peaks in barometric pressure occurred at the moon's first quarter, full and last quarters. The strongest peak was at the second quarter moon. Below is a summary of their research data.

Improve your Remote Viewing Accuracy Techniques using Quantum Microtubules

LUNAR POINTS.	Number of observations.	Mean heights of the barometer.	Reduction to millimetres.
		pou. lig.	m. m.
Mean general height, - - - -	6915	27 11.29	755.44
Conjunction, or new Moon, - - -	234	27 11.27	755.39
First Octant, - - - - - -	234	27 11.26	755.37
First quadrature, - - - - -	234	27 11.26	755.37
Second Octant, - - - - -	235	27 10.94	754.65
Opposition, or full Moon, - - -	234	27 11.20	755.23
Third Octant, - - - - - -	234	27 11.47	755.70
Second quadrature, - - - -	234	27 11.68	756.32
Fourth Octant, - - - - -	235	27 11.31	755.48
Northern Lunistice, - - - -	258	27 11.42	755.73
Southern Lunistice, - - - -	258	27 11.28	755.42
Moon in Perigee, (Parall. equa. 60' 24")	252	27 10.97	754.72
Moon in Apogee, (Parall equa. 54' 4" -	252	27 11.46	755.82

Summary

The reason ARV sessions are more accurate during perigee moons is due to the fact that earth's barometric air pressure is lower. Lower air pressure = the chance of more rainfall/water/moisture. The first quarter moon and full moons are times that more water moisture is in earth's atmosphere and water moisture exists in the brain's microtubules, which is most likely enhancing the resonance from signals received from earth's Schuman resonance.

The only benefit received from a higher barometric air pressure for associative remote viewing sessions, is the sun's 10.7cm radio flux is usually higher or steady, which is a favorable condition for associative remote viewing sessions.

While these conditions may not always occur at these moon phases, they act as a general guideline to follow when looking for the best atmospheric conditions to perform ARV sessions.

Seasonal Variation of Barometric Air Pressure According to Sunspots and Region
A research study found that during summer, **inland** regions experience **stronger lower** air **pressure** then average and the ocean coastal regions experience above **average high pressure**.

During winter the exact opposite effect occurs and that during solar maximum, this influence (*and possibly the equinoxes*) may possibly be enhanced. This would mean performing associative remote viewing sessions inland during wintertime and along the coast during summertime would be a favorable environment for associative remote viewing sessions.

The study also found that during **sunspot maximum** barometric air pressure occurred under north-east winds and the minimum under south-west winds. During **solar minimum** the maximum and minimum barometric air pressure occurred under north and south east winds[25].

High Air Pressure and Births

A study looking at births between January 1997 and December 2003 discovered a significant increase in the number of deliveries and ruptures during low barometric pressure (P < 0.01). During days larger changes in barometric pressure occurred (*regardless of increasing or decreasing*) deliveries increased. The relationship was found to be statistically significant (P < 0.01). Also a small relationship existed between ruptures of the fetal membranes, delivery and barometric pressure. The study concluded that low barometric air pressure induces rupture of the fetal membranes [26].

Further Reading

Spontaneous delivery is related to barometric pressure. O. Akutagawa et al. April 2007

Biodynamic Gardening and the Influence of the Moon's Forces

The science of Biodynamic Gardening utlizes lunar, solar and environmental forces to eliminate or greatly reduce the need for pesticides in farming. Biodynamics teaches that the soil is affected by these forces and uses the term **'soil tide'**.

Crops that produce food above ground are planted during waxing moons. During the full moon absorption of liquids into plant matter is maximized. Seed germination is faster during full moons, however the seeds become more prone to fungus attack especially during periods of warmer weather and higher humidity because of enhanced water moisture.

One of the 4 elements of the Tao most affected by water is the moon. For example sap in plants. At new Moons underground activity in the soil increases and above ground sap flow is weaker. Also the flow of water in plants growing above the ground (sap in plants) is much reduced. From new to full moons the sap flow greatly increases in the above ground regions of plants. This is commonly the time to plant leafy, fruiting and flowering biennials and annuals. Fertile days during this time provide an extra boost for the healing of germination and pruning. These are shown as *'best days'* in the new moon and first quarter phases of a lunar gardening calendar. During this time liquid fertilizers are much better absorbed and the sap flow is high[26].

Seasonal Variation

Apogee moons show higher air pressure and perigee moons show lower air pressure. During winter when rain is more common, ARV sessions may be enhanced when the moon is in perigee. Especially full moon super moons. During full moons there is more rain on average, hence a lower pressure system. Rain also occurs more often at first quarter moons.

Cosmic Rays and Water Moisture
Another interesting fact is rainfall is more common during solar minimum, or when solar activity is declining. This causes an increase in cosmic ray activity, which in turn causes an increase in clouds, contributing to more rain. This means that the period just before solar minimum ARV sessions would be more successful due to this increased water moisture in the air, especially during full moons and super moon full moons. I cover this in much greater detail in my article titled: Is Climate Change Responsible for the 2017 Hurricanes in Texas. Louisiana, Georgia and Florida? published at the address below: www.ez3dbiz.com/cosmic_rays_climate_change.html

The Moon's Influence on Nature
The moon's first quarter is when the influence of the Sun's energy is decreasing. An energetical switch occurs during this time where the yin and yang become balanced and the activities of organisms increase.

Metaphysically teaching, from the first quarter moon to the full moon the link with our unconscious mind is stronger and the actualization of our ideas takes place. This is the best moon phase for undertaking new journeys or to change jobs.

Please note this book includes general guidelines that enhance associative remote viewing. A more comprehensive step by step instruction method on how to perform an Associative Remote Viewing session can be found in our second book **Associative Remote Viewing Remote Viewing. The Complete User's Manual on Experiencing Future Consciousness**.

The Star Arcturus and Remote Viewing

The star Arcturus rises in the east at 5:30 p.m. at my location in Honolulu, Hawaii during late spring (April 28th). During this time, the second peak period of remote viewing (**8:45 LST**) takes place at approximately 6:30 p.m. During this time Sirius rises at 11 a.m. in the morning and is zenith (directly overhead) at 4:30 p.m.

Improve your Remote Viewing Accuracy Techniques using Quantum Microtubules

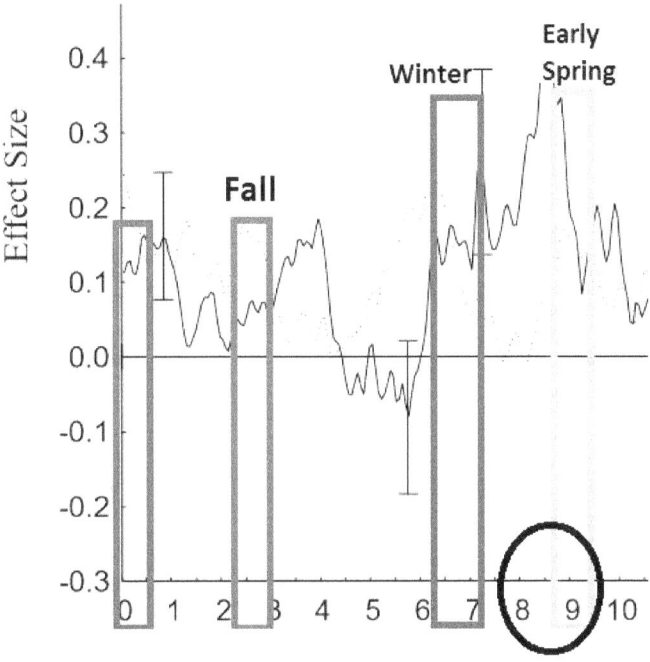

Reference
Apparent Association Between Effect Size In Free Response Anomalous Cognition Experiments And Local Sidereal Time. *Seasonal Emphasis by author.*

Chapter 38. How to Find Favorable Solar Weather Conditions to Enhance Remote Viewing Accuracy

We covered in an earlier chapter about how certain periods of solar weather enhance the success of remote viewing. Let's now review this in greater detail as there are a few other tips that can help plot favorable future solar weather conditions.

Because geomagnetic storms show a seasonal variation with peaks occurring in spring and October, the KP (*earth's geomagnetic energy*) entering the "*declining sweet spot*" is more likely to take place and be stronger after the spring equinox in march and after the fall equinox around September.

Once favorable solar weather conditions have been found, perform the ARV session. This gives you a maximum of 4 days of clear associative remote viewing conditions into the future. 4 is the number of maximum stability. The further out you see, the more unstable the timeline. Only very, very large scale events are able to be seen further out, much like weather forecasting can now accurately predict the path of a hurricane during hurricane season.

One of the clear advantages of looking out up to 4 days into the future is you can look back

Improve your Remote Viewing Accuracy Techniques
using Quantum Microtubules

on what the solar weather conditions were at the time the associative remote viewing session was conducted. This is because it takes up to 24 hours for NOAA's Solar Weather computers to process the previous' day's solar weather conditions such as KP numbers, the sun's 10.7cm radio flux levels and so on. If any of these were not favorable on the day of the ARV session, the ARV session may not be accurate. Also as stated earlier, a telling sign that the conditions are not favorable is it will be harder to go into heart math coherence. Another clue is the future date you are viewing

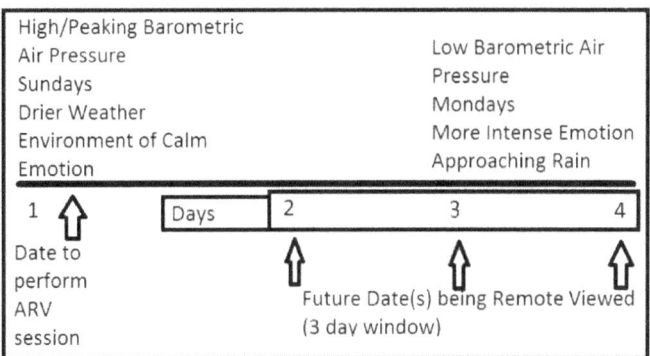

will be harder to "connect with" and may seem blurry or "out of phase" during the ARV session. The following illustration is a simplified summary of conditions favorable to an associative remote viewing session.

The 0.8 MEV Energetic Particles
Between 3 and 5 days before favorable solar weather conditions occur, there is usually a peak or increase in the 0.8 MEV particle photon flux. The 0.8 MEV particle numbers can be found at this address

http://legacy-www.swpc.noaa.gov/ftpdir/latest/DPD.txt

Another condition contributing to favorable solar wind speeds we mentioned earlier is the sun's solar wind speed will be declining towards the favorable period of 350. Favorable solar winds are between 370 and 350.

An LST Time Clock
Use the LST time clock to find your 8:45LST and 13:30LST times in your area.

http://www.wwu.edu/skywise/skymobile/skywatch.html

LST Seasonal Calendar
This next calendar we are about to show is the result of years of finding the peak time to perform associative remote viewing sessions based on season and region.

It is based on the original Spottiswoode calendar that reviewed thousands of remote viewing tests to find the best time of day to perform remote viewing. Our years of research

Improve your Remote Viewing Accuracy Techniques using Quantum Microtubules

discovered the daily peak of accuracy took place around midnight. When this is combined with the peak periods of remote viewing graph (*Spottiswoode calendar*), ARV sessions are conducted during this peak at midnight for maximum accuracy.

The peaks are seasonal with the main peak occurring around spring in North America (*13:30LST*). The following page lists the seasonal LST time slots along with a general guideline of favorable lunar conditions.

All ARV sessions are conducted between 11:20 p.m. and 12:30 a.m. (30 minutes after mid-night).

All ARV sessions are performed within 24 to 48 hours of the following moon phase (best to last)

The Half Moon heading towards the full moon (best time)

The Half Moon headed towards the new moon (2nd best time)

Full Moons (3rd best time)

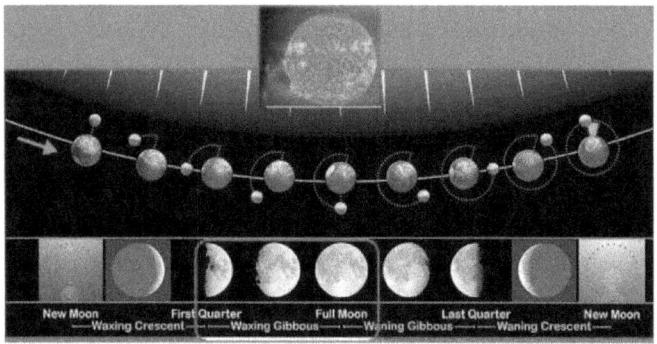

ARV Window - Begins 72 hours before the moon's first quarter. Ends up to 72 hours after the full moon. Solar Weather Conditions must be favorable during this window to locate the key ARV date.

How to Use the Calendar
After calculating to your time zone, match the times to the rectangular sections in the chart. These are the times remote viewing accuracy peaks. If they fall within range of higher values during that season, then performing an associative remote viewing session at midnight on that date will yield good results. If not, then wait until the date when the chart shows that remote viewing accuracy is higher in that rectangular section.

Peak Seasonal Remote Viewing Seasonal LST Accuracy Time Slots

Early Spring - 8:45LST to 9:30LST (best seasonal time). In Hawaii at Midnight on Feb 11th it is 08:29LST

Spring - 12:30LST to 14:30LST. In Hawaii at Midnight on April 11th it is 12:22LST

SUMMER - = ARV SESSION RESULTS ARE AT THEIR YEARLY LOW FOR ACCURACY IN THIS REGION

Late Summer - 20:30LST to 00:30LST. In Hawaii at Midnight on August 11th it is 20:23LST

Fall - 2:15LST to 3:00LST. In Hawaii at Midnight on November 11th it is 02:25LST

Winter - 6:15LST to 7:15LST. In Hawaii at Midnight on January 11th it is 06:25LST

The following page after the seasonal calendar shows the times in Hawaiian Time. Dates and Times are approximate.

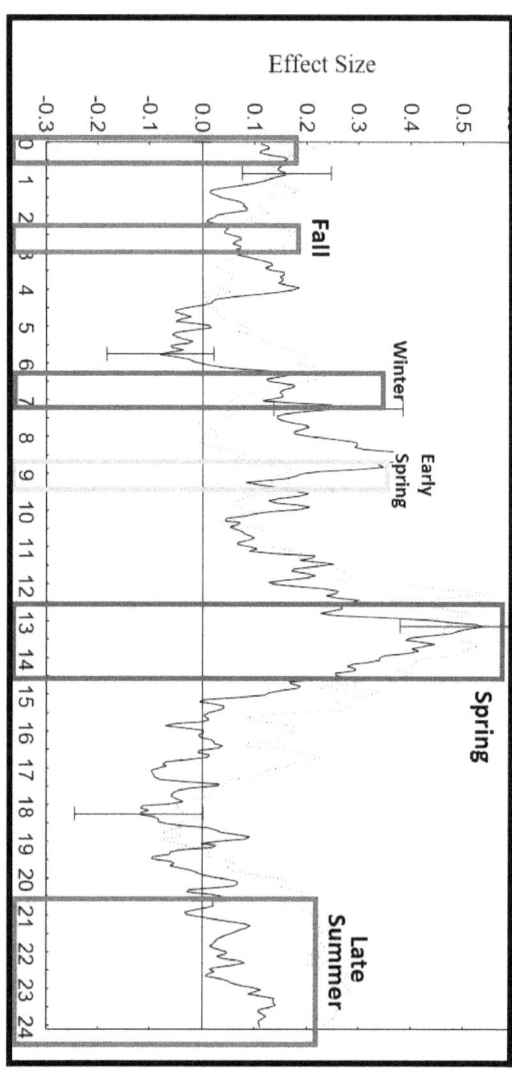

Improve your Remote Viewing Accuracy Techniques using Quantum Microtubules

Date ....LST Time	July 1, 17:42
January 1, 05:47	July 2, 17:46
January 2, 05:50	July 3, 17:49
January 3, 05:54	July 4, 17:53
January 4, 05:59	July 5, 17:57
January 5, 06:02	July 6, 18:01
January 6, 06:06	July 7, 18:05
January 7, 06:10	July 8, 18:08
January 8, 06:14	July 9, 18:12
January 9, 06:18	July 10, 18:16
January 10, 06:21	July 11, 18:20
January 11, 06:25	July 12, 18:24
January 12, 06:30	July 13, 18:28
January 13, 06:34	July 14, 18:32
January 14, 06:37	July 15, 18:36
January 15, 06:41	July 16, 18:40
January 16, 06:45	July 17, 18:44
January 17, 06:49	July 18, 18:48
January 18, 06:53	July 19, 18:52
January 19, 06:57	July 20, 18:56
January 20, 07:01	July 21, 18:57
January 21, 07:05	July 22, 19:04
January 22, 07:09	July 23, 19:08
January 23, 07:13	July 24, 19:12
January 24, 07:17	July 25, 19:16
January 25, 07:21	July 26, 19:20
January 26, 07:25	July 27, 19:24
January 27, 07:29	July 28, 19:28
January 28, 07:33	July 29, 19:32
January 29, 07:37	July 30, 19:36
January 30, 07:41	July 31, 19:40
January 31, 07:45	August 1, 19:44
February 1, 07:49	August 2, 19:48

February 2, 07:53	August 3, 19:52
February 3, 07:57	August 4, 19:55
February 4, 08:01	August 5, 19:59
February 5, 08:05	August 6, 20:03
February 6, 08:09	August 7, 20:07
February 7, 08:13	August 8, 20:11
February 8, 08:17	August 9, 20:15
February 9, 08:21	August 10, 20:19
February 10, 08:25	August 11, 20:23
February 11, 08:29	August 12, 20:27
February 12, 08:33	August 13, 20:31
February 13, 08:37	August 14, 20:35
February 14, 08:41	August 15, 20:39
February 15, 08:45	August 16, 20:43
February 16, 08:49	August 17, 20:47
February 17, 08:53	August 18, 20:51
February 18, 08:57	August 19, 20:55
February 19, 09:01	August 20, 20:59
February 20, 09:05	August 21, 21:03
February 21, 09:09	August 22, 21:07
February 22, 09:13	August 23, 21:11
February 23, 09:17	August 24, 21:15
February 24, 09:21	August 25, 21:19
February 25, 09:25	August 26, 21:23
February 26, 09:29	August 27, 21:27
February 27, 09:33	August 28, 21:31
February 28, 09:37	August 29, 21:35
March 1, 09:40	August 30, 21:39
March 2, 09:44	August 31, 21:43
March 3, 09:48	September 1, 21:47
March 4, 09:52	September 2, 21:50
March 5, 09:56	September 3, 21:54
March 6, 10:00	September 4, 21:58
March 7, 10:04	September 5, 22:02
March 8, 10:08	September 6, 22:06
March 9, 10:12	September 7, 22:09

Improve your Remote Viewing Accuracy Techniques using Quantum Microtubules

March 10, 10:16	September 8, 22:13
March 11, 10:20	September 9, 22:17
March 12, 10:24	September 10, 22:21
March 13, 10:28	September 11, 22:25
March 14, 10:31	September 12, 22:29
March 15, 10:35	September 13, 22:33
March 16, 10:40	September 14, 22:37
March 17, 10:43	September 15, 22:41
March 18, 10:47	September 16, 22:45
March 19, 10:51	September 17, 22:48
March 20, 10:55	September 18, 22:52
March 21, 10:59	September 19, 22:56
March 22, 11:03	September 20, 23:00
March 23, 11:07	September 21, 23:04
March 24, 11:11	September 22, 23:08
March 25, 11:15	September 23, 23:12
March 26, 11:19	September 24, 23:16
March 27, 11:23	September 25, 23:19
March 28, 11:27	September 26, 23:23
March 29, 11:31	September 27, 23:27
March 30, 11:35	September 28, 23:31
March 31, 11:39	September 29, 23:35
April 1, 11:43	September 30, 23:39
April 2, 11:47	October 1, 23:43
April 3, 11:50	October 2, 23:47
April 4, 11:54	October 3, 23:54
April 5, 11:58	October 4, 23:58
April 6, 12:01	October 5, 00:03
April 7, 12:06	October 6, 00:07
April 8, 12:10	October 7, 00:11
April 9, 12:14	October 8, 00:15
April 10, 12:18	October 9, 00:19
April 11, 12:22	October 10, 00:23
April 12, 12:26	October 11, 00:27
April 13, 12:30	October 12, 00:30
April 14, 12:34	October 13, 00:33

April 15, 12:38	October 14, 00:35
April 16, 12:42	October 15, 00:38
April 17, 12:46	October 16, 00:42
April 18, 12:50	October 17, 00:46
April 19, 12:54	October 18, 00:50
April 20, 12:58	October 19, 00:55
April 21, 13:01	October 20, 00:59
April 22, 13:05	October 21, 01:02
April 23, 13:09	October 22, 01:06
April 24, 13:13	October 23, 01:10
April 25, 13:17	October 24, 01:14
April 26, 13:21	October 25, 01:18
April 27, 13:25	October 26, 01:21
April 28, 13:29	October 27, 01:25
April 29, 13:33	October 28, 01:29
April 30, 13:37	October 29, 01:33
May 1, 13:41	October 30, 01:37
May 2, 13:44	October 31, 01:45
May 3, 13:48	November 1, 01:49
May 4, 13:52	November 2, 01:53
May 5, 13:56	November 3, 01:55
May 6, 14:00	November 4, 01:57
May 7, 14:06	November 5, 02:01
May 8, 14:10	Note: End 2016 Daylight
May 9, 14:14	Saving Time (-1 hour)
May 10, 14:18	November 6, 02:05
May 11, 14:22	November 7, 02:09
May 12, 14:24	November 8, 02:13
May 13, 14:28	November 9, 02:17
May 14, 14:32	November 10, 02:21
May 15, 14:36	November 11, 02:25
May 16, 14:40	November 12, 02:29
May 17, 14:44	November 13, 02:33
May 18, 14:48	November 14, 02:37
May 19, 14:52	November 15, 02:41
May 20, 14:56	November 16, 02:45

Improve your Remote Viewing Accuracy Techniques using Quantum Microtubules

May 21, 15:00	November 17, 02:48
May 22, 15:04	November 18, 02:52
May 23, 15:08	November 19, 02:56
May 24, 15:12	November 20, 03:00
May 25, 15:15	November 21, 03:03
May 26, 15:19	November 22, 03:07
May 27, 15:23	November 23, 03:11
May 28, 15:27	November 24, 03:15
May 29, 15:31	November 25, 03:19
May 30, 15:35	November 26, 03:23
May 31, 15:39	November 27, 03:27
June 1, 15:43	November 28, 03:31
June 2, 15:47	November 29, 03:35
June 3, 15:51	November 30, 03:40
June 4, 15:55	December 1, 03:44
June 5, 16:00	December 2, 03:48
June 6, 16:03	December 3, 03:52
June 7, 16:07	December 4, 03:56
June 8, 16:11	December 5, 04:00
June 9, 16:15	December 6, 04:03
June 10, 16:18	December 7, 04:07
June 11, 16:22	December 8, 04:11
June 12, 16:26	December 9, 04:15
June 13, 16:30	December 10, 04:19
June 14, 16:34	December 11, 04:23
June 15, 16:38	December 12, 04:27
June 16, 16:42	December 13, 04:33
June 17, 16:46	December 14, 04:37
June 18, 16:50	December 15, 04:41
June 19, 16:54	December 16, 04:46
June 20, 16:58	December 17, 04:50
June 21, 17:02	December 18, 04:54
June 22, 17:06	December 19, 04:58
June 23, 17:10	December 20, 05:02
June 24, 17:14	December 21, 05:06
June 25, 17:18	December 22, 05:10

June 26, 17:22 June 27, 17:26 June 28, 17:30 June 29, 17:34 June 30, 17:38	December 23, 05:14 December 24, 05:18 December 25, 05:25 December 26, 05:29 December 27, 05:33 December 28, 05:28 December 29, 05:33 December 30, 05:37 December 31, 05:41

Reasons for Failed ARV Sessions
A false sense of confidence can arise when the protocols are not fully carried out during an ARV session. This can result in errors in the data collected during the session. It is best to follow these 3 simple principles. 1 - Preparing the body's Parasympathetic Nervous System. 2 - Preparing for the ARV session ahead of time, including knowing the date favorable solar weather conditions will take place. 3 - Proper preparation of the tools and equipment before an ARV session.

A Remote Viewing Financial Markets Template
Use / make copies of the chart on the following page to use in your financial markets ARV sessions.

Improve your Remote Viewing Accuracy Techniques using Quantum Microtubules

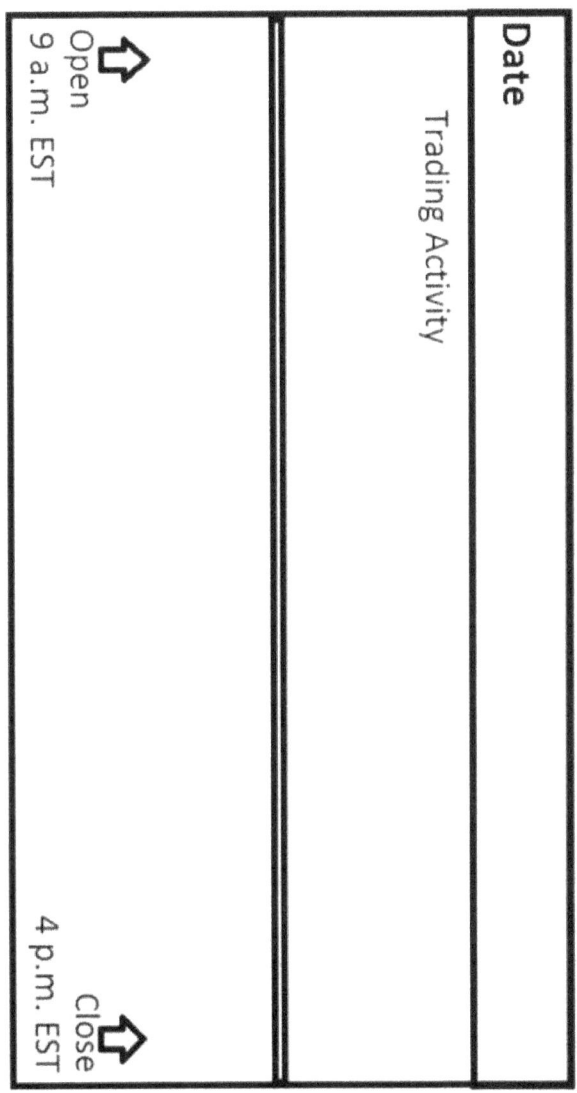

Scott Rauvers

Why Time Flows at Different Speeds According to its Mass

It is a fact that gravity affects time. It may be that "*time slows down*" or the mind is able to comprehend the rapid speed at which time and information are flowing during favorable solar weather conditions. This may be due to a form of "quantum time dilation" taking place. Because these effects occur at the quantum level, trained consciousness is able to pick up these subtle effects (information). Hence, If indeed time dilation is taking place at the quantum level, it may be occurring in the quantum foam. It has already been proven mathematically that time dilation occurs in the quantum foam. Below is a quote from the brilliant paper titled: Quantum Foam, Gravity And Gravitational Waves[27].

"*The quantum foam induces actual dynamical time dilations and length contractions. These are in agreement with the Lorentz interpretation of special relativistic effects*"

Summary
The rate at which time flows is never always constant, and this is especially more so at the quantum/nano level. For example large bodies orbiting earth such as the moon cause time

dilation. Hence, time flows slightly slower when the moon is full (*we weigh less when the moon is directly overhead*). It may be that the right equipment (*resonators*), the right solar weather conditions and the proper protocol used in associative remote viewing help enhance this effect to our advantage, allowing a clearer glimpse into future events.

We experience every-day time as a rapid sequence of events, like a river if you will. At certain times the flow of the river is slower, hence we are clearly able to see what is flowing down the river.

References. Chapter 37 & 38

1. INNAMORATI, M., SIGNORINI, P. (1980): Ritmineivegetali: rilevamentoedanalisi. G. Bot. Italiano 114 (3-4): 124.
2. SPRUYT, E., VERBELEN, J.-P., DE GREEF, J. A. (1987): Expression of Circaseptan and Circannual Rhythmicity in the Imbibition of Dry Stored Bean Seeds. Plant Physiol. 84: 707-710.
3. BROWN, F. A., Chow, C. S. (1973): Lunar-correlated Variations in Water Uptake by Bean Seeds. Biol. Bull. 145: 265-278.
4. Kolisko, E. and L. Kolisko. (1939 / 1953): Agriculture of Tomorrow, Stroud, Gloucester, England: Kolisko Archive, original publication, 1939, Bournemouth 1947 Trad. allem.:Landwirtschaft der Zukunft, Meier &Cie, Schaffhausen Switzerland, 1953.
5. Lunar-Correlated Variations in Water Uptake by Bean Seeds Frank A. Brown, Jr. and Carol S. Chow Biological Bulletin Vol. 145, No. 2 (Oct., 1973), pp. 265-278

6. SEMMENS, E. (1923): Effect of Moonlight on the germination of Seeds. Nature Vol. 111: 49-50.
7. POLARIZED LIGHT AND STARCH HYDROLYSIS. BY JOHN W. M. BUNKER AND EDMUND G. E. ANDERSON. (From the Laboratory of Biochemistry, Massachusetts Institute of Technology, Cambridge.) (Received for publication, February 21, 1928.)
8. Proceedings of the American Pharmaceutical Association at the ..., Volume 34 By American Pharmaceutical Association. Meeting
9. Linalool. Wikipedia.
10. Odorographia a natural history of raw materials and drugs. By John Chrles Sawer.
11. Canadian Pharmaceutical Journal, Volume 22
12. Semi-annual Report of Schimmel & Co. (Fritzsche Brothers)
13. J. Zurzycki. ACTA Societatis Botanicorum Poloniae. Vol XXIV #3. 1955
14. The History of Barbados By R.H. Schomburgk
15. The Eclectic, Volume 3

16. Report of the Annual Meeting, Volume 29 By British Association for the Advancement of Science
17. Front Cover 0 Reviews Write review The Civil Engineer and Architect's Journal, Volume 18 By William Laxton
18. Popular Lectures on Science and Art, Volume 1 By Dionysius Lardner
19. American Journal of Science, Volume 15
20. Spiess, H. (1994): Chronobiologische Untersuchungen mit besonderer Berucksichtigung lunarer Rhythmen im biologische-dynamischen Pflanzenbau, Darmstadt; 2 vols.
21. Relation of the weather and the lunar cycle with the incidence of trauma in the Groningen region over a 36-year period. W. Stomp et al. Nov 2009
22. Rainfall variations induced by the lunar gravitational atmospheric tide and their implications for the relationship between tropical rainfall and humidity Authors Tsubasa Kohyama, and John M. Wallace. January 2016
23. Lunar gravitational atmospheric tide, surface to 50 km in a global, gridded data set Authors Tsubasa

Kohyama, and John M. Wallace. Dec 2014
24. American Journal of Science, Volume 15. Page 174
25. The Encyclopaedia Britannica: A Dictionary of Arts, Sciences. Volume 16 Page187 Spontaneous delivery is related to barometric pressure. Akutagawa O et al. April 2007
26. BIO-DYNAMIC ASSOCIATION OF INDIA (BDAI) BDAI Secretariat, #20, 16th 'C' Main, 2nd 'A' Cross, 4th Block, Koramangala, Bangalore 560 034, India. he Planting Calendar Rhythms
27. QUANTUM FOAM, GRAVITY AND GRAVITATIONAL WAVES. Reginald T. Cahill. School of Chemistry,Physics and Earth Sciences. Flinders University. Adelaide 5001, Australia.

Closing Remarks / Final Summary

Both the effects of the moon (*gravity*) and earth's geomagnetic activity (*electromagnetism*) cannot be conventionally shielded, which is why both conditions that are not favorable increase the chance an associative remote viewing session will be inaccurate. However, knowing the favorable conditions (*sweet spot and right lunar phase*) enhance remote viewing. This proves that the right geomagnetic and lunar conditions are vital to successful remote viewing projects.

Everything around us consists of information which flows through us, being interpreted by our brains. This is what sets our moods and inspires us. Favorable solar weather conditions inspire us, generating new ideas and creativity.

Both bergamot and mistletoe generate biophotons (*which occur in the left side of the brain*) and both are powerful cancer fighters. The left brain rules logic. The Tao states excess intuition is balanced by more logic. One method to create balance may be to spend the mid to late afternoon of an ARV session listening to Alpha rhythm binaural beats or performing other exercises to enhance Alpha brainwaves while inhaling bergamot essential oil, burning incense (*generates alpha waves*) or performing other actions which encourage left brain responses to generate the necessary balance necessary for a successful remote viewing session later on.

Essential Oils and Creativity

Because creativity is related to intuition and essential oils lower blood pressure, it would make sense that certain essential oils would enhance/boost one's creativity. Let's take a look at some of the properties of essential oils and which ones are commonly used to enhance creativity. The most popular essential oil credited with enhancing creativity is Bergamot. Its scent is sweet and fruity. Bergamot keeps one alert and focused while reducing anxiety, stress and apathy.

The second most popular essential oil credited with enhancing creativity is Jasmine, which has a deliciously fragrant scent producing feelings of confidence, euphoria, calm and optimism. Basil essential oil (high in linalool) is said to the third eye and mind. Other essential oils credited with creativity include Neroli (citrus like scent) which creates feelings of personal contentment and peace while reducing stress and calming the nerves. Rose, Sandalwood and Frankincense also work well with sandalwood and are said to be good for writers who have writer's block. It instills self-awareness and self-confidence while creating a peaceful work environment. Others include Peppermint,

Scott Rauvers

Rosemary, Lavender, Eucalyptus and Patchouli and Petitgrain.

A List Of 6 Tea Recipes That Enhance Intuition

1. Calms the Mind and Enhances Intuition

1 tsp of chamomile buds
2 tsp of black tea
1 tsp of mugwort

Steep in approx two cups of hot water for one to two minutes. Sip before bed.

2. Meditation and Prophetic Dreams

1 tsp chamomile buds
2 tsp of white tea leaves
2 tsp of mugwort
A pinch of cinnamon
A pinch of nutmeg

Steep for between five and ten minutes in approximately 2 to 3 cups of hot water. Consume before a nightly mediation or before bed. Use a dream journal to write out any prophetic dreams received during the night.

3. Strengthening Clairvoyance

1 tsp lavender buds
1 tsp valerian root

2 tsp chamomile buds
A pinch of cinnamon
Steep mixture in a cup hot water for between one and four minutes before bed. Works well with meditations that open the third eye.

4. Receiving Psychic Messages and Astral Travel

2 tsp mugwort leaves
dash of honey (manuka is preferred)

Steep mugwort for ten minutes in 1 to 2 cups of hot water. Add honey and drink before bed.

5. Meditation and Visualisation Work

1 tsp peppermint leaves
1 tsp skullcap
2 tsp green tea leaves
1 tsp rose buds

Steep between 5 and 6 minutes in two cups of hot water. Consume before meditation or visualisation work. Also works well for enhancing vivid dreams or past life regression work.

6. Energy Healing while you Sleep

1 tsp passionflower
1 tsp lemon balm
2 tsp of chamomile buds
1 tsp of dried rosemary
1 tsp peppermint

Steep between 4 and 5 minutes in about two cups of hot water and sip before bed.

Scott Rauvers

Solar Eclipses and Wind Speed

A reserach study conducted in England during the 1999 solar eclipse found that the average wind speed across a region over southern England located in an inland cloud-free zone dropped by 0.7 metres per second. Also they noted that the wind's direction turned an average anticlockwise direction of 17°. Hence the eclipse caused the winds to flow in from the east. more easterly. Also temperatures fell by an average of 1°C.

Article Reference - Eclipses' effect on wind revealedMarch 29, 2012 by Tom Marshall. Phys.org

Techinical Reference - Diagnosing eclipse-induced wind changes. SL Gray and RG Harrison, Proc. R. Soc. A. doi: 10.1098/rspa.2012.0007

Heart Rate Variability (HRV) And Exercise

This discovery was made by Dr. Herbert Dardik who states that humans are waves of energy with everything in us oscillating. Dr. Dardik is Head of Vascular Surgery at Englewood Hospital, NJ.

It may be that this oscillating is affected by lunar cycles. Dr. Dardik has put together an exercise routine that is not too strenuous and that is good for the health of the body. Two clinical trials have proven that his regime does work. Here is his schedule using a cycle exercise machine.

- The week before the New Moon, exercises are done in the early morning between 6 and 9 a.m. The cycles should be very gentle with 3 to 4 moderate bursts of exertion and recovery. This should be done three times over the week. This stage is designed to kick start metabolism

- The following week after the New Moon the exercises are done a little later on.

The best time is between 9 and 12 noon. During this time exercise intensity is kicked up a notch with 3 to 4 bursts of exertion and recovery performed three times a week.

- The week before the Full Moon exercise is performed between the hours of 3 and 6 in the mid to late afternoon. Cycles are done at maximum intensity, reaching the maximum heart rate that is appropriate for the person.

- The week after a Full Moon is the recovery period. Next a week of rest takes place with the cycling beginning all over again.

I want to state here that over the years, we have found that doing mild exercise the afternoon before an ARV session, which consists of a short to medium bicycle ride, seems to add to the accuracy of the ARV session.

Clinical Trials of Dr. Dardik's Lunar Exercise Routine

Immune function and stress resistance improved and relief of anxiety improved. Participants slept better and had more energy. 9% increase in HRV, VO2 max increased 15.5% and a 7.5% drop in diastolic blood pressure (Goldsmith R, et al. Am J Med and Sports. 2002; 4: 135–141). The greater one's heart rate variability (HRV) the overall better heart health. A more limited HRV increases cardiovascular risk.

Anti-inflammatory signalling molecules, such as IL-10 and adrenocorticotropin became elevated (Cadet P. Int J Molecular Medicine. 2003; 12: 485–492).

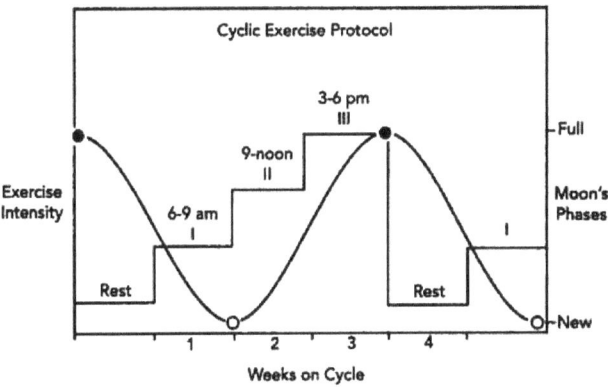

It is interesting to note that moving objects as well as the full moon cause objects to weigh less. Perhaps this combined with physical exercise is affecting HRV through changes in the heart. It is also interesting to note that in mythology, the star Arcturus is recommended as the star for travel or taking long journeys.

Further Reading

Can Gravitational Waves Predispose Neuro-Cardiovascular Circadian Rhythms?. Singh RB. Halberg Hospital and Research Institute, Moradabad, Uttar Pradesh, India), People's Friendship University of Russia, Moscow, Russia, Sergey Chibisov), (Department of Food science and Nutrition, Faculty of human Environmental Sciences, Mukagawa Women's University, Japan. Toru Takahashi. Graduate School of human Environment Medicine, Fukuoka, Women's University, Fukuoka, Japan.DOI: 10.15761/CRT.1000159

Improve your Remote Viewing Accuracy Techniques
using Quantum Microtubules

Materials References

Monoterpenes in Essential Oils

Grapefruit	93%
Silver Fir	92%
Bitter Orange	90%
Mandarin	90%
Orange	90%
Tangerine	90%
Balsam Fir	83%
Angelica	80%
Lemon	80%
Frankincense	78%
Celery Seed	77%
Cypress	76%
Parsley	75%
Fleabane	73%
Elemi	72%
Galbanum	72%
Douglas Fir	70%
Nutmeg	68%
White Fir	65%
Lime	62%
Bergamot	55%

Phenol Levels in Essential Oils

Wintergreen	97%
Anise	90%
Birch	90%
Clove	77%
Basil	76%
Tarragon	75%
Fennel	72%
Oregano	70%
Thyme	50%
Mountain Savory	49%
Peppermint	39%
Tea Tree	33%
Calamus	29%
Cinnamon Bark	26%
Moroccan Thyme	26%
Citronella	23%
Marjoram	19%
Nutmeg	18%
Lemon Eucalyptus	11%
Parsley	11%

Keytone Levels in Essential Oils

Camphor		a-Thujone	
White Camphor	45%	Cedar Bark	77%
Rosemary	18%	Cedar Leaf	72%
Juniper	14%	Thuja	45%
Sage	14%	Sage	23%
Blue Tansy	13%	Mugwort	6%
Yarrow	11%	Idaho Tansy	5%
Lavandin	8%		
Rosemary Verbenon	8%		
Idaho Tansy	6%		
Coriander	5%		

Fenchone		b-Thujone	
Thuja	11%	Idaho Tansy	67%
Fennel	10%	Thuja	12%
Cedar Leaf	5%	Sage	8%
		Cedar Bark	7%
		Cedar Leaf	7%
		Mugwort	6%

Monoterpenes in Essential Oils Chart #2

19.5%	of essential oils contain 5% or more of:	β–Pinene
19.5%	of essential oils contain 5% or more of:	l–limonene
14.2%	of essential oils contain 5% or more of:	d–limonene
13.3%	of essential oils contain 5% or more of:	Sabinene
12.4%	of essential oils contain 5% or more of:	γ–Terpinene
9.7%	of essential oils contain 5% or more of:	Camphene
8.8%	of essential oils contain 5% or more of:	Myrcene
6.2%	of essential oils contain 5% or more of:	p–Cymene
6.2%	of essential oils contain 5% or more of:	Ocimene
5.3%	of essential oils contain 5% or more of:	δ–3–Carene
5.3%	of essential oils contain 5% or more of:	α–Terpinene
3.5%	of essential oils contain 5% or more of:	α–Phellandrene
2.7%	of essential oils contain 5% or more of:	β–Phellandrene

Essential Oils that have the Most Popular Monoterpenes

Cypress	51%
Cistus	50%
Balsam Fir	38%
Frankincense	38%
Pine	33%
Douglas Fir	32%
Juniper	30%
Myrtle	26%
Rosemary Verbenon	25%
Silver Fir	24%
Angelica	22%
Nutmeg	22%
Eucalyptus	16%
Dill	15%

(*continued next page*)

Niaouli	11%
Rosemary	11%
E. Radiata	10%
Lemon	10%
Roman Chamomile	10%
White Fir	10%
Bay Laurel	7%
Ravensara	7%
Black Pepper	6%
Fennel	6%
Valerian	6%
Coriander	5%
Marijuana	5%
Tea Tree	5%
Vitex	5%

A list of Terpene alcohols
carveol, geraniol, piperitol, borneol, carveol, isoborneol, santalol, citronellol, thujanol, tetrahydromyrcenol, terpineol, nerol, terpineol, menthol, dihydromyrcenol, myrcenol, rhodinol, hydroxycitronerol, farnesol and perilla alcohol, linaloolmyrtenol, nerolidol,

A list of Cyclic ketones
ethylmaltol, cyclotene, dihydronootkatone, isomenthone, carvone, menthone, dihydrocarvone, pulegone, jasmone, sotolone,

damascone, cyclopentadione, piperitone, cyclohexene, damascenone, maltol, nootkatone, butylcyclohexanone, furaneol, methylionone, isomethylionone and camphor.

A list of Aromatic ketones
raspberry ketone, nisylacetone anisylideneacetone, acetonaphthone, acetophenone, methylacetophenone and methoxyacetophenone.

Nicotine hydrogen tartrate salt can be purchased online

Lithium and Potatoes

Certain locations contain more lithium then others. For example some mineral waters are high in lithium and the groundwater in northern Chile is one example. Plants of the nightshade family such as potatoes, peppers and tomatoes absorb lithium very well. Potato is a hygroscopic material, meaning it readily absorbs moisture.

Potato juice is used to naturally lighten the skin due to it containing the enzyme catecholase. Potatoes are also used for treating acne breakouts.

Lithium- and manganese-rich transition metal oxides have hexagonal patterns (*Unravelling structural ambiguities in lithium- and manganese-rich transition metal oxides. Alpesh Khushalchand Shukla et al. Nat Commun. 2015; 6: 8711.Published online 2015 Oct 29. doi: 10.1038/ncomms9711. PMCID: PMC4846316*).

Improve your Remote Viewing Accuracy Techniques
using Quantum Microtubules

Van Der Waals Radius of the elements

Text lists sorted by: Value | Atomic Number | Alphabetical
Plots: 3D Live | Shaded | Ball | Crossed Line | Scatter | Sorted Scat
Log scale plots: 3D Live | Shaded | Ball | Crossed Line | Scatter | Sorted Scat
Good for this property: Crossed Line

Hydrogen 120.pm	Aluminum N/A
Zinc 139.pm	Calcium N/A
Helium 140.pm	Scandium N/A
Copper 140.pm	Titanium N/A
Fluorine 147.pm	Vanadium N/A
Oxygen 152.pm	Chromium N/A
Neon 154.pm	Manganese N/A
Nitrogen 155.pm	Iron N/A
Mercury 155.pm	Cobalt N/A

Tellurium Dioxide and Lead Molybdate. Minerals used in Acousto-optic Devices

Tellurium Dioxide, also called TeO2 or Paratellurite (laser quality) makes an excellent piezo-optical material when the crystals are grown using the Czochralsky method. This allows it to be used to make tunable polarisation filters, radio frequency spectral analysers, imaging devices, splitters, deflectors, and other types of acousto-optoelectronic equipment used for laser radiation control. This is due to its high acousto-optical figure.

Lead Molybdate (PbMoO4 or PM) crystal is an efficient material used for acousto-optic devices. Its use is in deflectors, acousto-optic modulators and phase-shifters when grown by the Czochralsky-Kyropulos method. Lead Molybdate contains high optical homogeneity, low optical losses and is stable in laser radiation environments. Tungsten, lead molybdate and zinc can be extracted from Selenium and Tellurium (*Trace Analysis of Semiconductor Materials. J. Paul Cali et al*).

Piezoelectric Crystals that have Elastic Properties

Rochelle salt, sodium chlorate, sodium bromate, potassium dihydrogen phosphate, ammonium dihydrogen phosphate and quartz, are assembled. Non-piezoelectric crystals include lithium fluoride, sodium chloride and pyrite.

Egyptians Healing Rods and the Schumann Resonance

Did the ancient Pharaohs of Ancient Egypt know how to harness the Schumann Resonance and use it to improve their health and for longevity? Scattered throughout the ancient Egyptian ruins one can see statues holding Egyptian rods in their hands with their left foot forward.

Dr. Valery Uvarov of Russia discovered that the rods are hollow. The rod held in the left hand is made from Zinc and the rod held in the right hand is made from Copper. Various substances including charcoal powder (carbon) or quartz crystals can be placed in each rod. If one looks at a table of chemical elements one will see that Zinc, a dense element, is directly adjacent to the light and airy chemical Copper. By having rods of these elements held in each hand may be causing the Zinc to create low amplitudes in the Schumann Resonance while the Copper creates high amplitudes. Hence, high and low amplitudes of the Earth's Schuman resonance generates a series of spirals at the intersection point that is located in the center of the body when a person holds the rods in their palms. These spirals of energy rejuvenate and generate balance allowing one to experience high energy levels.

Tuning Forks for the Parasympathetic Nervous System

In our second book on remote viewing we mention the use of a C256 and a G384 tuning fork which are both struck at the same time just before an ARV session is conducted. A C256 and a G384 struck both together create what's called the "**interval of the fifth**". Placing each of these on each side of the head after they have been struck has been struck helps bring immediate balance to the body's sympathetic and parasympathetic nervous systems. This tone generates balance and creates harmony by unifying opposite energies by balancing the parasympathetic and sympathetic nervous systems.

Other Tuning Forks for the body's Parasympathetic Nervous system

C-64Hz - This tuning fork loosens sacral ligaments and stimulates balance between the sympathetic and parasympathetic nervous systems. It is especially useful for grounding and healing emotional trauma.

The Otto Tuners

The Otto Tuners resonate with the deeper tones of Earth. Otto 128 (as well as the C & G Tuners) increase nitric oxide within the body. It allows the body to gradually move into a Parasympathetic "rest and relax" dominance.

The Otto Tuner is the note of C in the diatonic scale. The C frequency project sound into the body's bones and replenish the tissue tone and bring re-alignment. Works great for relaxing bones and for sore muscles.

Otto Tuning Forks are available in 3 frequencies. This Tuner is at a frequency of 64hz and vibrates at a low pitch due. Most people begin at 128hz and then move to 64hz and then to the 32Hz tuning fork. It is very effective placed on the vertebra, sternum and meridian points. It is mostly primarily for grounding and relaxation.

Nutrients that calm the sympathetic system include selenium, manganese, zinc, calcium, magnesium and kelp.

Acupressure Points for Influencing Heart Rate Variability (HRV)

A research study conducted in Harbin, China, recorded participant's HRV activity during stimulation of the Shenmen point on the left ear (ear acupressure). The study found that HR decreased after and during the acupressure stimulation. The study also found that the effect did not occur after the first stimulation, but instead appeared in the phase following the second acupressure stimulation, which took place 10 minutes after the first stimulation. A maximum occurred after the third stimulation, which took place 20 min after the first stimulation. However the effect was not long-lasting. The study found that this may serve in treating insomnia.

It sounds like using this technique with essential oils that enhance sleep and valerian may yield very good results for those who can't get to sleep at night, as well as bring balance to HRV.

The acupoint **Shenmen** on the left ear is commonly used for healing pain, insomnia and skin itching. Its action is due to easing the mind, relieving pain and subduing inflammation. The acupressure procedure lasted 15 seconds each time, with two pressure movements taking place each second. This resulted in 30 pressure movements per stimulation. The Acupressure took place every 10 minutes, totaling of 3 sessions during the entire period. The maximum increase in

HRV took place after the third acupressure stimulation at the ear.

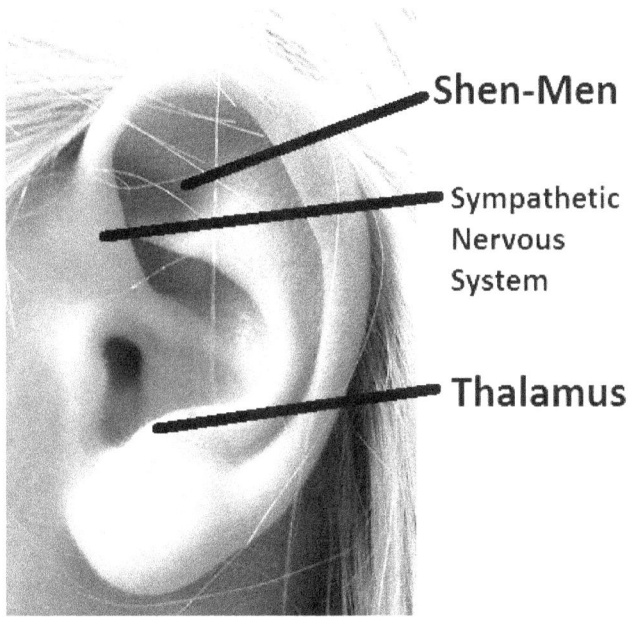

The study concluded the following
The participant's Heart rate decreased significantly after and during the acupressure stimulation following the second acupressure stimulation, which took place 10 minutes after the first stimulation. The participant's total HRV showed significant stimulation-dependent increases. These took place immediately after each acupressure stimulation cumulating to maximum after the third stimulation, which took place 20 minutes after the first stimulation. However no long-lasting effect took place.

Reference
Ear Acupressure, Heart Rate, and Heart Rate Variability in Patients with Insomnia. Evidence-Based Complementary and Alternative Medicine. Lu Wang et al. Stronach Research Unit for Complementary and Integrative Laser Medicine, Research Unit of Biomedical Engineering in Anesthesia and Intensive Care Medicine, TCM Research Center Graz, Medical University of Graz, Auenbruggerplatz 29, 8036 Graz, Austria
16 January 2013.

Research Study Number 2. Accupressure Points and Heart Rate Variability (HRV)

A research study showed that application of acupressure showed significant reductions in dysmenorrheic pain and also showed alterations in the body's sympathetic and parasympathetic nervous systems. The acupressure points used were the following -

SP6 - This is located on the inside leg and just above the ankle. SP6 has a strong influence on blood flow and helps stop pain in the abdomen.

LIV4 – This is located on the ankle 1 t-sun (1 inch) anterior to the medial malleolus, located in-between the tendon of the extensor hallusis longus muscle.

St41 - This is located on the ankle crease, located midway between the tips of the malleoli between the extensor digitorum longus and extensor hallicis longus tendon.

Note - These acupressure points can be easily located doing an online image search for any of the aforementioned points.

The acupressure was applied by applying pressure using a blunt probe in the regions mentioned above for 2 minutes. The results were high frequency (LF/HF) ratio and low frequency, was reduced significantly in dysmenorrhoeic women and the acupressure therapy increased the HF/LF ratio (group III). Measurement of the participant's heart rate variability showed significant decreases in the low frequency value in the pre therapy group. This indicates sympathetic activity and the acupressure therapy significantly returned it back to normal control (group III).

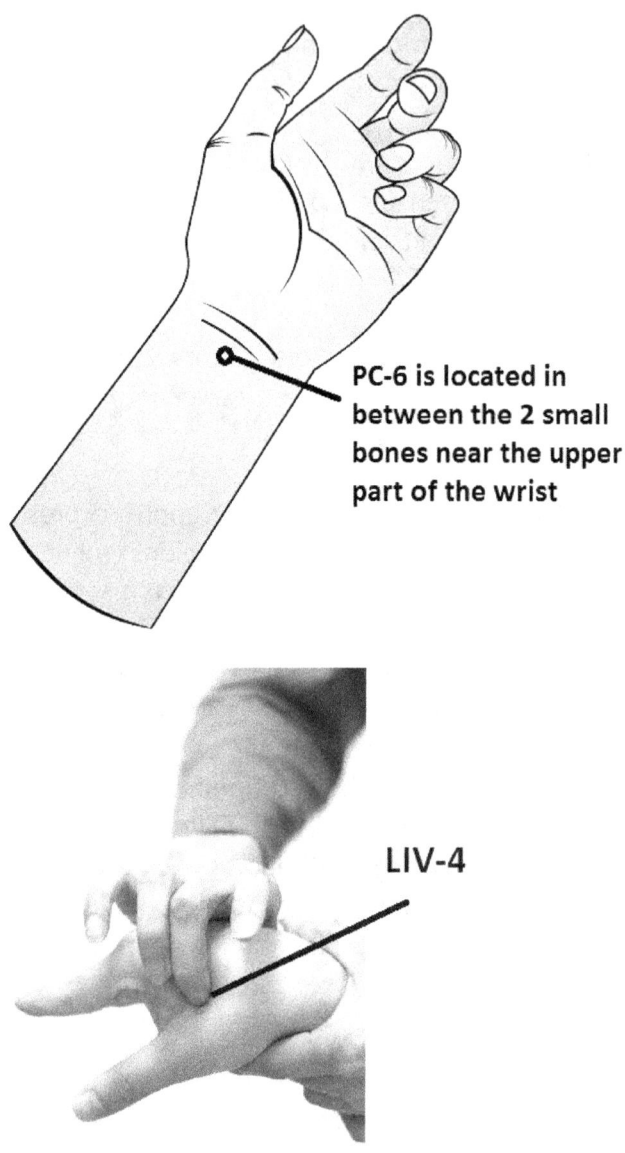

PC-6 is located in between the 2 small bones near the upper part of the wrist

LIV-4

Reference

Effect Of Acupressure And Changes In Heart Rate Variability In Dysmenorrhoea. R. Archana et al. Recent Research in Science and Technology 2011, 3(10): 01-06 ISSN: 2076-5061 www.recent-science.com. Department of Physiology, Saveetha Medical College, Thandalam, 933, Chennai-600101, India

Acupuncture Points for Influencing Heart Rate Variability (HRV)

A research study involving acupuncture points Hegu (**LI4**) and Neiguan (**PC6**) and its effect on heart rate variability in subjects under fatigue and non-fatigue states took place. The study found that stimulations of the acupuncture points showed a significant decrease in their heart rate (HR), low frequency (LF) power, HRV total power (TP), and the ratio of low frequency to high frequency (LF/HF). Also a significant increase in the HF power in normalized units (NU) during post stimulation period while in fatigue state was observed ($P<0.05$). Also stimulation of the acupuncture points showed a significant increase both in the HF power and LF power in absolute units (AU) ($P<0.05$). No significant change in NU occurred during the post stimulation period in non-fatigue state.

The study concluded the modulating effect of acupuncture on the participant's heart rate variability depended on the points of stimulation and also on the functional state of the participants (if the participant's were in a state of fatigue or not).

Reference
Effects of acupuncture on heart rate variability in normal subjects under fatigue and non-fatigue state. Eur J Li

Z et al. Appl Physiol. 2005 Aug;94(5-6):633-40. Epub 2005 May 20. Wang C, Mak AF, Chow DH.

Music and Exercise and its effects on the body's Autonomic Nervous System

A research study conducted by Professor Masachiro Kohzukiand and Assistant Professor Yoshiko Ogawa at Tohoku University Graduate School reported that relaxing music causes a decrease in parasympathetic nervous system activity only after the body has been exercising. Studies show that physical exercise increases the body's sympathetic nervous system while decreasing parasympathetic nervous system activity. This causes an increased heart rate. This increased heart rate declines rapidly after exercise.

The study involved 26 graduate students taking part in four sessions in random order for four days. The students were asked to exercise on a cycle ergometer and listen to their favorite music. The study found that High frequency power (HF) which is an index of the body's parasympathetic nervous system activity became significantly increased during the music session. The student's heart rate increased, and their high frequency power decreased during the bicycling session.

The study showed no significant difference existed in their high frequency power after or before bicycling with music (although their heart rate was significantly increased).

The study concluded that music increased the body's parasympathetic activity and attenuated the exercise-induced decrease in their parasympathetic nervous system and that music play an effective role for improving post-exercise parasympathetic reactivation. This can result in faster recovery and assist in a reduction in cardiac stress after exercise.

Reference
Music Attenuated a Decrease in Parasympathetic Nervous System Activity after Exercise"
Assistant professo Yoshiko Ogawa, Professor Masachiro Kohzuki. Department of Internal Medicine and Rehabilitation Science, Tohoku University Graduate School of Medicine.

Improve your Remote Viewing Accuracy Techniques
using Quantum Microtubules

Scott Rauvers

ABOUT THE AUTHOR

Scott Rauvers is the founder of the **Solar Institute**, a think tank that studies solar weather and its effects on behavior, the economy and human health. Additional topics are also researched as special projects when necessary.

Discoveries made by the Solar Institute over the past decade have ended in solutions that utilize less resources, significantly improve health via non-invasive methods, are non-polluting and enhance productivity, all saving time, energy and resources in the process. These benefits are the future, one in where our planet's resources are preserved and maintained and where future energy needs are met. Health and healing follow this same tradition, only that awareness is greater and understanding plays a role.

Scott Rauvers

Founder of the Institute for Solar Studies on Behavior and Human Health.

www.ez3dbiz.com

Improve your Remote Viewing Accuracy Techniques using Quantum Microtubules

The 10 Rules of Discovery Science

1) Substances / Tools used to measure / induce an effect can also influence the object(s) being measured / studied

2) If the same information you are researching keeps popping up, it is true

3) Follow the data. After recording the data, assemble the theory. This creates reproducible results

4) Create a summary of your facts and conclusions

5) The more passive one is before nature, the more nature will reveal its secrets

6) Nature never uses a language we can understand. However, nature is talking to us all the time in the form of energy and information. The key is to listen and identify what it is saying.

7) To have the real facts revealed, turn yourself into an opponent of all you study. Be sure you thoroughly assess all the main and marginal parts, then oppose them from every possible angle, including all its hidden aspects. Sticking faithfully to this course of action will clearly yield all the facts.

8) Review your data, then mold the data into proper context by simplifying the data by connecting the relevant dots

9) Constant and Reproducible results - all final data must be 100% reproducible

10) Great minds discuss ideas, average minds discuss events and small minds discuss people.

Further Reading

Arnette, J. (1999). The theory of essence. III. Neuroanatomical and neurophysiological aspects of interactionism. Journal of Near-Death Studies, 14, 73-101.

Aspect, A., Dalibard, J., and Roger, G. (1982). Experimental test of Bell inequalities using time-varying analyzers. Physical Review Letters, 49, 1804.

Audain, L. (1999). Near-death experiences and the theory of the extraneuronal hyper space. Journal of Near-Death Studies, 18, 103-115.

Becker, C. B. (1995). A philosopher's view of near-death research. Journal of Near-Death Studies, 14, 17-28.

Begley, S. (1996, July 15). You must remember this. Newsweek, p. 64.

Bensimon, G., and Chermat, R. (1991). Microtubule disruption and cognitive defects: Effect of colchicine on learning behavior in rats. Pharmacology, Biochemistry, and Behavior, 38, 141-145.

Colli, J. E. (2001). Angels and aliens: Encounters with both near-death and UFOs. Pre sented at the International Association of Near-Death Studies 2001 International Con ference, Seattle, WA.

Cotton, P. (1994). Medical news and perspectives: Biology enters repressed memory fray. Journal of American Medical Association, 272, 1725-1726.

Fabiani, M., Stadler, M. A., and Wessels, P. M. (2000). True but not false memories pro duce a sensory signature in human lateralized brain potentials. Journal of Cognitive Neuroscience, 12, 941-949.

Frohlich, H. (1988). Theoretical physics and biology. In H. Frohlich (ed.), Biological coherence and response to external stimuli (pp. 1-24). Berlin, Germany: Springer Verlag.

Gauthier, I., Hayward, W. G., Tarr, M. J., Anderson, A. W., Skudlarski, P., and Gore, J. C. (2002). BOLD activity during mental rotation and viewpoint-dependent object recognition. Neuron, 34, 161-

Goswami, A. (1993). The self-aware universe: How consciousness creates the material world. New York, NY: Tarcher/Putnam.

Greene, F. G. (1981). A glimpse behind the life review. Journal of Religion and Psychical Research, 4, 113-130.

Hameroff, S. (2001). Feasibility of macroscopic quantum mechanisms in the brain. Retrieved April 18, 2002, from the University of Arizona course "Consciousness at the Millennium: Quantum Approaches to Understanding the Mind" Web site: http://www.consciousness.arizona.edu/quantum/week7a.htm

Hameroff, S. (1994). Quantum coherence in microtubules: A neural basis for emergent consciousness? Journal of Consciousness Studies, 1, 91-118.

Hirano, I., and Hirai, N. (1986). Holography in the single-photon region. Applied Optics, 25, 1741-1742. Ho, M-W. (1999, October 2). Coherent Energy, Liquid Crystallinity and Acupuncture [Talk presented to the British Acupuncture Society]. Retrieved April 18, 2002, from the Institute of Science in Society Web

Ho, M-W. (1998a). Organism and psyche in a participatory universe. In D. Loye (ed.), The evolutionary outrider. The impact of the human agent on evolution: Essays in honour of Ervin Laszlo (pp. 49-65).

Westport, CT: Praeger. Ho, M-W. (1998b). The rainbow and the worm: The physics of organisms. Singapore: World Scientific Publishing. Koruga, D. (1995). Information physics: In search of a scientific basis of consciousness. In D. Koruga and D. Rakovic (eds.), Consciousness: Scientific challenge of the 21st century (pp. 243-261). Belgrade, Serbia: European Centre for Peace and Development (ECPD) of the United Nations University for

Peace. Koruga, D., Hameroff, S., Withers, J., Loutfy, R., and Sundareshan, M. (1993). Fullerene C60: History, physics, nanobiology, nanotechnology. New York, NY: North-Holland.

Krishnan, V. (1996). Misidentified flying objects. [Letter]. Journal of Near-Death Studies, 14, 287-290. Lamoreaux, S. (1997, January 6). Demonstration of the Casimir force in the 0.6 to 6 micron range. Physical Review Letters, 78(1), 5-8.

Laszlo, E. (1995). The interconnected universe:

Conceptual foundations of transdisciplinary unified theory. Singapore: World Scientific Publishing.

Loftus, E. F. (1997, September). Creating false memories. Scientific American, 277(3), 70-75.

Loftus, E. F., and Hoffman, H. G. (1989). Misinformation and memory: The creation of new memories. Journal of Experimental Psychology: General, 118, 100-104.

Loftus, E. F., and Loftus, G. R. (1980). On the permanence of stored information in the human brain. American Psychologist, 35, 409-420.

Lorimer, D. (1990). Whole in one: The near-death experience and the ethic of interconnectedness. London, England: Arkana.

MacNeill, J. (2001, May). Holographic memory: Laser beams store data in three dimensions. Technology Review, pp. 96-97.

Marcer, P. J., and Schempp, W. (1996). A mathematically specified template for DNA and the genetic code in terms of the physically realizable processes of quantum holography.

In A. M. Fedorec and P. J. Marcer (eds.), Proceedings of the Greenwich Symposium on Living Computers (pp. 45-62). Wiltshire, England: British Computer Society.

Marcer, P. J., and Schempp, W. (1997). Model of the neuron working by quantum holography, Informatica, 21, 519-534.

Marcer, P. J., and Schempp, W. (1999). Quantum holography: The paradigm of quantum entanglement. In D. M. Dubois (ed.), Computing anticipatory systems: CASYS '98, Second International Conference (pp. 461-467). College Park, MD: American Institute of Physics.

Marshall, I. (1989). Consciousness and Bose-Einstein condensates. New Ideas in Psychology, 7(1), 73-83.

Mitchell, E. (1999). Nature's mind: The quantum hologram. Retrieved April 18, 2002 from the National Institute for Discovery Science Web site: http://www.nidsci.org/articles/naturesmind-qh.html

Moody, R. (1975). Life after life. Covington, GA: Mockingbird Books. Morse, M., and Perry, P. (1992). Transformed by the light: The powerful effect of near-death experiences on people's lives. New York, NY: Villard.

Nimtz, G. (1998). Superluminal signal velocity. Annals of Physics (Leipzig), 7, 618-624.

Nimtz, G. (1999). Evanescent modes are not necessarily Einstein causal. European Phys ical Journal B, 7, 523-525.

Noyes, R., and Kletti, R. (1977). Panoramic memory: A response to the threat of death, Omega, 8, 181-194.

Penfield, W. (1969). Consciousness, memory, and man's conditioned reflexes. In K. H. Pribram (ed.), On the biology of learning (pp. 127-168). New York, NY: Harcourt, Brace and World.

Penfield, W. (1975). The mystery of the mind: A critical study of consciousness and the human brain. Princeton, NJ: Princeton University Press.

Popp, F., Li, K., Nagl, W., and Klima, H. (1983). Indications of optical coherence in biological systems and its possible significance.

In H. Frohlich and Kremer (Eds.), Coherent excitations in biological systems (pp. 117-122). New York: Springer-Verlag.

Pribram, K. (1969, January). The neurophysiology of remembering. Scientific American, 220(1), 73-86.

Puthoff, H. E. (1996). CIA initiated remote viewing program at Stanford Research Institute. Journal of Scientific Exploration 10, 63-76.

Radin, D. (1997). The conscious universe: The scientific truth of psychic phenomena. San Francisco, CA: Harper San Francisco.

Ring, K., and Valarino, E. E. (1998). Lessons from the light: What we can learn from the near-death experience. New York, NY: Plenum/Insight.

Rosenthal, M. H., and Larson, C. P. (1978). Protection of the brain from progressive ischemia. Western Journal of Medicine, 128, 145.

Sabom, M. (1998). Life and death: One doctor's fascinating account of near-death experiences. Grand Rapids, MI: Zondervan.

Schacter, D. L., Reiman, E., Curran, T., Yun, L. S., Bandy, D., McDermott, K. B., and Roediger, H. L. (1996). Neuroanatomical correlates of veridical and illusory recognition memory: Evidence from positron emission tomography. Neuron, 17, 267-274.

Sokolov, I. (1996). Consistency of memory for combat-related traumatic events in veterans of Operation Desert Storm. American Journal of Psychiatry, 154, 173-177.

Stapp, H. P. (1994). Theoretical model of a purported empirical violation of the predictions of quantum theory. Physical Review A, 50(1), 18-22.

Stuchebrukhov, A. A. (1996). Tunneling currents in electron transfer reaction in proteins. II. Calculation of electronic super exchange matrix element and tunnelling currents using nonorthogonal basis sets. Journal of Chemical Physics, 105, 10819-10829.

Van der Kolk, B. A., and Fisler, R. (1995). Dissociation and the fragmentary nature of traumatic memories: Review and experimental confirmation. Journal of Traumatic Stress, 8, 505-525.

Van der Kolk, B. A., McFarlane, A. C., and Weisaeth, L. (Eds.). (1996). Traumatic stress: The effects of overwhelming experience on mind, body and society. New York, NY: Guilford Press.

Van der Kolk, B. A., and van der Hart, O. (1991). The intrusive past: The flexibility of memory and the engraving of trauma. American Imago, 48, 425-454.

Wade, J. (1996). Changes of mind: A holonomic theory of the evolution of consciousness. Albany, NY: State University of New York Press. Wagenknecht, H. A., Rajski, S. R., Pascaly, M., Stemp, E. D., and Barton, J. K. (2001). Direct observation of radical intermediates in protein-dependent DNA charge trans port. Journal of the American Chemical Society, 123, 4400-4407.

Stuart Hameroff, Meyer-Overton Meets Quantum Physics: Consciousness, Memory and Anesthetic Binding in Tubulin Hydrophobic Channels

Travis Craddock, Volatile anesthetic interactions with tubulin and coherent energy transfer

Hirano, I., and Hirai, N. (1986). Holography in the single-photon region. Applied Optics, 25, 1741-1742. Ho, M-W. (1999, October 2). Coherent Energy, Liquid Crystallinity and Acupunc ture [Talk presented to the British Acupuncture Society]. from the Institute of Science in Society

Ho, M-W. (1998a). Organism and psyche in a participatory universe.

In D. Loye (ed.), The evolutionary outrider. The impact of the human agent on evolution: Essays in honour of Ervin Laszlo (pp. 49-65). Westport, CT: Praeger.

Ho, M-W. (1998b). The rainbow and the worm: The physics of organisms. Singapore: World Scientific Publishing.

Koruga, D. (1995). Information physics: In search of a

scientific basis of consciousness.

In D. Koruga and D. Rakovic (eds.), Consciousness: Scientific challenge of the 21st century (pp. 243-261). Belgrade, Serbia: European Centre for Peace and Development (ECPD) of the United Nations University for Peace.

Loftus, E. F., and Hoffman, H. G. (1989). Misinformation and memory: The creation of new memories. Journal of Experimental Psychology: General, 118, 100-104.

Loftus, E. F., and Loftus, G. R. (1980). On the permanence of stored information in the human brain. American Psychologist, 35, 409-420.

Lorimer, D. (1990). Whole in one: The near-death experience and the ethic of interconnectedness. London, England:

Arkana. MacNeill, J. (2001, May). Holographic memory: Laser beams store data in three dimensions. Technology Review, pp. 96-97.

Marcer, P. J., and Schempp, W. (1996). A mathematically specified template for DNA and the genetic code in terms of the physically realizable processes of quantum holography.

In A. M. Fedorec and P. J. Marcer (eds.), Proceedings of the Greenwich Symposium on Living Computers (pp. 45-62). Wiltshire, England: British Computer Society.

Marcer, P. J., and Schempp, W. (1997). Model of the

neuron working by quantum holography, Informatica, 21, 519-534.

Marcer, P. J., and Schempp, W. (1999). Quantum holography: The paradigm of quantum entanglement.

In D. M. Dubois (ed.), Computing anticipatory systems: CASYS '98, Second International Conference (pp. 461-467). College Park, MD: American Institute of Physics. Marshall, I. (1989). Consciousness and Bose-Einstein condensates. New Ideas in Psychology, 7(1), 73-83.

Mitchell, E. (1999). Nature's mind: The quantum hologram. Retrieved April 18, 2002 from the National Institute for Discovery Science

Moody, R. (1975). Life after life. Covington, GA: Mockingbird Books. Morse, M., and Perry, P. (1992). Transformed by the light: The powerful effect of near-death experiences on people's lives. New York, NY: Villard.

Nimtz, G. (1998). Superluminal signal velocity. Annals of Physics (Leipzig), 7, 618-624.

Nimtz, G. (1999). Evanescent modes are not necessarily Einstein causal. European Physical Journal B, 7, 523-525.

Aspect, A., Grangier, P. & Roger, G. 1982 Experimental realization of Einstein–Podolsky– Rosen–Bohm Gedanken Experiment: a new violation of Bell's inequalities. Phys. Rev. Lett. 48, 91–94.

Beck, F. & Eccles, J. C. 1992 Quantum aspects of brain activity and the role of consciousness. Proc. Natn. Acad. Sci. USA 89, 11357–11361.

Conrad, M. 1994 Ampli?cation of superpositional effects through electronic–conformational interactions. Chaos Solitons Fractals 4, 423–438.

Dennett, D. 1991 Consciousness explained. Boston, MA: Little, Brown and Company

Engelborghs, Y. 1992 Dynamic aspects of the conformational states of tubulin and microtubules. Nanobiol. 1, 97–105

Franks, N. P. & Lieb, W. R. 1982 Molecular mechanisms of general anaesthesia. Nature 300, 487–493.
Franks,N.P.&Lieb,W.R.1985 Mapping of general anaesthetic target sites provides a molecular basis for effects. Nature 316, 349–351.

Fr¨ohlich, H. 1968 Long-range coherence and energy storage in biological systems. Int. J. Quantum Chem. 2, 641–649.

Fr¨ohlich, H. 1970 Long range coherence and the actions of enzymes. Nature 228, 1093.

Gray, J. A. 1998 Creeping up on the hard question of consciousness. In Toward a science of consciousness. II. The second Tucson discussions and debates (ed. S. Hamero?, A. Kaszniak & A. Scott), pp. 279–291. Cambridge, MA: MIT Press.

Grundler, W. & Keilmann, F. 1983 Sharp resonances in yeast growth prove nonthermal sensitivity to microwaves. Phys. Rev. Lett. 51, 1214–1216

Halsey, M. J. 1989 Molecular mechanisms of anaesthesia. In General anaesthesia (ed. J. F. Nunn, J. E. Utting & B. R. Brown Jr), 5th edn, pp. 19–29. London: Butterworths.

Hamero?, S. R. 1994 Quantum coherence in microtubules: a neural basis for emergent consciousness. J. Conscious. Stud. 1, 91–118

Grundler, W. & Keilmann, F. 1983 Sharp resonances in yeast growth prove nonthermal sensitivity to microwaves. Phys. Rev. Lett. 51, 1214–1216.

Hamero?, S. 1998d 'Funda-Mentality'. Is the conscious mind subtly connected to a basic level of the universe? Trends Cognitive Sci. 2, 119–127

Hamero?, S. R. & Penrose, R. 1996b Conscious events as orchestrated space-time selections. J. Conscious. Stud. 3, 36–53.

Hamero?, S. R. & Watt, R. C. 1982 Information processing in microtubules. J. Theor. Biol. 98, 549–561.

Hamero?, S. R. & Watt, R. C. 1983 Do anesthetics act by altering electron mobility? Anesth. Analg. 62, 936–94

Jibu, M., Hagan, S., Hamero?, S. R., Pribram, K. H. & Yasue, K. 1994 Quantum optical coherence in cytoskeletal microtubules: implications for brain function. Biosystems 32, 195–209.

Jibu, M., Pribram, K. H. & Yasue, K. 1996 From conscious experience to memory storage and retrieval: the role of quantum brain dynamics and boson condensation of evanescent photons. Int. J. Mod. Phys. B10, 1735–1754.

Joliot, M., Ribary, U. & Llinas, R. 1994 Human oscillatory brain activity near 40 Hz coexists with cognitive temporal binding. Proc. Natn. Acad. Sci. USA 91, 11748–11751.

Neubauer, C., Phelan, A. M., Keus, H. & Lange, D. G. 1990 Microwave irradiation of rats at 2.45 GHz activates pinocytotic-like uptake of tracer by capillary endothelial cells of cerebral cortex. Bioelectromag. 11, 261–268

Penrose,R.1987.Newton,quantumtheoryandreality.In3 00 years of gravity (ed.S.W.Hawking & W. Israel). Cambridge University Press

Penrose, R. 1996 On gravity's role in quantum state reduction. Gen. Relativ. Grav. 28, 581–600

Penrose, O. & Onsager, L. 1956 Bose–Einstein condensation and liquid helium. Phys. Rev. 104, 576–584

Tejada, J., Garg, A., Gider, S., Awschalom, D. D., DiVincenzo, D. P. & Loss, D. 1996 Does macroscopic quantum coherence occur in ferritin? Science 272, 424–426.

Tuszy'nski, J., Hamero?, S., Sataric, M. V., Trpisova, B. & Nip, M. L. A. 1995 Ferroelectric behavior in microtubule dipole lattices; implications for information processing, signalling and assembly/disassembly. J. Theor. Biol. 174, 371–380

W. Meissner, R. Ochsenfeld, "Ein neur effect bei eintritt der supraleitfahigkeit", Naturwissenschaften, vol. 21, n. 44, pp. 787-788, 1933.

F. London, H. London, "The electromagnetic equations of the superconductor", Proceedings of the Royal Society of A: Mathematical, Physical and Engineeering Sciences, vol. 149, 1935.

V. L. Ginzburg, L. Landau, Zh. Eksp. Teor. Fiz., 20(1064), 1950.

J. Barden, L. N. Cooper, J. R. Schrieffer, "Theory of superconductivity", Phys. Rev. vol. 106, pp. 162-164, 1957.

I. N. Marshall, "Consciousness and Bose-Einstein condensates", New Ideas Psychol., vol. 7, pp. 73-83, 1989.

F. Crick, C. Koch, "Towards a neurobiological theory of consciousness", Semin. Neurosci, vol. 2, pp. 263-275, 1990.

W. Singer, "Synchronization of cortical activity and its putative role in information processing and learning", Annu. Rev. Physiol., vol. 55, pp. 349-374, 1993.

J. S. Bell, "On the Einstein Podolsky Rosen paradox", Physics, 1, 195-200 (1964), reprinted in Speakable and Unspeakable in Quantum Mechanics, Cambridge University Press, 1987.

(various authors), "Interpretations of quantum mechanics", list, discussion, and references in Wikipedia

Y. Aharonov, P. G. Bergmann, and J. L. Lebowitz, "Time Symmetry in the Quantum Process of Measurement", Phys. Rev. 134, B1410–B1416, 1964.

O. Costa de Beauregard, "Timelike Nonseparability and Retrocausation", arXiv:quantph/9804069v1,
1998.

H. Price, Time's Arrow and Archimedes Point, Oxford University Press, 1996.

A. C. Elitzur and L. Vaidman, "Quantum Mechanical Interaction-Free Measurements", Found. Phys. 23, 987-97 (1993), arXiv:hep-th/9305002.

S. P. Walborn et al, "A double-slit quantum eraser", Phys. Rev. A 65, 033818 (2002), arXiv:quantph/0106078v1.

L. S. Schulman, Techniques and Applications of Path Integration, John Wiley & Sons, New York,
1981.

J. G. Cramer, Reviews of Modern Physics 58, 647-687 (1986); www.npl.washington.edu/TI.

J. G. Cramer, "Reverse Causation and the Transactional Interpretation of Quantum Mechanics", in Frontiers of Time: Retrocausation -- Experiment and Theory, D. P. Sheehan editor, AIP Conference. Proceedings 863, American Institute of Physics, 2006.

C. Rovelli, "Relational Quantum Mechanics", Intl J Theoretical Physics 35, 8: 1637 (1996), arXiv:quant-ph/9609002.

W. H. Zurek, "Decoherence and the transition from quantum to classical - revisited", arXiv:quantph/0306072, 2003.

J. A. Wheeler and R. P. Feynman, "Interaction with the Absorber as the Mechanism of Radiation", Reviews of Modern Physics, 17, 157–161 (1945)

A. Suarez, "Time and nonlocal realism: Consequences of the before-before experiment", arXiv:quant-ph/0708.1997v1, 2007.

A. Suarez, "Does Quantum Mechanics imply influences acting backward in time in impact series experiments?", arXiv:quant-ph/9801061v1, 1998.

J. Dunningham and V. Vedral, "Nonlocality of a single particle", Phys. Rev. Lett. 99, 180404 (2007), arXiv: quant-ph/0705.0322v1.

A. C. Elitzur, S. Dolev, and A. Zeilinger, "Time-Reversed EPR and the Choice of Histories in Quantum Mechanics", Quantum Computers and Computing, 2002; and in the Proceedings of XXII Solvay Conference in Physics, New York, World Scientific, 2002; quant-ph/0205182.

R. Hanbury Brown and R. Q. Twiss, "Interferometry of the intensity fluctuations in light. I. Basic theory: the correlation between photons in coherent beams of radiation", Proc of the Royal Society of London A 242

(1957): 300–324.

R. Hanbury Brown and R. Q. Twiss, "Interferometry of the intensity fluctuations in light. II. An experimental test of the theory for partially coherent light", Proc of the Royal Society of London A 243 (1958): 291–319.

D. I. Radin, "Experiments Testing Models of Mind-Machine Interaction", J Scientific Exploration, 20, 3, pp 375-401 (2006), www.deanradin.com/papers/MarkovModels.pdf

H. Schmidt, "Observation of a Psychokinetic Effect Under Highly Controlled Conditions", J. Parapsychology, 57 (1993), also available at www.fourmilab.ch/rpkp.

R. G. Jahn and B. J. Dunne, "The PEAR Proposition", J. Scientific Exploration, 19, 2 (2005), pp. 195-246; also available at www.princeton.edu/~pear.

R. D. Nelson, D. I. Radin, R. Shoup, and P. A. Bancel, "Correlations of Continuous Random Data with Major World Events", Foundations of Physics Letters, 15, 6 (2002), see www.boundary.org/bi/randomness.htm.

D. J. Bierman and D. I. Radin, "Anomalous unconscious emotional responses: Evidence for a reversal of the arrow of time", in Tuscon III: Towards a Science of Consciousness, MIT Press, 1998.

D. I. Radin, "Unconscious perception of future emotions: An experiment in presentiment", J. Scientific Exploration, 11 (2), 1997.

D. J. Bierman, "Empirical research on the radical

subjective solution of the measurement problem. Does time get its direction through conscious observation?", in Frontiers of Time: Retrocausation --Experiment and Theory, D. P. Sheehan editor, AIP Conference Proceedings 863, American Institute of Physics, 2006.

D. I. Radin, "Psychophysiological and perceptual tests of possible retrocausal effects in humans", in Frontiers of Time: Retrocausation -- Experiment and Theory, D. P. Sheehan editor, AIP Conference Proceedings 863, American Institute of Physics, 2006.

R. Nelson, "Anomalous anticipatory responses in networked random data", in Frontiers of Time: Retrocausation -- Experiment and Theory, D. P. Sheehan editor, AIP Conference Proceedings 863, American Institute of Physics, 2006.

R. Shoup, and T. Etter, "Proposal: The RetroComm Experiment - Using Quantum Randomness to **Send a Message Back in Time**", Boundary Institute (2004), www.boundary.org/bi/articles/RetroComm_exp.pdf.

R. Shoup, "Anomalies and Constraints", J. Scientific Exploration, 16, 1 (2002), also available at www.boundary.org/bi/causality.htm.

D. Bem, "Feeling the Future: Experimental Evidence for Anomalous Retroactive Influences on Cognition and Affect", J Personality and Social Psychology, 100, 407-425 (2011), preprint and discussion at http://dbem.ws.

D. I. Radin, The Conscious Universe, HarperEdge, 1997.

D. I. Radin, Entangled Minds, Paraview, 2006.

D. I. Radin, "Testing Nonlocal Observation as a Source of Intuitive Knowledge", Explore Journal, 4, 1, 25-35 (Jan 2008).

R. Wiseman and M. Schlitz, "Experimenter effects and the remote detection of staring", J Parapsychology, 61, 3, 197-208 (1998), also at www.richardwiseman.com/resources/staring1.pdf

J. Palmer, "ESP research findings: the process approach" in Foundations of Parapsychology, Routledge & Kegan Paul: London, 1986, pp 184-222.

H. Schmidt, "Can an Effect Precede Its Cause? A Model of a Noncausal World", Foundations of Physics, 8 (5/6), 1978; also at www.fourmilab.ch/rpkp.

Comscire Corp., "Design Principles and Testing of the QNG Model J1000KU", white paper (2003), see www.comscire.com.

id Quantique Corp., "Quantis Quantum Random Number Generator", white paper (2004), see www.idquantique.com.

Y. Aharonov, et al, "Multiple-time states and multiple-time measurements in quantum mechanics", Phys. Rev. A 79, 052110 (2009), arXiv:quant-ph/0712.0320v1.

J. S. Lundeen and A. M. Steinberg, "Experimental Joint Weak Measurement on a Photon Pair as a Probe of Hardy's Paradox", Phys. Rev. Lett. 102, 020404 (2009), arXiv:quant-ph/0810.4229.

A. Zeilinger, "Three challenges from John Archibald Wheeler", in Science and Ultimate Reality, Cambridge University Press, 2004

J. A. Wheeler, "The 'Past' and the Delayed-Choice Double-Slit Experiment", in Mathematical Foundations of Quantum Theory, A. R. Marlow editor, Academic Press, 1978.

V. Jacques et al, "Experimental Realization of Wheeler's Delayed-Choice Gedanken Experiment". Science 315: 966-968 (2007), arXiv:quant-ph/0610241v1.

W. K. Wootters and W. H. Zurek, "A single quantum cannot be cloned", Nature 299, 802 (1982).

M. D. Westmoreland and B. Schumacher, "Quantum Entanglement and the Nonexistence of Superluminal Signals", arXiv:quant-ph/9801014v2, 1998.

S. J. van Enk, "No-cloning and superluminal signaling", arXiv:quant-ph/9803030v1, 1998.

R. Omnes, The Interpretation of Quantum Mechanics, Princeton University Press (1994).

J. von Neumann, Mathematical Foundations of Quantum Mechanics, Princeton University Press. (1932).

J. Jenkins et al, "Analysis of Experiments Exhibiting Time-Varying Nuclear Decay Rates: Systematic Effects or New Physics?", arXiv: nucl-ex/ 1106.1678v1, 2011.

Bem D. J. (2011). Feeling the future: experimental

evidence for anomalous retroactive influences on cognition and affect. J. Pers. Soc. Psychol. 100, 407–425 10.1037/a0021524 [PubMed] [Cross Ref]

Bem D., Tressoldi P. E., Rabeyron T., Duggan M. (2014). Feeling the Future: A Meta-Analysis of 90 Experiments on the Anomalous Anticipation of Random Future Events (SSRN Scholarly Paper No. ID 2423692). Rochester, NY: Social Science Research Network; Available onlie at: http://papers.ssrn.com/abstract=2423692

Bierman D. J. (2008). Consciousness induced restoration of time symmetry (CIRTS): a psychophysical theoretical perspective, in Proceedings of the 51st Parapsychological Association Annual Meeting, 33–49

Bierman D. J., Radin D. I. (1997). Anomalous anticipatory response on randomized future conditions. Percept. Mot. Skills 84, 689–690 [PubMed]

Bierman D. J., Scholte H. S. (2002). Anomalous anticipatory brain activation preceding exposure to emotional and neutral pictures, in Paper Presented at The Parapsychological Association, 45th Annual Convention (Paris:).

Cardeña E. (2014). A call for an open, informed study of all aspects of consciousness. Front. Hum. Neurosci. 8:17 10.3389/fnhum.2014.00017 [PMC free article] [PubMed] [Cross Ref]

Cramer J. G. (1986). The transactional interpretation of quantum mechanics. Rev. Mod. Phys. 58, 647–687

Cramer J. G. (2007). A test of Quantum Nonlocal Communication. CENPA Annual Report. Seattle, WA: University of Washington, 52

Cramer J. G. (2014). Status of Nonlocal Quantum Communication Test. CENPA Annual Report. Seattle, WA: University of Washington, 114

Ferguson C. J. (2014). Comment: why meta-analyses rarely resolve ideological debates. Emot. Rev. 6, 251–252 10.1177/1754073914523046 [Cross Ref]

Franklin M. S. (2007). Can practice effects extend backwards in time?, in A Talk Presented at the 26th Annual Society for Scientific Exploration Meeting (Lansing, MI: Michigan State University;).

Honorton C., Ferrari D. C. (1989). Future telling: A meta-analysis of forced-choice precognition experiments, 1935–1987. J. Parapsychol. 53, 1–308

Jacques V., Wu E., Grosshans F., Treussart F., Grangier P., Aspect A., et al. (2007). Experimental realization of wheeler's delayed-choice Gedanken experiment. Science 315, 966–968 10.1126/science.1136303 [PubMed] [Cross Ref]

Jolij J. (2014). Social priming and psi. PISCesLAB. Available online at: http://www.jolij.com/?p=188

Mossbridge J., Tressoldi P. E., Utts J. (2012). Predictive physiological anticipation preceding seemingly unpredictable stimuli: a meta-analysis. Front. Percept. Sci. 3:390 10.3389/fpsyg.2012.00390 [PMC free article] [PubMed] [Cross Ref]

Mossbridge J., Tressoldi P. E., Utts J., Ives J., Radin D., Jonas W. (2014). Predicting the unpredictable: critical analysis and practical implications of predictive anticipatory activity. Front. Hum. Neurosci. 8:146 P 10.3389/fnhum.2014.00146 [PMC free article] [PubMed] [Cross Ref]

Puthoff H. E. (1984). ARV (Associational Remote Viewing) applications, in Research in Parapsychology 1984, eds White R., Solfvin J., editors. (Metuchen, NJ: Scarecrow Press;), 121

Rabeyron T. (2014). Retro-priming, priming, and double testing: psi and replication in a test–retest design. Front. Hum. Neurosci. 8:154 10.3389/fnhum.2014.00154 [PMC free article] [PubMed] [Cross Ref]

Radin D. (1997). Unconscious perception of future emotions: an experiment in presentiment. J. Sci. Explor. 11, 163–180

Radin D. (2011). Electrocortical activity prior to unpredictable stimuli in meditators and nonmeditators. Explore 7, 286–299 10.1016/j.explore.2011.06.004 [PubMed] [Cross Ref]

Rhine J. B. (1938). Experiments bearing on the precognition hypothesis: I.Pre-shuffling card calling. J. Parapsychol. 2, 38–54

Rouder J. N., Morey R. D. (2011). A Bayes factor meta-analysis of Bem's ESP claim. Psychon. Bull. Rev. 18, 682–689 10.3758/s13423-011-0088-7 [PubMed] [Cross Ref]

Schwarzkopf D. S. (2014). We should have seen this coming. Front. Hum. Neurosci. 8:332 10.3389/fnhum.2014.00332 [PMC free article] [PubMed] [Cross Ref]

Smith C. C., Laham D., Moddel J. (2014). Stock market prediction using associative remote viewing by inexperienced remote viewers. J. Sci. Explor. 28, 7–16

Spottiswoode S. J. P., May E. C. (2003). Skin conductance prestimulus response: analyses, artifacts and a pilot study. J. Sci. Explor. 17 617–641

Sternberg S. (1969). The discovery of processing stages: extensions of Donders' method. Acta Psychol. 30, 276–315 10.1016/0001-6918(69)90055-9 [Cross Ref]

Wagenmakers E., Wetzels R., Borsboom D., van der Maas H. L. (2011). Why psychologists must change the way they analyze their data: the case of psi: Comment on Bem (2011). J. Pers. Soc. Psychol. 100, 426–432 10.1037/a0022790 [PubMed] [Cross Ref]

Wagner M. W., Monnet M. (1979). Attitudes of college professors towards extrasensory perception. Zetetic Scholar 6, 7–17

Wheeler J. A. (1984). In Quantum Theory and Measurement, ed Zurek J. A., editor. (Princeton, NJ: Princeto University Press;), 182–213

Wheeler J. A., Feynman R. P. (1945). Interaction with the absorber as the mechanism of radiation. Rev. Mod. Phys. 17, 157–181

Wilson A. (2013). Social Priming: of Course it Only Kind of Works. Notes from Two Scientific Psychologists. Available online at: http://psychsciencenotes.blogspot.c

M. Bizzarri, A. Pasqualato, A. Cucina and V. Pasta, "Physical forces and non linear dynamics mould fractal shape. Quantitative morphological parameters and cell phenotype," Histology and Histopathology, Vol. 28, pp. 155-174, 2013.

W. Bras, J. Torbet, G. P. Diakum, G. L.J. A. Rikken and J. F. Diaz, "The diamagnetic susceptibility of the tubulin dimer," Journal of Biophysics, Vol. 2014, 5 pages, 2014 Article ID 985082.   ISSN: 2153-8212 Journal of Consciousness Exploration & Research Published by QuantumDream, Inc.
www.JCER.com

M. Rahnama, J. A. Tusynski, I. Bokkon, M. Cifra, P. Sardar and V. Salari, "Emission of mitochondrial biophotons and their effect on electrical activity of membrane via microtubules," Journal of Integrative Neuroscience, Vol. 10, 65-88, 2011.

W. Bras, G. P. Diakun, J. F. Diaz et al., "The susceptibility of pure tubulin to high magnetic fields: a magnetic birefringence and X-ray fiber diffraction study," Biophysical Journal, Vol. 74, pp. 15091521, 1988.

P. M. Vassilev, R. T. Dronzine, M. P. Vassileva and G. A. Georgiev, "Parallel arrays of microtubules formed in electric and magnetic fields," Bioscience Reports, Vol. 2, pp. 1025-1029, 1982

N. J. Murugan, B. T. Dotta, L. M. Karbowski, and M. A. Persinger, "Conspicuous bursts of photon emissions in malignant cell cultures following injections of morphine: implications for cancer treatment", International Journal of Current Research, Vol. 6, pp. 10588-10592, 2014

B. T. Dotta, C. A. Buckner, R. M. Lafrenie and M. A. Persinger, "Photon emissions from human brain and cell culture exposed to rotating magnetic fields shared by separate light-stimulated brains and cells," Brain Research, Vol 388, pp. 77-88, 2011. 21.

F-A. Popp, "Photon storage in biological system," Electromagnetic Information, Urban and Schwarzenberg, 1979, pp. 123-149.

L. Yu. Berzhanskaya, O. Yu. Beloplotova and V. N. Berzhansky, "Electromagnetic field effect on luminescent bacteria," IEEE Transactions on Magnetics, Vol. 31, pp. 4274-4275, 1995

B. T. Dotta, R. M. Lafrenie, L. M. Karbowski and M. A. Persinger, "Photon emission from melanoma cells during brief stimulation by patterned magnetic fields: is the source coupled to rotational diffusion within the membrane?," General Physiology and Biophysics, Vol. 33, pp. 63-73, 2014.

Q. H. Mach and M. A. Persinger, "Behavioral changes with brief exposures to weak magnetic fields patterned to stimulate long-term potentiation," Brain Research, Vol. 1261, pp. 45-53, 2009. 36.

K. S. Saroka and M. A. Persinger, "Quantitative evidence for direct effects between earth-ionosphere

Schumann Resonances and human cerebral cortical activity," International Letters of Chemistry, Physics and Astronomy Vol. 20(2) pp. 166-194, 201

S. Tanahshi, "Detection of line splitting of Schumann resonances from ordinal data," Journal of Atmospheric and Terrestrial Physics, Vol. 38, pp. 135-142, 1976.

T. Harmony, "The functional significance of delta oscillations in cognitive processing," Frontiers in Integrative Neuroscience, Vol. 7, Dec, 2013, Article 83.

A. Lisi, A. Foletti, M. Ledda, F. De Carlo, L. Giuilani, E. D'Emilia, and S. Grimaldi, "Resonance as a tool to transfer information to living systems: the effect of 7 Hz calcium ion energy resonance on human epithelial cells (HaCaT) differentiation," PIERS Proceedings, Cambridge, U.SA., July 2-6, 2008.

G. Benitez-King, G. Ramirez-Rodriguez, and L. Ortiz-Lopez, "Altered microtubule associated proteins in schizophrenia," NeuroQuantology, Vol. 5, pp. 58-61, 2007.

Dmitreva I., Khabarova O., Obridko V., Ragulskaja M., Reznikov A. Experimental confirmations of bioeffective effect of magnetic storms. Astron. Astrophys. Trans. 2000;19:67–77. doi: 10.1080/10556790008241351. [Cross Ref]

Khabarova O. Investigation of the tchizhevsky-velhover effect. Biophysics. 2004;49:S60.

Dimitrova S., Stoilova I., Cholakov I. Influence of local geomagnetic storms on arterial blood pressure.

Bioelectromagnetics. 2004;25:408–414. doi: 10.1002/bem.20009. [PubMed] [Cross Ref]

Samsonov S., Sokolov V., Strekalovskaya A., Petrova P. On the Relationship of Cardiovascular Disease Exacerbation to Helio-Geophysical Disturbances; Proceedings of the XXVII Annual Seminar (Physics of Auroral Phenomena); Apatity, Russian. 2004; [(accessed on 3 July 2017)]. pp. 134–137. Available online: https://www.researchgate.net/profile/S_Samsonov/publication/228907582_On_the_relationship_of_cardiovascular_disease_exacerbation_to_heliogeophysical_disturbances/links/0912f508f089a6118c000000.pdf.

Azcárate T., Mendoza B., de la Peña S.S., Martínez J. Temporal variation of the arterial pressure in healthy young people and its relation to geomagnetic activity in Mexico. Adv. Space Res. 2012;50:1310–1315. doi: 10.1016/j.asr.2012.06.015. [Cross Ref]

Dimitrova S., Mustafa F., Stoilova I., Babayev E., Kazimov E. Possible influence of solar extreme events and related geomagnetic disturbances on human cardiovascular state: Results of collaborative Bulgarian–Azerbaijani studies. Adv. Space Res. 2009;43:641–648. doi: 10.1016/j.asr.2008.09.006. [Cross Ref]

Additional References

Can gravitational waves predispose neuro-cardiovascular circadian rhythms?. Singh RB. Halberg Hospital and Research Institute, Moradabad, Uttar Pradesh, India), People's Friendship University of Russia, Moscow, Russia, Sergey Chibisov), (Department of Food science and Nutrition, Faculty of human Environmental

Sciences, Mukagawa Women's University, Japan. Toru Takahashi. Graduate School of human Environment Medicine, Fukuoka, Women's University, Fukuoka, Japan.DOI: 10.15761/CRT.1000159

Zadeh SS, Pahlevanlo A, Rad NG, Jalili M, Singh RB, et al. (2011) Can diet and lifestyle factors and geomagnetic forces predispose aggression. The Open Nutr J 4: 176-179.

Breus TK, Baevskii RM, Chernikova AG (2012) Effects of geomagnetic disturbances on humans functional state in space flight. JBiSE 5: 341-355.

Breus TK, Halberg G, Cornelissen G (1995) Effect of solar activity on the physiological rhythms of biological systems. Biofizika 40: 737-748. [Crossref]

Breus TK, Cornelissen G, Halberg F, Levitin AE (1995) Temporal associations of life with solar and geophysical activity. Annales Geophysicae 13: 1211- 1222.

Syutkina EV, Cornelissen G, Grigoriev AE, Mitish MD, Turti T, et al. (1997) Neonatal intensive care may consider associations of cardiovascular rhythms with local magnetic disturbance. Scripta Medica 70: 217-226.

Chibisov SM, Breus TK, Levitin AE, Drogova GM (1995) Biological effects of a global magnetic storm. Biofizika 40: 959-968.

Cornelissen G, Halberg F, Breus TK, Syutkina EV, Baevskii RM, et al. (2002) Nonphotic solar associations of heart rate variability and myocardial infarction. Journal of Atmospheric and Solar-Terrestrial Physics 64: 707-728.

Halberg F, Cornélissen G, Regal P (2004) Chronoastrobiology: Proposal, nine conferences, heliogeomagnetics, transyears, near-weeks, near-decades, phylogenetic and ontogenetic memories. Biomed Pharmacother 58: 150-187.

Singh RB, Fedacko J, Sharma JP, Vargova V, Sharma R, et al. (2010) Association of inflammation, heavy meals, magnesium, nitrite, and coenzyme Q10 deficiency and circadian rhythms with risk of acute coronary syndromes. World Heart Journal 2: 219-228.

Hammoudeh AJ, Alhaddad IA (2009) Triggers and the onset of acute myocardial infarction. Cardiol Rev 17: 270-274. [Crossref]

Singh RB, Pella D, Neki NS, Chandel JP, Rastogi S, et al. (2004) Mechanisms of acute myocardial infarction study (MAMIS). Biomed Pharmacother 58 Suppl 1: S111-115. [Crossref]

Singh RB, Shastun S, Hristova K, Fedacko J, Joshi P, et al. (2016) Blood pressure and blood glucose variability, the silent killers, in subjects with diabetes mellitus, flying blue. World Heart J 8: 109-120.

Cornelissen G, Siegelova J, Otsuka K (2016) Circadian disruption of the blood pressure rhythm as predictor of adverse cardiovascular outcome and overall mortality. World Heart J 8: 5-10.

Scott Rauvers

Index

10.7cm Radio Flux_____63, 64, 65, 66, 67, 72, 570, 609, 613, 621

10Hz_____90, 316, 453, 462, 463, 464

Alfvén waves_____69

Algae   253, 254, 255, 256, 302, 393, 396, 397, 399, 410, 411, 421, 422, 541, 542, 543, 603

Alpha Brain Waves_____87, 88, 90, 93, 97, 98, 285, 286, 308, 309, 310, 311, 312, 313, 314, 340, 361, 423, 462, 465, 472, 540, 543, 640

Alternate Universe (see parallel universe)

Anesthetics_____252, 346, 347, 348, 349, 350, 351, 354, 371, 372, 373, 476, 511, 664, 681, 685

Arcturus_____11, 148, 149, 618, 649

Aspirin_____244, 269, 356, 375, 386, 563

Back to the Future_____64, 151, 216

Barometric Air Pressure_____103, 438, 531, 532, 570, 608, 613, 614, 615

Bergamot_____79. 193, 251, 303, 304, 305, 411, 415, 416, 428, 533, 640, 641

Blue Light_____100, 144, 145, 150, 151, 176, 592

Bose-Einstein Condensation_____122, 180, 442, 444, 580, 678, 687

Brilliant Blue_____231, 265, 280, 327, 518, 523

Carbon nano_____164, 307, 387, 442

Improve your Remote Viewing Accuracy Techniques
using Quantum Microtubules

Catecholamine_____243, 263, 266, 277, 278, 279, 325, 326, 351, 532

Chiral / Chirality_____156, 157, 321, 322, 603

Chizhevsky_____66

Circularly Polarlized Light_____102, 210, 211, 212, 254, 274, 603

Colchicine_____86, 178, 351, 357, 358, 377, 674

Cordless Telephones_____135

Creatine   233, 234, 266

Cyanobacteria_____142, 143, 396, 409, 525, 536, 542, 544, 545

Cytoskeleton_____106, 201, 202, 208, 380

Dice_____114, 116, 119

Dogon Tribe_____146, 147, 148

Electromagnetic fields_____365, 507

EMwave 2_____35

Fenchyl_____605

First Quarter Moon_____102, 105, 145, 401, 550, 593, 607, 613, 617

Fish Oil_____100, 237, 241, 242, 243, 268, 276, 278, 325, 385, 392

Full Moon_____45, 47, 48, 83 to 87, 103, 366, 551, 571, 550, 591, 592, 597, 599, 600, 601, 606, 608, 613, 615, 616, 617, 623, 648, 649

Gamma waves_____465, 466

Gene (see PON1 Gene)

Geosmin_____525, 540, 541, 542, 544, 545

Geraniol_____229, 230, 281, 282, 341, 342, 343, 344 to 347, 395, 408, 411, 418, 419, 420, 427, 437, 447, 524, 525, 527, 528, 530, 535, 537, 603, 654

Gingko_____204, 229, 301, 306, 604

Glutamate_____98, 142, 173, 174, 175, 280, 303, 326, 327, 380, 396, 398, 408, 412, 428, 543, 545

Glutamic Acid_____239, 281

Gold_____162, 164, 389

Half Moon (see first quarter moon)

Heart Math_____35, 36, 46, 47, 74, 80, 102 to 104, 219, 230, 293, 309, 312, 440, 481, 517, 567, 572, 573, 621

Hemispheric Balancing_____93, 144, 175, 190, 236, 286, 362, 469, 479, 487

High Pressure_____ (see barometric air pressure)

HRV_____ (see heart math)

Incense (induces alpha brainwaves) _____423, 434, 640, 641

Jasmine_____251, 252, 253, 318, 319, 320, 423, 540, 641

Juniper_____78, 110, 305, 605

Kinesin_____348, 349, 372, 378

Lanthanum_____386 to 388, 445, 448, 553, 554

L-DOPA_____98, 235 to 237, 247

Lead Molybdate_____157, 658

Lemon_____211, 229, 275, 318, 340, 345, 370, 399, 409,

Improve your Remote Viewing Accuracy Techniques using Quantum Microtubules

423, 527, 541, 567, 602, 605, 645

Limonene_____75, 76, 193, 304, 321, 340, 342, 343, 350, 370, 398, 399, 416, 418, 419, 424, 429, 527, 601

Linalool_____222, 227, 258, 350, 398, 412 to 420, 428, 437, 527, 529, 530, 531, 533, 538, 540, 544, 567, 601, 603

Low Frequency_____40, 51, 52, 69, 93, 294, 395, 365, 405, 463, 466, 479, 487, 665, 668

Luminescence_____150, 162, 395, 479, 480

Madagascar Periwinkle_____352

Magnometer Activity_____571, 572

Manganese_____156, 161, 388, 414, 415, 480, 482, 483, 562, 564, 565, 591, 596, 656, 661

Mantis Shrimp_____273, 274

Melatonin_____141, 157, 174, 178, 239

Middle Latitude Fredericksburg K-Indices_____65, 130, 568

Midnight_____145, 152, 210, 336, 604, 623, 624, 625

Mind/Heart Coherence_____ (see heart math)

Mucuna Pruiens_____236, 237

Nerol_____258, 342, 345, 532, 529, 603, 641, 654

Neurotransmission _____ 100, 243, 269, 272, 284, 301, 323, 372

Neutrinos_____39, 294, 295, 496 to 499, 506 to 508

Nicotine_____288

Nummo _____ 147

Pacific Yew Tree _____ 204, 353, 380 to 383

Parahippocampal Gyrus _____ 190, 192, 193, 194, 559

Parallel Universe _____ 490, 495, 500 to 505, 506, 610

Patchouli _____ 78

Peppermint _____ 305, 319, 320, 418, 605, 641, 644, 645

Polymers _____ 444, 552

PON1 Gene _____ 242, 243

Pupil Dilation _____ 40, 49

Pyrroloquinoline _____ 276, 277, 325

Quantum Coherence _____ 109, 170, 180, 185, 253, 254, 339, 384, 400, 402 to 405, 410, 442, 444, 451, 457, 460, 501, 508, 512, 514, 515, 518, 586, 696, 675, 685, 686

Quantum Computer _____ 184

Quantum Dots _____ 161, 162, 163, 167

Resveratrol _____ 199, 231, 239, 242, 264, 269, 278, 325, 432

Reward Effect _____ 216, 261

Roulette _____ 36, 38, 44, 45, 48

Russian Telepathic Experiments _____ 287

Selenium _____ 159, 161, 167, 170, 232, 234, 235, 265 to 267, 305, 306, 355, 364, 376, 388, 551, 658, 661

Sensory Deprivation _____ 361

Sesquiterpenes _____ 304, 398, 420, 421, 423, 432, 473, 474,

Signal to Noise Ratio_____329, 332, 333, 334, 337, 461, 472

Silver_____164, 173, 388, 480

Sirius _____ 145 to 148, 618

Skin Temperature_____321, 361

Solar Wind_____48 to 54, 62 to 68, 570, 609, 622

Spinal Fluid_____238, 239, 268, 280, 281, 282, 324, 327, 344, 518, 523

Stochastic Resonance_____313 to 316, 470 to 472, 477, 478

Super fluidity_____441, 442

Synchronization_____52, 60, 63, 68, 70, 73, 311, 315, 316, 450, 467, 687

Taxol_____203, 204, 209, 349, 351, 356, 357, 360, 381, 382, 383, 384, 390, 391, 469, 477

Theanine_____279, 281, 310, 326 to 328, 343, 370

Theta Brain Waves_____88 to 94, 98, 101, 215, 284 to 285, 308 to 310, 423, 437, 469, 559, 573, 603, 604

Thyroid_____234, 238, 239, 266, 280, 577

TMD_____158, 173, 549, 550

Transthyretin_____10, 237, 238 to 242, 267, 268, 280

Tritium_____123, 543, 553

Tungsten_____153 to 157, 159 to 162, 387, 388, 417, 430, 445, 549, 658

Tungsten Disulfide_____156, 158, 159, 161, 179, 550, 551,

552

TXP Formula_____76, 230, 346, 415, 418, 436, 526, 567, 601, 602

Tyrosine_____222, 228, 324, 235, 266, 278, 325, 339, 417, 564

USSR PSI Labs_____297

UV emissions_____66

Vagus Nerve_____43

Valerian_____81, 282, 283, 284, 422, 423, 424, 425, 433, 643, 662

Van der Waals_____159, 179, 380, 549, 552, 553

Vinca Plant_____ (see Madagascar Periwinkle)

Wulfenite_____157, 158, 168, 169, 205

Ylang-Ylang_____320, 321, 437, 447, 529 to 533 538, 605

Zacks Functional MRI Experiment_____215

Zinc_____206, 242, 273, 276, 388, 480, 481, 482, 576, 659, 651

Zinc Oxide_____161

Improve your Remote Viewing Accuracy Techniques using Quantum Microtubules

Notes

Scott Rauvers

Notes

www.ingramcontent.com/pod-product-compliance
Lightning Source LLC
Chambersburg PA
CBHW050044230526
45470CB00004B/1399